Solid State
Pulse Circuits

Third Edition

Solid State
Pulse Circuits

David A. Bell

Lambton College
of Applied Arts and Technology
Sarnia, Ontario, Canada

A RESTON BOOK
Prentice Hall
Englewood Cliffs, NJ 07632

Library of Congress Cataloging-in-Publication Data

Bell, David A.
  Solid state pulse circuits.

  ''A Reston book.''
  Includes index.
  1. Pulse circuits.   2. Digital electronics.
3. Semiconductors.   I. Title.
TK7868.P8B44   1988   621.3815'34   87-11546
ISBN 0-8359-7052-3

Editorial/production supervision and
   interior design: Linda Zuk, WordCrafters Editorial Services, Inc.
Cover design: Lundgren Graphics, Ltd.
Manufacturing buyer: Carol Bystrom
Cover photo courtesy of The Stock Market, Christopher Springmann

Printed in the United States of America

10   9   8   7   6   5   4

ISBN 0-8359-7052-3   025

Prentice-Hall International (UK) Limited, *London*
Prentice-Hall of Australia Pty. Limited, *Sydney*
Prentice-Hall Canada Inc., *Toronto*
Prentice-Hall Hispanoamericana, S.A., *Mexico*
Prentice-Hall of India Private Limited, *New Delhi*
Prentice-Hall of Japan, Inc., *Tokyo*
Prentice-Hall of Southeast Asia Pte. Ltd., *Singapore*
Editora Prentice-Hall do Basil, Ltda., *Rio de Janeiro*

*Dedicated to
my mother and father*

Contents

Contents

Preface

This book is intended for use in electronics technology courses in colleges and universities. It should also be useful as a reference text for practicing electronics technicians, technologists, and engineers.

The first few chapters explain the characteristics of *pulse waveforms* and *capacitive-resistive circuits*, which must be understood before any study of pulse circuitry can commence. The operation of *diodes*, *transistors*, *FETs*, and *IC op-amps* in switching circuits is studied next. This leads to the design and analysis of *inverters*, *Schmitt trigger circuits*, *multivibrators*, *ramp generators*, and *function generators*, including their application to integrated circuits. *Logic gates*, *logic circuits*, and *IC logic families* are also studied.

For circuit design, the approach is a simple step-by-step procedure in which the reader knows exactly how each component value is arrived at. Many design examples are included in the text; device data sheets in the appendices are referred to where appropriate, and standard value components are selected.

Once individual circuits and gates are understood, they are used as building blocks to describe *digital counting*, *digital frequency meters*, *digital voltmeters*, *A-to-D and D-to-A conversion*, *pulse modulation*, and *time division multiplexing*.

Major changes from the second edition include new chapters on "*Basic Logic Gates and Logic Functions*" and "*Logic Circuits*," as well as an updated chapter on "*Integrated Circuit Logic Gates*." There is also additional material on *IC op-amps in switching circuits*, *ADCs*, *DACs*, and *IC function generators*.

I am grateful to those who made suggestions for improvements to the first and second editions. Comments on the third edition are also welcome.

David A. Bell

Chapter 1

Pulse Fundamentals

INTRODUCTION

The term *pulse waveform* is normally applied only to approximately rectangular waveshapes. However, many different types of waveforms are involved in the study of pulse circuits. Waveforms are defined in terms of amplitude and time interval measurements. Each of the various waveforms can be shown to contain a fundamental-frequency sine wave and many higher frequency sinusoidal components, known as *harmonics*. A study of the fundamental frequency and its harmonics shows a definite relationship between the bandwidth of a circuit and the distortion produced in a square wave output from the circuit.

1-1 TYPES OF WAVEFORMS

Repetitive Waveforms and Transients

When one quantity varies in relation to another quantity, the relationship can be represented by plotting a graph. Thus, graphs may be plotted to show how certain quantities vary with respect to time. A plot of dc voltage or current *versus* time normally produces a straight line graph, as in Figure 1-1(a) and (b). An alternating voltage, as its name implies, increases and decreases with respect to time, alternately swinging positive and negative.

When the instantaneous voltage levels v are plotted against time t, the graph that results is called the *waveform* of the voltage. In Figure 1-1(c) the instantaneous values of a sinusoidal alternating voltage are plotted against a base of time. It is seen that the voltage increases positively to a peak value, decreases through zero to a negative peak

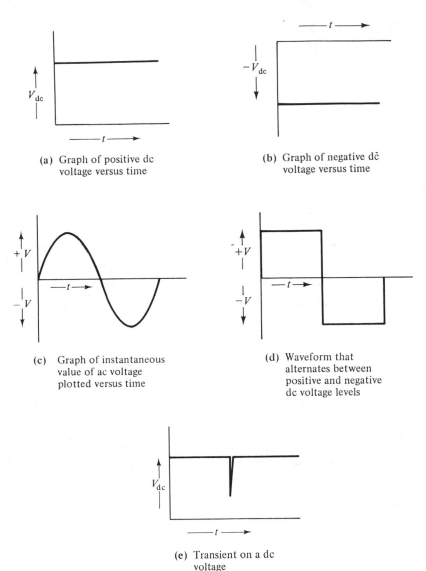

(a) Graph of positive dc
voltage versus time

(b) Graph of negative dc
voltage versus time

(c) Graph of instantaneous
value of ac voltage
plotted versus time

(d) Waveform that
alternates between
positive and negative
dc voltage levels

(e) Transient on a dc
voltage

Figure 1-1   Graphs of voltage (or current) versus time show the waveform of a quantity.

value, then returns to zero; then the cycle recommences. The sine wave is a repeating cycle of voltage (or current) with a sinusoidal relationship to time. All waveforms which are composed of identical cycles that keep repeating are termed *repetitive waveforms* or *periodic waveforms*. It is necessary to study only one cycle of such a waveform to gain an understanding of the behavior of the voltage or current involved. When successive cycles of an alternating voltage are *not* identical, the waveform is described as *aperiodic*. Figure 1-1(d) shows a wave that alternates between fixed positive and

negative direct voltage levels. This is termed a rectangular wave, and, like the sine
wave, it has a repetitive alternating waveform.

Sometimes a direct voltage suddenly decreases (or increases) for a brief instant
and then returns to its normal level [Figure 1-1(e)]. This may happen, for example,
when a load is suddenly switched onto a power supply. Such brief *nonrepetitive* waveforms are termed *transients*.

Display Methods

The *cathode-ray oscilloscope* is widely used in the study of electrical waveforms. In
the oscilloscope, an electron beam striking a fluorescent screen produces a tiny spot of

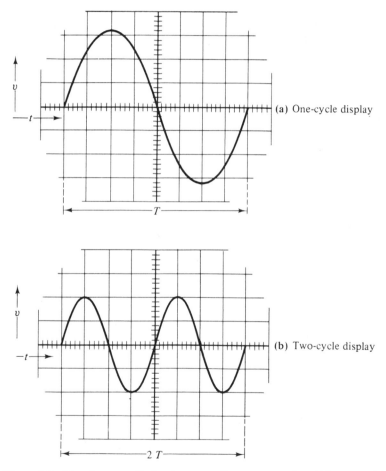

(a) One-cycle display

(b) Two-cycle display

Figure 1-2   Waveforms are displayed on an oscilloscope screen by a light spot which
is deflected vertically by the instantaneous voltage and horizontally in proportion to
time.

light. The light spot becomes a line when the electron beam is deflected vertically by the voltage to be displayed and horizontally in proportion to time (Figure 1-2). The light spot starts at the left-hand side of the screen and moves horizontally to the right as it is deflected vertically. At the end of one (or more) cycles of the waveform, the light spot returns almost instantaneously to the left-hand side of the screen and then recommences the sweep from left to right. Thus, a repetitive waveform is traced on the screen again and again. When a permanent record of the waveform is required, a camera is used to photograph the display on the oscilloscope. A camera can be employed also to obtain photographs of any transient waveform that might be displayed briefly on the oscilloscope screen.

Miscellaneous Waveforms

Sinusoidal.   The most common electrical waveform is the sine wave, shown in Figure 1-2. Half-wave rectification removes the negative (or positive) half-cycles of a sine wave [Figure 1-3(a)], while full-wave rectification produces a train of unidirectional half-sine waves, as shown in Figure 1-3(b).

Rectangular.   When a dc voltage suddenly changes from one level to another, the change is referred to as a *step change*. The change might be positive or negative, as shown in Figure 1-4(a). A *rectangular waveform* consists simply of successive cycles of positive step changes followed by negative step changes. When the time duration t_1 for the upper dc level is equal to the time duration t_2 for the lower level, the waveform is termed a *square wave* [Figure 1-4(b)]. When t_1 and t_2 are unequal, as illustrated in Figure 1-4(c), the wave is usually referred to as a *pulse waveform*.

Ramp.   A voltage that increases or decreases at a constant rate with respect to time has a graph that is a positive or negative *ramp* [Figure 1-5(a)]. A repetitive cycle

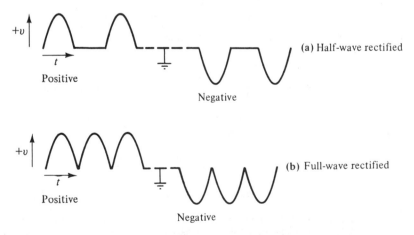

Figure 1-3   Positive and negative, half-wave and full-wave rectified sine waves.

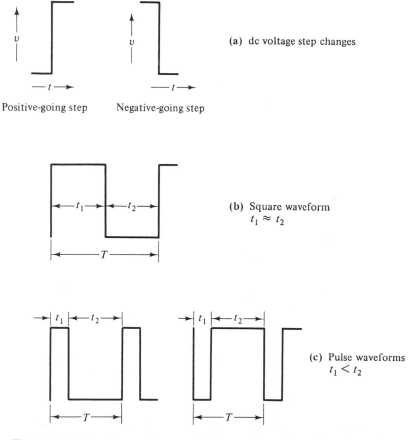

(a) dc voltage step changes

Positive-going step   Negative-going step

(b) Square waveform
$t_1 \approx t_2$

(c) Pulse waveforms
$t_1 < t_2$

Figure 1-4 Rectangular waveforms consist of successive positive-going and negative-going steps. When $t_1 = t_2$, the waveform is termed a square wave. When $t_1 \neq t_2$, it is referred to as a pulse wave.

of a positive ramp followed by a negative ramp is known as a *triangular waveform* [Figure 1-5(b)]. When one ramp is much steeper than the other, as illustrated in Figure 1-5(c), the waveform is usually termed a *sawtooth waveform*.

Exponential.   In this case, the voltage varies with respect to time according to the equation $V = E(1 - \epsilon^{-kt})$ or $V = E\epsilon^{-kt}$, where t is time, k is a constant, ϵ is the *exponential constant* ($\epsilon = 2.718$), and E is a constant voltage. The resultant graphs of voltage *versus* time are of the form shown in Figure 1-6(a). Repetitive cycles of positive and negative exponentials produce an *exponential waveform* [Figure 1-6(b)]. An exponential change followed by a step change gives the waveforms shown in Figure 1-6(c). Introduction of a gap results in the *spike waveforms* of Figure 1-6(d). Obviously, a great variety of waveforms can be produced by combining two or more of the various voltage changes discussed above.

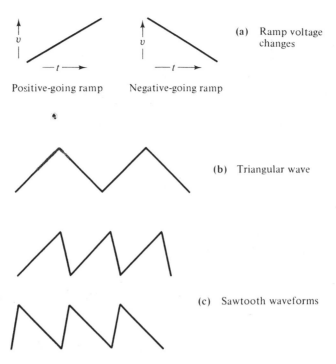

Positive-going ramp   Negative-going ramp

Figure 1-5   A triangular waveform consists of successive positive-going and negative-going ramp changes. When one ramp is steeper than the other, a sawtooth waveform results.

1-2 CHARACTERISTICS OF PULSE WAVEFORMS

Ideal Pulse Waveform

Consider the *ideal pulse waveform* shown in Figure 1-7. In this particular case the pulses are positive with respect to ground. The pulse amplitude is simply the voltage level of the top of the pulse measured from ground. The first edge of the pulse (at $t = 0$) is referred to as the *leading edge*, *rising edge*, or *positive-going edge*. The second edge is termed the *trailing edge*, *lagging edge*, *falling edge*, or *negative-going edge*.

The *time period T* is the time measured from the leading edge of one pulse to the leading edge of the next pulse. If $T = 1$ sec, then the *pulse repetition frequency* (PRF) is 1 cycle/sec or 1 pulse per sec (pps), so that PRF = $1/T$ pps. Instead of pulse repetition frequency, the term *pulse repetition rate* (PRR) is sometimes used.

The time measured from the leading edge to the trailing edge of one pulse is known as the *pulse width* (PW), the *pulse duration* (PD), or sometimes as the *mark length*. The time between pulses is simply referred to as the *space width*. The proportion of the time period occupied by the pulse is defined as the *duty cycle*, or the *mark-to-space* (M/S) *ratio*:

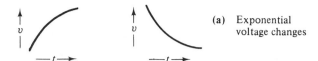

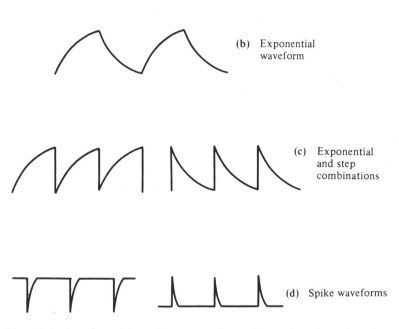

Positive-going exponential Negative-going exponential

(a) Exponential
voltage changes

(b) Exponential
waveform

(c) Exponential
and step
combinations

(d) Spike waveforms

Figure 1-6 An exponential waveform occurs when positive-going and negative-going exponential changes follow each other. A combination of an exponential change and a step change gives a spike waveform.

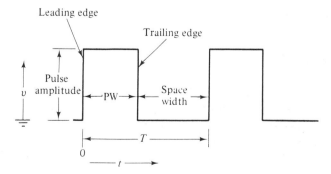

Figure 1-7 An ideal pulse waveform has perfectly vertical leading and lagging edges and perfectly flat tops and bottoms.

$$\text{Duty cycle} = (PW/T) \times 100\% \qquad\qquad (1\text{-}1)$$

and
$$\text{M/S ratio} = PW/(\text{space width}) \qquad\qquad (1\text{-}2)$$

The duty cycle usually is expressed as a percentage, while the mark-to-space ratio is normally stated simply as a ratio.

EXAMPLE 1-1

For the pulse waveform displayed in Figure 1-8, determine the pulse amplitude, PRF, PW, duty cycle, and M/S ratio. The vertical scale is 1 V per division, and the horizontal scale is 0.1 ms per division.

Solution

$$\text{Pulse amplitude} = (3.5 \text{ divisions}) \times (1\text{V/division})$$

$$= 3.5 \text{ V}$$

$$T = (6 \text{ divisions}) \times (0.1 \text{ ms/division})$$

$$= 0.6 \text{ ms}$$

$$\text{PRF} = 1/T = 1/0.6 \text{ ms} = 1667 \text{ pps}$$

$$\text{PW} = (2.5 \text{ divisions}) \times (0.1 \text{ ms/division})$$

$$= 0.25 \text{ ms}$$

$$\text{Space width} = 3.5 \times 0.1 \text{ ms} = 0.35 \text{ ms}$$

From Eq. 1-1, $\quad \text{Duty cycle} = \dfrac{PW}{T} \times 100\%$

$$= \frac{0.25 \text{ ms}}{0.6 \text{ ms}} \times 100\% = 41.7\%$$

From Eq. 1-2, $\quad \text{M/S ratio} = \dfrac{PW}{\text{Space width}} = \dfrac{0.25 \text{ ms}}{0.35 \text{ ms}}$

$$= 0.71$$

Rise Time, Fall Time, and Tilt

The pulse displayed in Figure 1-8 appears to have a perfectly flat top and perfectly vertical sides. When pulses are examined very carefully, however, it is found that the top is never perfectly flat. The amplitude of the lagging edge normally is less than that of the leading edge. In many cases, the slope at the top of the pulse may be so small that it cannot be easily measured or even observed. In other cases, as in Figure 1-9,

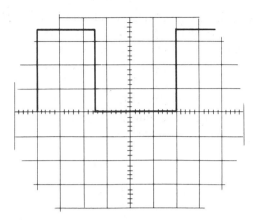

Figure 1-8 Pulse waveform displayed on an oscilloscope. The amplitude of the wave can be measured in terms of the vertical deflection in volts/division. The pulse width, space width, and time period are determined according to the horizontal time/division deflection.

the slope may be very obvious. The pulse voltage does not go from zero to its maximum level instantaneously, and from maximum to zero instantaneously. In fact, there is a definite *rise time* t_r and *fall time* t_f at the leading and lagging edges of the pulse. This is illustrated in the figure.

If the pulse width (PW) is measured near the top of the pulse in Figure 1-9, it would be quite different from the PW as measured close to the bottom of the pulse.

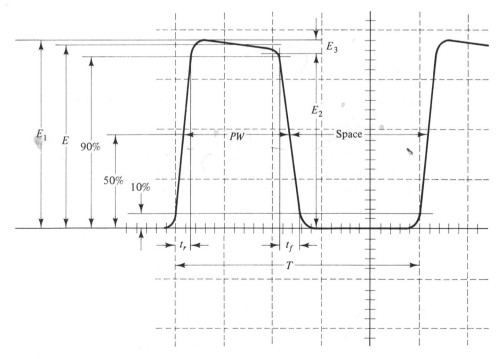

Figure 1-9   Practical pulse waveform, showing rise time t_r, fall time t_f, and tilt E_3.

Therefore, the PW is defined as the *average* pulse width and is normally measured at half the average amplitude. (See the figure.) The space width (SW) is measured at the same amplitude as the pulse width. The time period T is usually measured from the beginning of one pulse to the beginning of the next pulse, but it is also the sum of the pulse width and space width,

$$PW + SW = T$$

Again in Figure 1-9, E_1 is the maximum pulse amplitude, E_2 is the minimum amplitude, and E is the average pulse amplitude:

$$E = \frac{E_1 + E_2}{2}$$

The rise time is defined as the time required for the voltage to go from 10% to 90% of the average amplitude. Similarly, the fall time is the time required for the pulse to fall from 90% to 10% of the average amplitude (see Fig. 1-9). For a pulse with severe slope, the fall time may be measured from 90% of E_2. The *slope* or *tilt* (also termed *droop*) at the top of the waveform is defined in terms of the average amplitude. Usually, the tilt is expressed as a percentage. Sometimes, however, it is employed as a fraction, in which case it is termed the *fractional tilt*.

$$\text{Tilt} = \frac{E_3}{E} \times 100\% \qquad\qquad (1\text{-}3)$$

$$= \frac{E_1 - E_2}{E} \times 100\%$$

In Figure 1-10 a waveform is shown which is symmetrical above and below ground level. Once again, the tilt is determined by expressing the tilt voltage E_3 as a percentage of the average amplitude E.

EXAMPLE 1-2

(a) For the waveform displayed in Figure 1-9, determine the pulse amplitude, tilt, t_r, t_f, PW, PRF, mark-to-space ratio, and duty cycle. (b) For the square waveform in Figure 1-10, determine the tilt. The vertical scale is 100 mV/division, and the horizontal scale is 100 μs/division in each case.

Solution (a)

$$\text{Average pulse amplitude, } E = \frac{E_1 + E_2}{2} = \frac{380\,\text{mV} + 350\,\text{mV}}{2}$$

$$= 365\,\text{mV}$$

From Eq. 1-3,
$$\text{Tilt} = \frac{E_1 - E_2}{E} \times 100\%$$

$$= \frac{380\,\text{mV} - 350\,\text{mV}}{365\,\text{mV}} \times 100\% = 8.2\%$$

$$t_r = (0.3 \text{ division}) \times (100 \text{ μs/division})$$

$$= 30 \text{ μs}$$

$$t_f = (0.4 \text{ division}) \times (100 \text{ μs/division})$$

$$= 40 \text{ μs}$$

$$T = (5 \text{ divisions}) \times (100 \text{ μs/division})$$

$$= 500 \text{ μs}$$

$$\text{PRF} = 1/T = 1/500 \text{ μs} = 2000 \text{ pps}$$

$$\text{PW} \simeq (2.2 \text{ divisions}) \times (100 \text{ μs/division})$$

$$= 220 \text{ μs}$$

$$\text{Space width} \simeq (2.8 \text{ divisions}) \times (100 \text{ μs/division})$$

$$= 280 \text{ μs}$$

$$\text{M/S ratio} = \frac{220 \text{ μs}}{280 \text{ μs}} \approx 0.79$$

$$\text{Duty cycle} = \frac{220 \text{ μs}}{500 \text{ μs}} \times 100\% = 44\%$$

Solution (b)

$$E_3 \simeq (0.5 \text{ division}) \times (100 \text{ mV/division}) = 50 \text{ mV}$$

$$E \simeq (4.5 \text{ divisions}) \times (100 \text{ mV/division}) = 450 \text{ mV}$$

$$\text{Tilt} = \frac{E_3}{E} \times 100\% = \frac{50 \text{ mV}}{450 \text{ mV}} \times 100\%$$

$$= 11.1\%$$

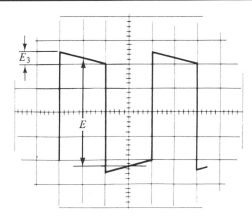

Figure 1-10 Square waveform with tilt on top and bottom.

The square waveform shown in Figure 1-11(a) is symmetrical above and below ground level. The positive and negative peaks are of equal amplitudes and equal widths (i.e., $t_1 = t_2$). This means that the average value of one cycle of the waveform is zero. If this waveform were applied to a dc voltmeter, the instrument would indicate zero. The average value of the waveform is found simply by summing the positive and negative areas enclosed by one cycle and dividing by the time period.

$$\text{Average voltage} = V_{av} = \frac{(V_1 \times t_1) + (V_2 \times t_2)}{T} \tag{1-4}$$

For Figure 1-11(a):

$$\text{Average voltage} = \frac{(6 \text{ V} \times 1 \text{ ms}) + (-6 \text{ V} \times 1 \text{ ms})}{2 \text{ ms}} = 0 \text{ V}$$

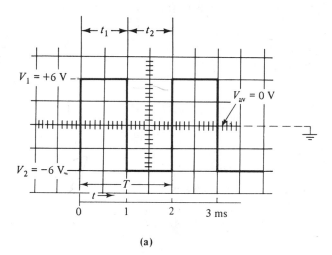

(a)

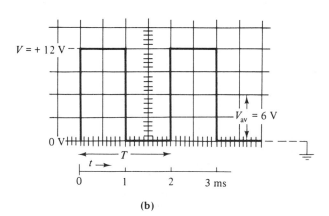

(b)

Figure 1-11 Two square waves which have equal peak-to-peak amplitudes, but different average values.

The waveform of Figure 1-11(b) has no negative portion, so its average voltage level is

$$V_{av} = \frac{(12 \text{ V} \times 1 \text{ ms}) - (0)}{2 \text{ ms}} = 6 \text{ V}$$

EXAMPLE 1-3

Determine the average voltage level of the pulse waveform in Fig. 1-12. The vertical scale is 2 V/div, and the horizontal scale is 1 ms/div.

Solution

From Eq. 1-4, $V_{av} = \dfrac{(V_1 \times t_1) + (V_2 \times t_2)}{T}$

$$= \frac{(8 \text{ V} \times 0.8 \text{ ms}) + (-1 \text{ V} \times 2.2 \text{ ms})}{3 \text{ ms}}$$

$$= 1.4 \text{ V}$$

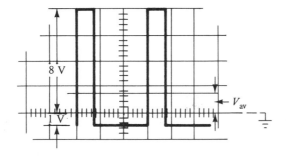

Figure 1-12 Pulse waveform with a pulse amplitude of 9 V, starting from a level of −1 V.

1-3 HARMONIC CONTENT OF WAVEFORMS

Frequency Synthesis

In Figure 1-13(a) two signal generators are shown connected in series. Generator A is producing a sinusoidal output waveform as illustrated. The output of generator B is also a sine wave, but its frequency is three times the frequency of signal generator A. The amplitude of the output from B is also less than that from A. The waveform produced by the two generators in series is the larger amplitude (and lower frequency) signal, with the smaller amplitude signal superimposed. It is seen that the combination approximately resembles a square wave with its peaks dented. Figure 1-13(b) shows a third generator connected in series with A and B. The output from generator C is smaller in amplitude than that from B, and the frequency of this third signal is five

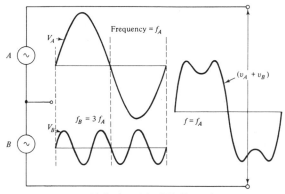

(a) Fundamental and third harmonic

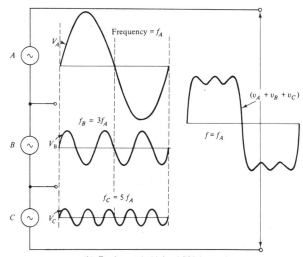

(b) Fundamental, third and fifth harmonic

Figure 1-13   A combination of a large-amplitude sinusoidal waveform with a number of smaller amplitude higher frequency sine waves can produce a square wave. The process, known as frequency synthesis, demonstrates that nonsinusoidal waveforms are made up of many sine wave components.

times the frequency of the output of generator A. The waveform produced by the three generators in series now more closely resembles a square wave. It is important to note that the resultant waveforms shown in Figures 1-13(a) and (b) are produced only when *the generators are synchronized,* i.e., when all component waves commence in a positive-going direction at the same instant.

The construction of the approximately square waveform is easily seen by referring to Figure 1-14, where the instantaneous amplitudes of waveforms A and B are added. At time t_1, for example, the amplitude of waveform A is 6.5 V, and that of B is approximately 2.5 V. Therefore, the amplitude of the resultant waveform at time t_1 is 9 V (point 1). At time t_2, the amplitude of B is zero, so the resultant amplitude is 8.5 V, that is, the amplitude of A at t_2 (point 2). At t_3, the amplitudes of B, -3 V, and of A, 10 V, are added together to produce an amplitude of 7 V. When this process is continued, it is seen how the final waveform is constructed.

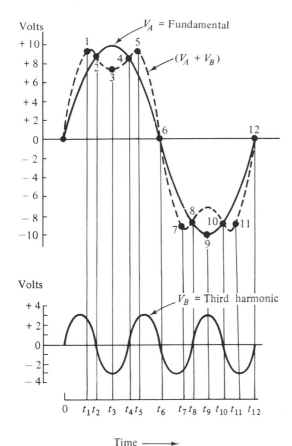

Figure 1-14 Addition of instantaneous levels of a fundamental-frequency sine wave and a third harmonic.

The process of building up a particular waveform by combining several sine waves of different frequencies and amplitudes is referred to as *frequency synthesis*. If the process were continued and appropriate higher frequency waveforms were added, the resultant each time would more closely resemble a square wave.

Harmonic Analysis

The converse of frequency synthesis is *harmonic analysis*. In this process, a waveform is analyzed to discover the sine wave frequencies it contains. By harmonic analysis, it can be shown that periodic nonsinusoidal waveforms are composed of combinations of pure sine waves. Some waveforms can also have dc components. One major component, a large-amplitude sine wave of the same frequency as the periodic wave under consideration, is termed the *fundamental*. The other components of a periodic waveform are sine waves with frequencies which are exact multiples of the frequency of the fundamental. These waves, referred to as *harmonics*, are numbered according to the ratio of their frequencies to that of the fundamental. For example, a harmonic with a frequency exactly double that of the fundamental is called the *second harmonic*. The frequency of the *third harmonic*, obviously, is three times the fundamental frequency.

By the mathematical operation known as *Fourier analysis*, waveforms can be analyzed to determine their harmonic content. The amplitude of each harmonic and its phase relationship to the fundamental can be found. Also, the amplitude of any dc component can be calculated. A perfect square wave which is symmetrical above and below ground can be shown, by Fourier analysis, to have a fundamental component and odd-numbered harmonics, but no even-numbered harmonics and no dc component. A pulse waveform is found to contain both odd- and even-numbered harmonics and (usually) a dc component. Sawtooth waveforms, triangular waveforms, and rectified sine waves are made up of more complicated combinations of odd- and even-numbered harmonics. In all cases, the harmonic content actually goes to infinity, but the amplitudes of the harmonics decrease as their frequencies increase. Thus, the higher frequency components are the least important.

Information which can be derived by harmonic analysis becomes very important when considering the circuitry through which various waveforms are processed. Suppose that a square wave with a frequency of 1 kHz is applied to an amplifier with an upper frequency limit of 15 kHz. In this case, the amplifier will not reproduce sine waveforms with frequencies greater than 15 kHz. Thus, the amplifier will not pass harmonics of 1 kHz greater than the fifteenth. If the square wave applied to the same amplifier had a frequency of 5 kHz, only the first, second, and third harmonics would be passed. If a 5 kHz square wave were to be amplified, and if all harmonics up to the thirty-third were to be reproduced, then the amplifier must have an upper frequency limit greater than 33×5 kHz.

1-4 WAVEFORM DISTORTION

Distortion and Frequency Response

If a square wave is applied to circuitry that does not pass all the necessary frequency components, the resultant output is a distorted square wave. The type of distortion depends upon whether the circuitry has poor low-frequency response or poor high-frequency response. In Figure 1-15(a) the long rise and fall times of the square wave show that the high-frequency harmonics are attenuated, and thus the circuit has poor high-frequency response. Figure 1-15(b) shows the output from a circuit which has good high-frequency response but poor low-frequency response. The *tilt* on the top and bottom of the square wave results because the low-frequency components were attenuated and phase shifted by the circuit. The waveform in Figure 1-15(c) shows both long rise times and tilt. This result is obtained when the involved circuitry has neither a low enough nor a high enough frequency response for the applied square wave. When circuits overemphasize some of the high-frequency harmonics, *overshoots*

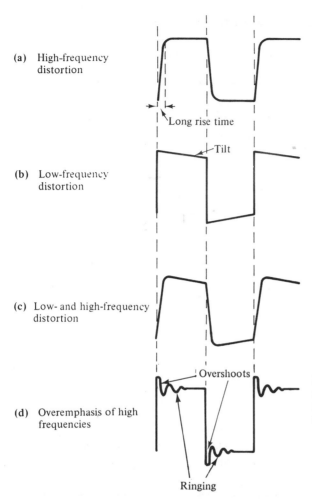

(a) High-frequency
     distortion

Long rise time

Tilt

(b) Low-frequency
     distortion

(c) Low- and high-frequency
     distortion

Overshoots

(d) Overemphasis of high
     frequencies

Ringing

Figure 1-15 Distortion of a square wave
results when the waveform is processed by
circuits with unsatisfactory frequency
response. Inadequate high-frequency
response results in increased rise and fall
times. Poor low-frequency response
produces tilt. Overemphasis of high
frequencies generates overshoots and
ringing.

are produced, as shown in Figure 1-15(d). *Ringing* occurs when a circuit oscillates for
a short time due to the presence of stray inductance and capacitance.

Rise Time and Upper Cutoff Frequency

When a pulse wave with no measurable rise and fall times is applied to a circuit, all
harmonic components with frequencies above the upper cutoff frequency of the circuit
are attenuated. Thus, the rise time and fall time of the output wave are limited by the
upper cutoff frequency of the circuit. In Chapter 2, the relationship between rise and
fall times, t_r and t_f, and upper cutoff frequency f_H is shown to be

$$t_r = t_f = \frac{0.35}{f_H} \tag{1-5}$$

By Eq. 1-5, the output rise time can be predicted when the upper cutoff frequency of the circuitry is known. The equation also affords a fast means of pulse testing to determine the cutoff frequency of any circuit or device.

EXAMPLE 1-4

The output waveform from an amplifier under pulse test has a rise time of 1 μs. Determine the upper 3 dB frequency of the amplifier.

Solution
From Eq. 1-5,

$$f_H = \frac{0.35}{t_r} = \frac{0.35}{1 \text{ μs}}$$

$$= 350 \text{ kHz}$$

It is possible to decide the upper cutoff frequency for a circuit that must pass pulse waveforms with an *acceptable* amount of high-frequency distortion. The question that arises, of course, is just what is an acceptable amount of distortion. Refer to Figure 1-15(a). The rise and fall times shown are approximately one-tenth of the pulse width. If the rise and fall times were much greater than PW/10, the pulse would be severely distorted. Therefore, in many cases $t_r \simeq$ PW/10 might be used as a guide for acceptable high-frequency distortion. In some other circumstance there may be a requirement for less distortion. In digital applications, where only the presence or absence of a pulse is important, quite large rise and fall times may be acceptable.

It should be noted that in circuits that use electronics devices the longest output rise and fall times occur when the pulse waveform has a maximum amplitude. If the pulse amplitude is halved, then the rise and fall times due to the circuitry are also usually halved.

EXAMPLE 1-5

A pulse waveform has a PRF of 1.5 kHz and a duty cycle of 3%. (a) Determine the minimum upper cutoff frequency of processing circuitry for acceptable reproduction of the waveform. (b) If the 1.5 kHz pulse is to be amplified by equipment with a high-frequency limit of 1 MHz, calculate the minimum pulse width and duty cycle that can be reproduced accurately.

Solution (a)   For a duty cycle of 3%:

$$\text{PW} = 0.03 \times T = 0.03 \times \frac{1}{f}$$

$$= 0.03 \times 1/1.5 \text{ kHz}$$

$$= 20 \text{ μs}$$

For a $t_r = 10\%$ of PW:

$$t_r = 0.1 \times 20 \ \mu s$$

$$= 2 \ \mu s$$

From Eq. 1-5, $$f_H = \frac{0.35}{t_r}$$

$$= \frac{0.35}{2 \ \mu s}$$

$$= 175 \text{ kHz}$$

Solution (b)   From Eq. 1-5,

$$t_r = \frac{0.35}{f_H}$$

$$= \frac{0.35}{1 \text{ MHz}}$$

$$= 0.35 \ \mu s$$

For $t_r = 10\%$ of PW:

$$\text{Minimum PW} = 10 \times t_r$$

$$= 10 \times 0.35 \ \mu s$$

$$= 3.5 \ \mu s$$

From Eq. 1-1, $$\text{Duty cycle} = \frac{\text{PW}}{T} \times 100\%$$

$$= \text{PW} \times f \times 100\%$$

$$= 3.5 \ \mu s \times 1.5 \text{ kHz} \times 100\%$$

$$\approx 0.5\%$$

As already discussed, there are always rise and fall times present in a pulse wave-form. Usually, a requirement of the circuitry that processes the waveform is that it does not add significantly to the input rise and fall times. For example, every oscilloscope has an upper cutoff frequency that produces rise and fall times in apparently perfect input pulses. Thus, t_r for the oscilloscope, as calculated from Eq. 1-5, should be much smaller than the rise and fall times of the input pulse. If the circuit-produced rise time t_{rc} is less than one-tenth of the signal rise time t_{rs}, the output rise time t_{ro} is not significantly affected by the circuitry. A circuit rise time which is one-third of the signal rise time will add only 5% to the output rise time, and this is usually acceptable. The equation relating these quantities is

$$t_{ro} = \sqrt{(t_{rs})^2 + (t_{rc})^2} \qquad (1\text{-}6)$$

EXAMPLE 1-6

(a) A pulse wave with a rise time of 10 μs is applied to an amplifier which has an upper cutoff frequency of 350 kHz. Calculate the rise time in the output waveform. (b) Determine the minimum upper cutoff frequency for an oscilloscope which is to display pulse waveforms with rise times of 100 ns. Calculate the displayed rise time.

Solution (a)   From Eq. 1-5,

$$t_{rc} = \frac{0.35}{f_H} = \frac{0.35}{350 \text{ kHz}}$$

$$= 1 \text{ μs}$$

From Eq. 1-6, $t_{ro} = \sqrt{(t_{rs})^2 + (t_{rc})^2}$

$$= \sqrt{(10 \text{ μs})^2 + (1 \text{ μs})^2}$$

$$\approx 10.05 \text{ μs}$$

Solution (b)

$$t_{rc} \approx \frac{t_{rs}}{3} = \frac{100 \text{ ns}}{3}$$

$$= 33 \text{ ns}$$

From Eq. 1-5, $f_H = \dfrac{0.35}{t_{rc}} = \dfrac{0.35}{33 \text{ ns}}$

$$= 10.6 \text{ MHz}$$

From Eq. 1-6, $t_{ro} = \sqrt{(100 \text{ ns})^2 + (33 \text{ ns})^2}$

$$= 105 \text{ ns}$$

Tilt and Lower Cutoff Frequency

If a perfect pulse waveform were applied to the input of a circuit with a lower cutoff frequency of $f_L = 0$, the top of the output pulse would be perfectly flat. When f_L is greater than zero, however, tilt is present on the output pulse, as illustrated in Fig. 1-15(b). In Chapter 2 it is demonstrated that, when the tilt is 10% or less, it is directly proportional to the pulse width and the lower cutoff frequency:

$$\text{Fractional tilt} = 2\pi f_L \text{ PW} \qquad (1\text{-}7)$$

For a square wave, $T = 2 \text{ PW} = \dfrac{1}{f}$, giving

$$\text{Fractional tilt} = \frac{\pi f_L}{f} \qquad (1\text{-}8)$$

Equations 1-7 and 1-8 can be used to determine the tilt that will be produced by a circuit with a known lower cutoff frequency. Alternatively, they can be employed to estimate the lower cutoff frequency of a circuit being pulse tested.

EXAMPLE 1-7

An amplifier with a lower cutoff frequency of 10 Hz is to be employed for amplification of square waves. For the tilt on the output waveform not to exceed 2%, calculate the lowest input frequency that can be amplified.

Solution   From Eq. 1-8,

$$f = \frac{\pi f_L}{\text{Fractional tilt}}$$

$$= \frac{\pi \times 10 \text{ Hz}}{0.02}$$

$$= 1.57 \text{ kHz}$$

EXAMPLE 1-8

Determine the upper and lower cutoff frequencies of a circuit which is to amplify a 1 kHz square wave, if the rise time of the output is not to exceed 200 ns and 3% tilt is acceptable.

Solution   From Eq. 1-5,

$$f_H = \frac{0.35}{t_r} = \frac{0.35}{200 \text{ ns}} = 1.75 \text{ MHz}$$

From Eq. 1-8,

$$f_L = \frac{f \times \text{Fractional tilt}}{\pi} = \frac{1 \text{ kH} \times 0.03}{\pi}$$

$$= 9.5 \text{ Hz}$$

EXAMPLE 1-9

Determine the upper and lower 3 dB frequencies of the circuitry which produced the output waveform shown in Figure 1-9, assuming that the input has a perfect waveform.

Solution   From Example 1-2,

$t_r = 30\ \mu s$,   PRF = 2000 pps,   Tilt = 8.2%,   PW = 220 μs

From Eq. 1-5,

$$f_H = \frac{0.35}{t_r} = \frac{0.35}{30\ \mu s} = 11.67\ kHz$$

From Eq. 1-7,

$$f_L = \frac{(Fractional\ tilt)}{2\pi \times PW} = \frac{0.082}{2\pi \times 220\ \mu s} = 59.3\ Hz$$

REVIEW QUESTIONS AND PROBLEMS

1-1. Define: repetitive waveforms, periodic waveforms, aperiodic waveforms, transients.

1-2. Draw sketches to show the shapes of the following waveforms: square, pulse, triangular, sawtooth, exponential.

1-3. For a pulse waveform, define: leading edge, lagging edge, trailing edge, T, PRF, PRR, PW, PD, M/S ratio, duty cycle.

1-4. For the pulse waveform illustrated in Figure 1-16, determine: pulse amplitude, PRF, PW, duty cycle, M/S ratio. The vertical scale is 0.1 V per division, and the horizontal scale is 1 ms per division.

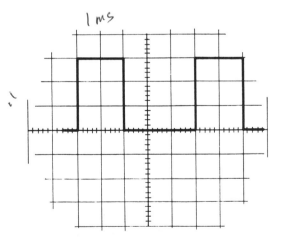

Figure 1-16   Problem 1-4.

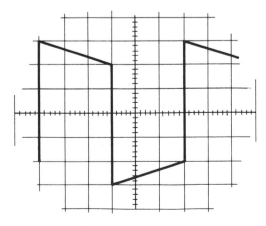

Figure 1-17   Problem 1-5.

1-5. (a) Define rise time, fall time, and tilt. (b) Determine the percentage tilt on the square wave shown in Figure 1-17.

1-6. For the waveform displayed in Figure 1-18, determine: pulse amplitude, tilt, t_r, t_f, PW, PRF, M/S ratio, and duty cycle. The vertical scale is 1 V/division, and the horizontal scale is 10 μs/division.

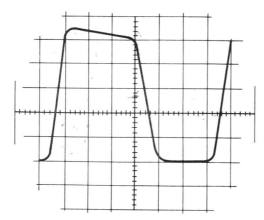

Figure 1-18 Problem 1-6.

1-7. If the pulse waveforms shown in Figure 1-19 were applied to a dc voltmeter, determine the voltages that would be indicated in each case.

1-8. (a) Define the following terms: fundamental, harmonic, frequency synthesis, harmonic analysis. (b) Determine the highest harmonic at the output of an amplifier which has an upper cutoff frequency of 1 MHz (i) when a 10 kHz square wave is applied to it? and (ii) when the input is a 150 kHz square wave.

1-9. (a) A 12 kHz pulse waveform is amplified by a circuit having a high-frequency limit of 1 MHz. Determine the minimum pulse width that can be reproduced accurately. (b) If the duty cycle of the 12 kHz pulse waveform becomes 0.5%, determine the approximate upper cutoff frequency of a circuit that will reproduce the waveform accurately.

1-10. Sketch a square wave that is amplified by equipment which (a) has poor low-frequency response; (b) has poor high-frequency response; (c) overemphasizes high frequencies; (d) has a combination of poor low-frequency and poor high-frequency responses.

1-11. A 1 kHz square wave output from an amplifier has t_r = 350 ns and tilt = 5%. Determine the upper and lower 3 dB frequencies of the amplifier.

1-12. Calculate the rise time and tilt that may be expected on the square wave output of an amplifier with a bandwidth extending from 10 Hz to 500 kHz. The applied square wave has a frequency of 5 kHz.

1-13. Determine the bandwidth of the circuitry that produced the output waveform shown in Figure 1-18 (Problem 1-6).

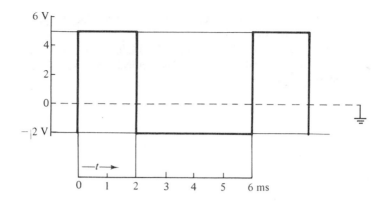

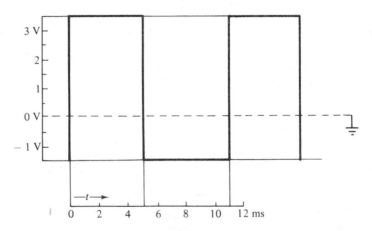

Figure 1-19   Problem 1-7.

1-14. Construct the waveform that results when the fundamental and harmonics shown in Figure 1-20 are added together.

1-15. Determine the lowest square wave frequency that can be passed by an amplifier with a lower cutoff frequency of 1 Hz if the output tilt is not to exceed 1%.

1-16. An oscilloscope has an upper cutoff frequency of 10 MHz. (a) Calculate the shortest pulse-rise time that can be displayed accurately. (b) Determine the displayed rise time when the input pulse has $t_r = 30$ ns.

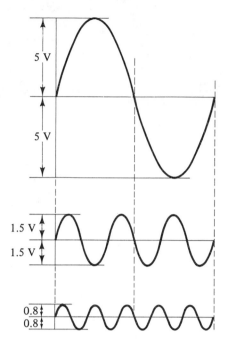

Figure 1-20 Problem 1-14.

Chapter 2

Capacitive Resistive (CR) Circuits

INTRODUCTION

When a capacitor is charged from a dc voltage source through a resistor, the instantaneous level of capacitor voltage may be calculated at any given time. There is a definite relationship between the **time constant** of a CR circuit and the times required for the capacitor to charge to approximately 63% and 99% of the input voltage. Also, an important relationship exists between the time constant of a circuit and the rise time of the output voltage from the circuit. Depending upon the arrangement of the CR circuit, it may be employed as an *integrator* or a *differentiator*.

2-1 CR CIRCUIT OPERATION

Capacitor Charging

Consider the circuit and graph shown in Figure 2-1. If the charge on capacitor C is zero at the instant that switch S is closed, then the voltage across R at $t = 0$ is

$$V_R = E - e_c$$

where E is the supply voltage and e_c is the capacitor voltage. The current through R at $t = 0$ is

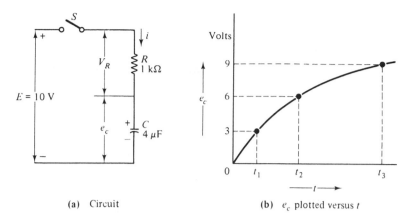

(a) Circuit   (b) e_c plotted versus t

Figure 2-1   Capacitor charged through a resistor from a constant voltage source. When
the capacitor voltage e_c is low, the voltage drop V_R across the resistor is high; consequently, the charging current is high and e_c increases rapidly. As e_c grows, V_R decreases
and the rate of charge decreases.

$$i_c = \frac{V_R}{R}$$

$$= \frac{E - e_c}{R} \qquad\qquad (2\text{-}1)$$

$$= \frac{10\,\text{V} - 0}{1\,\text{k}\Omega}$$

$$= 10\,\text{mA}$$

This current causes capacitor C to charge with the polarity illustrated, so that at
some time t_1 the capacitor voltage e_c might be 3 V. [See Figure 2-1(b).] This alters V_R:

$$V_R = E - e_c$$

$$= 10\,\text{V} - 3\,\text{V} = 7\,\text{V}$$

Now, $$i_c = (10\,\text{V} - 3\,\text{V})/1\,\text{k}\Omega = 7\,\text{mA}$$

Because C has accumulated some charge, e_c is increased and the voltage across
R is reduced; thus, the charging current through R is reduced. Since the current is
reduced, C is being charged at a lower rate than before. After some longer time period,
e_c increases to 6 V (t_2 on the graph). Now,

$$V_R = 10\,\text{V} - 6\,\text{V} = 4\,\text{V}$$

and $$i_c = \frac{4\,\text{V}}{1\,\text{k}\Omega} = 4\,\text{mA}$$

The charging current has now been further reduced. Consequently, an even longer time period is required to charge C by another 3 V.

The capacitor does not receive its charge at a constant rate. Instead, e_c is continuously increasing, so that the voltage across R is continuously decreasing, and consequently the charging current is decreasing. This means that C is charged at a rapid rate initially, and then the rate decreases as the capacitor voltage grows.

Voltage and Current Levels

It can be shown that the capacitor voltage follows the exponential equation

$$e_c = E - (E - E_o)\epsilon^{-t/CR} \tag{2-2}$$

where

e_c = capacitor voltage at instant t

E = charging voltage

E_o = initial charge on the capacitor

ϵ = exponential constant ≈ 2.718

t = time from commencement of charge (seconds)

C = capacitance of capacitor being charged (Farads)

R = charging resistance (ohms)

When there is no initial charge on the capacitor,

$$E_o = 0$$

and

$$e_c = E - [E - 0]\epsilon^{-t/CR}$$

or

$$e_c = E(1 - \epsilon^{-t/CR}) \tag{2-3}$$

Thus, since

$$i_c = \frac{E - e_c}{R}$$

$$= \frac{E - E(1 - \epsilon^{-t/CR})}{R}$$

it follows that

$$i_c = \frac{E}{R}\epsilon^{-t/CR}$$

and

$$i_c = I\epsilon^{-t/CR} \tag{2-4}$$

where $I = E/R$ is the initial level of charging current when $t = 0$.

EXAMPLE 2-1

Calculate the levels of capacitor voltage e_c in the circuit of Figure 2-1(a) at 2 ms intervals from the instant when switch S is closed. Plot a graph of e_c *versus* time. $E = 10$ V, $C = 4$ μF, and $R = 1$ kΩ.

Solution   Since $E_o = 0$, Equation 2-3 may be used to calculate e_c.

At $t = 0$,

$$e_c = E(1 - \epsilon^0) = 0 \text{ V}$$

point 1
(in Fig. 2-2)

At $t = 2$ ms,

$$e_c = 10 \text{ V}(1 - \epsilon^{-2\,\text{ms}/(4\,\mu\text{F}\times 1\,\text{k}\Omega)}) = 3.93 \text{ V}$$

point 2

At $t = 4$ ms,

$$e_c = 10 \text{ V}(1 - \epsilon^{-4\,\text{ms}/(4\,\mu\text{F}\times 1\,\text{k}\Omega)}) = 6.32 \text{ V}$$

point 3

At $t = 6$ ms,

$$e_c = 7.77 \text{ V}$$

point 4

At $t = 8$ ms,

$$e_c = 8.65 \text{ V}$$

point 5

At $t = 10$ ms,

$$e_c = 9.18 \text{ V}$$

point 6

At $t = 12$ ms,

$$e_c = 9.5 \text{ V}$$

point 7

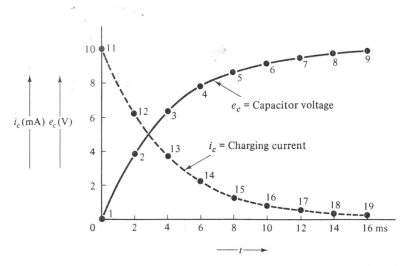

Figure 2-2   Capacitor voltage and current plotted *versus* time for a *CR* circuit. Charging current i_c is initially a maximum when capacitor voltage e_c is zero. As e_c increases exponentially, i_c decreases exponentially.

At $t = 14$ ms,

$$e_c = 9.7 \text{ V}$$ point 8

At $t = 16$ ms,

$$e_c = 9.82 \text{ V}$$ point 9

EXAMPLE 2-2

Determine the instantaneous levels of charging current in the circuit of Figure 2-1(a) at 2 ms time intervals from the instant that switch S is closed. Plot a graph showing i_c *versus* time.

Solution

Eq. 2-1,
$$i_c = \frac{E - e_c}{R}$$

At $t = 0$, $i_c = \dfrac{10 \text{ V} - 0}{1 \text{ k}\Omega} = 10 \text{ mA}$ point 11
                                                                                    (in Fig. 2-2)

At $t = 2$ ms, $e_c = 3.93$ V (from Example 2-1)

$$i_c = \frac{10 \text{ V} - 3.93 \text{ V}}{1 \text{ k}\Omega} = 6.07 \text{ mA}$$ point 12

At $t = 4$ ms,

$$i_c = \frac{10 \text{ V} - 6.32 \text{ V}}{1 \text{ k}\Omega} = 3.68 \text{ mA}$$ point 13

At $t = 6$ ms,

$$i_c = \frac{10 \text{ V} - 7.77 \text{ V}}{1 \text{ k}\Omega} = 2.23 \text{ mA}$$ point 14

At $t = 8$ ms,

$$i_c = \frac{10 \text{ V} - 8.65 \text{ V}}{1 \text{ k}\Omega} = 1.35 \text{ mA}$$ point 15

At $t = 10$ ms,

$$i_c = \frac{10 \text{ V} - 9.18 \text{ V}}{1 \text{ k}\Omega} = 0.82 \text{ mA}$$ point 16

At $t = 12$ ms,

$$i_c = \frac{10 \text{ V} - 9.5 \text{ V}}{1 \text{ k}\Omega} = 0.5 \text{ mA}$$ point 17

At $t = 14$ ms,

$$i_c = \frac{10\text{ V} - 9.7\text{ V}}{1\text{ k}\Omega} = 0.3\text{ mA} \qquad\qquad \text{point 18}$$

At $t = 16$ ms,

$$i_c = \frac{10\text{ V} - 9.82\text{ V}}{1\text{ k}\Omega} = 0.18\text{ mA} \qquad\qquad \text{point 19}$$

EXAMPLE 2-3

A 1 μF capacitor is charged from a 6 V source through a 10 kΩ resistor. If the capacitor has an initial voltage of -3 V, calculate its voltage after 8 ms.

Solution
From Eq. 2-2, $\qquad\qquad e_c = E - (E - E_o)\epsilon^{-t/CR}$

$$e_c = 6\text{ V} - [6\text{ V} - (-3\text{ V})]\epsilon^{-8\text{ ms}/(1\,\mu\text{F} \times 10\,\text{k}\Omega)}$$

$$= 6\text{ V} - [9\text{ V}]\epsilon^{-0.8} = 1.96\text{ V}$$

Refer again to Figure 2-1(a). Suppose that the capacitor becomes completely charged to the level of the supply voltage. Now assume that the input voltage is instantaneously reduced to zero, while the switch S is still closed. The result is that the capacitor discharges through resistor R. Equation 2-2 can be used to calculate the capacitor voltage at any time during discharge. The initial charge on the capacitor is E (i.e., the level of supply voltage before it went to zero). But during discharge E becomes zero. Therefore, Equation 2-2 can be simplified to

$$e_c = 0 - (0 - E)\epsilon^{-t/CR}$$

or $\qquad\qquad\qquad\qquad e_c = E\epsilon^{-t/CR}$ \hfill (2-5)

When a capacitor voltage discharge curve is plotted [using Equation 2-5], it is found to be similar in shape to the charging current graph in Figure 2-2.

2-2 NORMALIZED CHARGING GRAPH

In Figure 2-3, *normalized* charge and discharge curves are presented. These can be employed to graphically solve many problems. The normalized curves are plotted for the case of $E = 1$ V, $C = 1$ F, and $R = 1$ Ω. For these values, the capacitor voltage can be determined at any given time t after commencement of charge or discharge.

When the supply voltage is not 1 V, the capacitor voltage at any given time can be found simply by multiplying the voltage from the graph by the value of E. For example, when $t = 2$ s on the charge curve, $e_c = 0.86$ V. If, instead of 1 V, $E = 5$ V then

$$e_c = 0.86 \times 5\text{ V} = 4.3\text{ V}$$

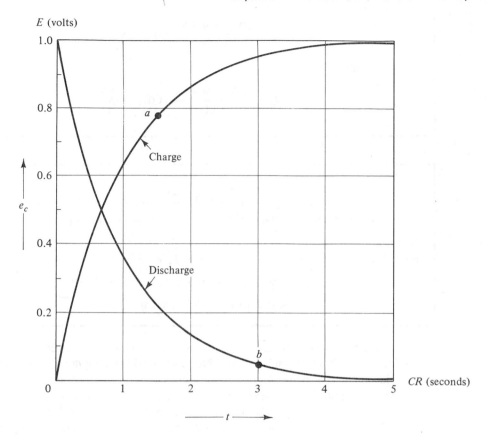

Figure 2-3 Normalized charge and discharge curve for a CR circuit. To use the curve for circuit calculations, the vertical scale is multiplied by the supply voltage, and the horizontal scale is multiplied by the circuit time constant CR.

Similarly, when C and R are other than 1 F and 1 Ω, respectively, the time at any instant is multiplied by $C \times R$. As an example of this, take the case of $C = 1$ μF and $R = 1$ kΩ when $e_c = 0.5$ V.

From the curve, the time for e_c to reach 0.5 V is

$$t = 0.7 \text{ s} \times 1 \text{ }\mu\text{F} \times 1 \text{ k}\Omega$$

$$= 0.7 \text{ ms}$$

EXAMPLE 2-4

Using the normalized charge and discharge curves in Figure 2-3, determine
(a) e_c at 1.5 ms starting from $e_c = 0$ V, when $R = 1$ kΩ, $C = 1$ μF, and $E = 10$ V.
(b) e_c at 6 ms from full charge when $R = 20$ kΩ, $C = 0.1$ μF, and $E = 12$ V.

Solution (a)   Each second on the time scale becomes

$$t = CR = 1 \ \mu F \times 1 \ k\Omega = 1 \ ms$$

and each volt on the voltage scale is

$$e = E \times 1 \ V = 10 \ V$$

At $t = 1.5$ ms (point a on the charge curve),

$$e_c = 10 \ V \times 0.78$$
$$= 7.8 \ V$$

Solution (b)   Each second on the time scale becomes

$$t = CR = 0.1 \ \mu F \times 20 \ k\Omega = 2 \ ms$$

and each volt on the voltage scale is

$$e = E \times 1 \ V = 12 \ V$$

At $t = 6$ ms (point b on the discharge curve),

$$e_c = 12 \ V \times 0.05$$
$$= 0.6 \ V$$

2-3 CR CIRCUIT EQUATIONS

Circuit Time Constant

Refer to the graphs in Figure 2-4, where charging current and capacitor voltage are plotted *versus* time for the circuit of Fig. 2-1(a). It is seen that when $t = 4$ ms, e_c is 6.32 V. In this case 4 ms is equal to the product of capacitance and resistance. That is,

$$t = C \times R = 4 \ \mu F \times 1 \ k\Omega = 4 \times 10^{-3}$$

Therefore, when $t = CR$, e_c is 6.32 V, or 63.2% of E.

Now consider Equation (2-3) once again:

$$e_c = E(1 - \epsilon^{-t/CR})$$

When $t = CR$,

$$e_c = E(1 - \epsilon^{-CR/CR})$$
$$= E(1 - \epsilon^{-1})$$
$$= E \times 0.632 \tag{2-6}$$

Thus, when $t = CR$, $e_c = 63.2\%$ of E, regardless of the value of E, C, or R.

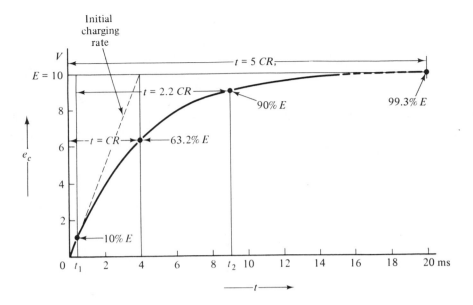

(a) *CR* and *t* relationships to e_c

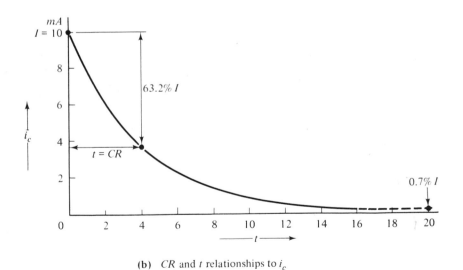

(b) *CR* and *t* relationships to i_c

Figure 2-4   Relationships between (a) time constant *CR*, time *t*, and capacitor voltage e_c, and (b) *CR*, *t*, and charging current i_c.

The product CR is termed the *time constant* of a circuit. As will be seen, the time constant is a very important quantity. The Greek letter τ is frequently employed as the symbol for the time constant. For a resistive capacitive circuit, $\tau = CR$. Note that instead of using time directly, the horizontal scale in Figure 2-4(a) could be plotted as τ, 2τ, etc.

Now consider the equation for the instantaneous charging current [Equation 2-4],

$$i_c = I\epsilon^{-t/CR}$$

and again let $t = CR$.

Then

$$i_c = I\epsilon^{-CR/CR} = I\epsilon^{-1}$$

$$= I \times 0.368$$

Thus, when $t = CR$,

$$i_c = I(1 - 0.632) \tag{2-7}$$

Thus, after time $t = CR$, the charging current is reduced by 63.2% of its initial value. [See Figure 2-4(b).] Once again, a time-constant scale could be used instead of a time scale.

Constant-Rate Charging

If the charging current were to remain constant at its initial level, the quantity of charge contained in the capacitor would be

$$Q = I \times t \text{ coulombs}$$

Also

$$Q = C \times V \text{ coulombs}$$

where C is the capacitance (in farads) and V is the capacitor voltage (in volts).

Therefore,

$$It = CV$$

or

$$V = \frac{It}{C} \tag{2-8}$$

It is important to note that *Equation 2-8 applies only to circuits in which the charging current is held at a constant level. It does not apply to the circuit of Figure 2-1(a).* However, this equation can be employed to learn a little more about the time constant CR.

The initial level of charging current in an ordinary resistive capacitive circuit [as in Figure 2-1(a)] is $I = E/R$. If this were to remain constant, then the time for the capacitor to become completely charged could be calculated from Equation 2-8.

$$t = \frac{CV}{I}$$

and, for $V = E$ and $I = E/R$,

$$t = \frac{C \times E}{E/R} = CR$$

As illustrated by the broken line in Figure 2-4(a), with the initial charging current maintained constant, the capacitor would be completely charged in a time period $t = CR$.

Time for Complete Charge

Once again, refer to Figure 2-4(a). It is seen that even at 16 ms, the capacitor is not completely charged to the level of the supply voltage. Theoretically, because the charging current continuously decreases, the capacitor cannot ever become completely charged to the supply voltage level. However, at a period of five time constants, that is, $t = 5 \times CR$, the capacitor is more than 99% charged and, for all practical purposes, can be regarded as completely charged. For the circuit of Figure 2-1(a), after $t = 5 \times CR$ the capacitor voltage is

$$e_c = 10 \text{ V}(1 - \epsilon^{-5CR/CR}) = 9.93 \text{ V}$$

Similarly, it can be shown that i_c reduces to less than 1% of its initial level (I) after a time period of $5 CR$. [See Figure 2-4(b).]

Rise Time and Fall Time

It was pointed out in Chapter 1 that the rise time of a pulse output from a circuit is defined as the time taken for the output to go from 10% to 90% of maximum output level. The CR circuit shown in Figure 2-1(a) has a step input of 10 V applied to it when switch S is closed. It has been shown that the output eventually approaches the maximum level of the step input. The rise time of the output can then be calculated as

$$t_r = (t \text{ at } e_c = 90\% \ E) - (t \text{ at } e_c = 10\% \ E)$$

An expression for t at a given level of e_c can be derived from Eq. 2-2:

$$e_c = E - (E - E_o)\epsilon^{-t/CR}$$

$$(E - E_o)\epsilon^{-t/CR} = E - e_c$$

$$\epsilon^{-t/CR} = \left(\frac{E - e_c}{E - E_o}\right)$$

$$\epsilon^{t/CR} = \left(\frac{E - E_o}{E - e_c}\right)$$

$$\frac{t}{CR} = \ln\left(\frac{E - E_o}{E - e_c}\right)$$

$$t = CR\ln\left(\frac{E - E_o}{E - e_c}\right) \tag{2-9}$$

For e_c = 90% of E, [t_2 on Figure 2-4(a)],

$$t_2 = CR\ln\left(\frac{E-0}{E-0.9\,E}\right)$$

$$\simeq 2.3\;CR \qquad\qquad (2\text{-}10)$$

For e_c = 10% of E, [t_1 on Figure 2-4(a)],

$$t_1 = CR\ln\left(\frac{E-0}{E-0.1\,E}\right)$$

$$t_1 \simeq 0.1\;CR \qquad\qquad (2\text{-}11)$$

Therefore, the rise time of the output is

$$t_r = (t_2 - t_1) = CR(2.3 - 0.1)$$

so that

$$\boxed{t_r = 2.2\;CR} \qquad\qquad (2\text{-}12)$$

When the time is calculated for the capacitor voltage to *fall* from the 90% level to the 10% level, the result is t_f = 2.2 CR. Thus, Eq. 2-12 is applicable to both rise time and fall time.

Equation 2-12 can be applied to any CR circuit when the time constant is known. Also, many circuits and individual devices can be defined as having a particular circuit time constant. Using this circuit time constant, Eq. 2-12 can be employed to calculate the output rise and fall times when a pulse input is applied.

Relationship between f_H and t_r

Consider Figure 2-5(a), which shows the complete equivalent circuit at the input of an amplifier. The components are the signal source resistance R_s, coupling capacitor C_c, amplifier input resistance R_i, and amplifier input capacitance C_i. The voltages are the signal voltage v_s and the circuit output v_o.

When the signal frequency is high, the impedance of the coupling capacitor is very much smaller than the source resistance. Also, the impedance of the input capacitance is very much smaller than the input resistance. Thus,

$$X_{Cc} \ll R_s \text{ and } X_{Ci} \ll R_i$$

Consequently, C_c and R_i can be omitted from the high-frequency equivalent circuit, giving the *low-pass circuit* of Figure 2-5(b).

As long as X_{Ci} remains very much greater than R_s in the high-frequency equivalent circuit, there is very little potential division of the input voltage, and v_o remains approximately equal to v_s. As the impedance of C_i falls with increasing signal frequency, the input voltage is potentially divided across R_s and X_{Ci}, to give an output v_o lower than v_s. The equation for v_o is,

$$v_o = v_s \times \frac{X_{Ci}}{R_s - j\,X_{Ci}}$$

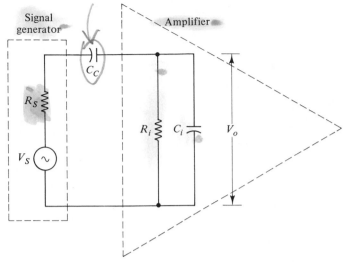

(a) Complete equivalent circuit
of a signal generator capacitor
coupled to an amplifier

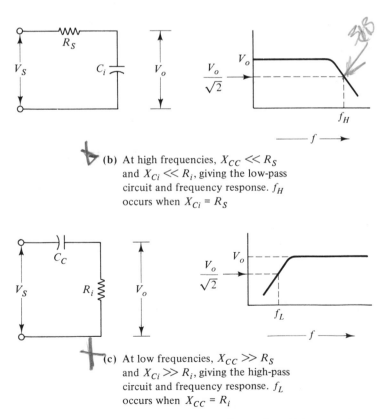

(b) At high frequencies, $X_{CC} \ll R_S$
and $X_{Ci} \ll R_i$, giving the low-pass
circuit and frequency response. f_H
occurs when $X_{Ci} = R_S$

(c) At low frequencies, $X_{CC} \gg R_S$
and $X_{Ci} \gg R_i$, giving the high-pass
circuit and frequency response. f_L
occurs when $X_{CC} = R_i$

Figure 2-5 Input and source capacitance and resistance form high-pass or low-pass
circuits, depending upon the signal frequency. In each case, the cutoff frequency occurs
when $X_C = R$.

or

$$|v_o| = v_s \times \frac{X_{Ci}}{\sqrt{R_s^2 + X_{Ci}^2}}$$

When $X_{Ci} = R_s$,

$$v_o = v_s \times \frac{1}{\sqrt{2}}$$

or

$$v_o = v_s - 3 \text{ dB}$$

Thus, as illustrated by the voltage/frequency graph in Figure 2-5(b), the circuit upper cutoff frequency f_H occurs when $X_{Ci} = R_s$.

Since

$$X_{c_i} = \frac{1}{(2\pi f C_i)}$$

$$R_s = \frac{1}{(2\pi f_H C_i)}$$

or

$$f_H = \frac{1}{2\pi C_i R_S} \qquad (2\text{-}13)$$

As discussed in Chapter 1, the upper cutoff frequency is responsible for rise and fall times in a circuit. Thus, rewriting Eq. 2-13,

$$C_i R_s = \frac{1}{2\pi f_H}$$

Now, substituting for CR in Eq. 2-12,

$$t_r = 2.2 \times \frac{1}{2\pi f_H} = \frac{0.35}{f_H}$$

which is a proof of Eq. 1-5.

Relationship between f_L and Tilt

Now consider the low-frequency equivalent circuit for Figure 2-5(a), as illustrated in Figure 2-5(c). At low frequencies,

$$X_{Cc} \gg R_s \text{ and } X_{Ci} \gg R_i$$

Consequently, C_i and R_s can be omitted at low frequencies, giving the *high-pass circuit* of Figure 2-5(c). In this case,

$$|v_o| = v_s \times \frac{R_i}{\sqrt{R_i^2 + X_{Cc}^2}}$$

When $X_{Cc} = R_i$, $$v_o = v_s \times \frac{1}{\sqrt{2}}$$

or $$v_o = v_s - 3 \text{ dB}$$

As illustrated by the voltage/frequency graph in Figure 2-5(c), the circuit lower cutoff frequency f_L occurs when $X_{Cc} = R_i$, or

$$R_i = \frac{1}{(2\pi f_L C_c)}$$

giving $$f_L = \frac{1}{2\pi C_c R_i} \qquad (2\text{-}14)$$

In Chapter 1, it was explained that the lower cutoff frequency of a circuit produces tilt on pulse waveforms. Referring to Eq. 2-11, it is seen that for a 10% change in capacitor voltage,

$$t = 0.1 \, CR$$

Note that, because the capacitor voltage change does not exceed 10%, the voltage across the resistor remains approximately constant, and consequently, the charging current can be taken as a constant quantity. Equation 2-11 also applies to a 10% drop in voltage from a maximum level. In other words, for a 10% *tilt*, the time, or pulse width, involved is

$$\text{PW} = 0.1 \, CR$$

This can be rewritten as

$$\frac{\text{PW}}{CR} = \frac{0.1 \, E}{E} = \frac{E_3}{E}$$

(see Figure 1-9), or

$$\text{Fractional tilt} = \frac{\text{PW}}{CR} \qquad (2\text{-}15)$$

From Eq. 2-14, $$C_c R_i = \frac{1}{2\pi f_L}$$

Thus, substituting for $C_c R_i$ in the equation for fractional tilt yields

$$\text{Fractional tilt} = 2\pi f_L \text{PW}$$

which is a proof of Eq. 1-7. It should be noted that this equation applies only when the tilt is 10% or less. When the tilt exceeds 10%, the falling voltage is exponential instead of linear. Consequently, Eq. 2-2 would have to be applied.

EXAMPLE 2-5

A 5 V step is switched *on* to a 39 kΩ resistor in series with a 500 pF capacitor. Calculate the rise time of the capacitor voltage, the time for the capacitor to charge to 63.2% of its maximum voltage, and the time for the capacitor to become completely charged.

Solution

From Eq. 2-12, $\qquad t_r = 2.2\, CR$

$$= 2.2 \times 500 \text{ pF} \times 39 \text{ k}\Omega$$

$$= 42.9 \text{ μs}$$

$e_c = 0.632\, E$ at $t = CR$,

so, $\qquad\qquad\qquad t = 500 \text{ pF} \times 39 \text{ k}\Omega$

$$= 19.5 \text{ μs}$$

The capacitor is 99.3% charged at

$$t = 5\, CR = 97.5 \text{ μs}$$

EXAMPLE 2-6

An oscilloscope has a coupling capacitor $C_c = 1$ μF when the input is set to ac, and an input resistance $R_i = 1$ MΩ. Determine the minimum square wave frequency that can be displayed if the tilt is not to exceed 1%.

Solution

From Eq. 2-15, $\qquad \text{PW} = \text{Fractional tilt} \times C_c R_i$

$$= 0.01 \times 1 \text{ μF} \times 1 \text{ M}\Omega$$

$$= 0.01 \text{ s}$$

$$f = \frac{1}{T} = \frac{1}{2\,\text{PW}} = \frac{1}{2 \times 0.01 \text{ s}}$$

$$= 50 \text{ Hz}$$

EXAMPLE 2-7

A pulse generator with an output resistance of $R_s = 600$ Ω is connected to an oscilloscope with an input capacitance of $C_i = 30$ pF. Determine the fastest rise time that can be displayed.

Solution

Eq. 2-12, $t_r = 2.2\, C_i R_s$

$= 2.2 \times 30\ \text{pF} \times 600\ \Omega$

$= 39.6\ \text{ns}$

2-4 CR CIRCUIT RESPONSE TO SQUARE WAVES

A resistive capacitive circuit with an input square wave is shown in Figure 2-6(a). The capacitor voltage first increases from zero to a level e_1 at time t_1. [See Figure 2-6(b).] Between t_1 and t_2 the applied voltage is zero, so the capacitor discharges to e_2 volts. Then the capacitor charges to a new level e_3 at time t_3. To determine the level of e_c at any time greater than t_2, it is necessary first to calculate e_1 at time t_1. Then e_2 must be calculated by using e_1 as the initial voltage on the capacitor and noting that the input voltage is zero from t_1 to t_2. Between t_2 and t_3, the initial voltage is e_2 volts, and the input voltage is again greater than zero.

EXAMPLE 2-8

Calculate the capacitor voltage in the circuit of Figure 2-6(a) at 14 ms from $t = 0$ when the square wave in Fig. 2-6(b) is applied as input.

Solution

Eq. 2-2, $e_c = E - (E - E_o)\epsilon^{-t/CR}$

At $t = 4$ ms,

$$e_c = 20\ \text{V} - (20\ \text{V} - 0)\epsilon^{-4\,\text{ms}/1\,\mu\text{F}\times3.3\,\text{k}\Omega}$$

$$= 14.05\ \text{V}\ [e_1\ \text{in Figure 2-6(b)}]$$

Thus, from $t = 4$ ms to $t = 8$ ms, $E = 0$ V and $E_o = 14.05$ V.
At $t = 8$ ms,

$$e_c = 0 - (0 - 14.05\ \text{V})\epsilon^{-4\,\text{ms}/1\,\mu\text{F}\times3.3\,\text{k}\Omega}$$

$$= 4.18\ \text{V}\ [e_2\ \text{in Figure 2-6(b)}]$$

Thus, from $t = 8$ ms to $t = 12$ ms, $E = 20$ V and $E_o = 4.18$ V.
At $t = 12$ ms,

$$e_c = 20\ \text{V} - (20\ \text{V} - 4.18\ \text{V})\epsilon^{-4\,\text{ms}/1\,\mu\text{F}\times3.3\,\text{k}\Omega}$$

$$= 15.29\ \text{V}\ [e_3\ \text{in Figure 2-6(b)}]$$

Thus, from $t = 12$ ms to $t = 16$ ms, $E = 0$ and $E_o = 15.29$ V. Finally, at $t = 14$ ms,

$$e_c = 0 - (0 - 15.29 \text{ V})\epsilon^{-2\,\text{ms}/1\,\mu\text{F}\times3.3\,\text{k}\Omega}$$

$$= 8.34 \text{ V} \ [e_4 \text{ in Figure 2-6(b)}]$$

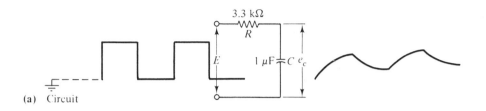

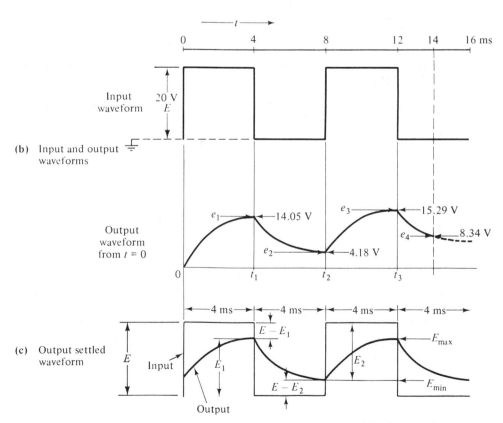

Figure 2-6   CR circuit response to a positive square wave. The capacitor voltage builds up from zero over several cycles of the input, finally reaching maximum and minimum settled levels.

After several intervals of charging, partially discharging, and recharging, the capacitor maximum and minimum voltages will eventually arrive at a settled condition. When this occurs, the capacitor always charges to a maximum voltage level E_{max} and discharges to a minimum level E_{min}, as shown in Figure 2-6(c). These final levels occur when the charging and discharging voltages are equal. Thus, in Figure 2-6(c) $E_1 = E_2$. Also, $E_{max} = E_1$ and $E_{min} = (E - E_{max})$.

Calculating E_{min}, starting from $E_o = E_{max}$ and $E = 0$ V,

$$E_{min} = e_c = 0 - (0 - E_{max})\epsilon^{-t/CR}$$

$$= E_{max}\epsilon^{-t/CR}$$

and since $E_{min} = (E - E_{max})$,

$$E - E_{max} = E_{max}\epsilon^{-t/CR}$$

$$E = E_{max}\epsilon^{-t/CR} + E_{max}$$

$$= E_{max}(\epsilon^{-t/CR} + 1)$$

and

$$E_{max} = \frac{E}{1 + \epsilon^{-t/CR}} \tag{2-16}$$

EXAMPLE 2-9

For the circuit of Figure 2-6(a), determine the maximum and minimum levels at which the capacitor voltage will settle.

Solution

Eq. 2-16,

$$E_{max} = \frac{E}{1 + \epsilon^{-t/CR}}$$

$$= \frac{20 \text{ V}}{1 + \epsilon^{-4\,\text{ms}/(1\,\mu\text{F} \times 3.3\,\text{k}\Omega)}}$$

$$= 15.41 \text{ V}$$

$$E_{min} = E - E_{max}$$

$$= 20 - 15.41 = 4.59 \text{ V}$$

The charging current for the circuit of Figure 2-6(a) may be most easily calculated at any instant as

$$i_c = \frac{E - e_c}{R}$$

Note that during the time intervals when $E = 0$, i_c is a negative quantity: because the capacitor is discharging, the current through R is reversed. The charging current is further discussed in Section 2-6.

2-5 INTEGRATING CIRCUITS

Integration of Pulse Waveforms

Figure 2-7 shows a CR circuit with a square wave input and with the output voltage taken across the capacitor. The shape of the output (capacitor) voltage waveform is dependent upon the relationship between the time constant (CR) and the pulse width (PW).

Consider the case where CR is smaller than PW, [waveform (a) in Figure 2-7]. In Section 2-3, it was demonstrated that the capacitor is charged to 99.3% of the input voltage after time $t = 5\,CR$. Accordingly, let

$$CR = PW/10$$

Then
$$e_c = 99.3\% \text{ of } E \text{ at } t = 5\,(PW/10)$$

That is,
$$e_c \simeq E \text{ at } t = \tfrac{1}{2}\,PW$$

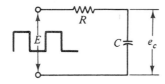

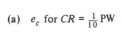

(a) e_c for $CR = \frac{1}{10}\,PW$

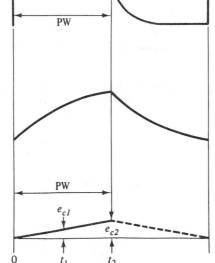

(b) e_c for $CR = PW$

(c) e_c for $CR = 10\,PW$

Figure 2-7 Integrating circuit and output waveforms for various CR and pulse-width relationships. The circuit functions as an integrator only when CR is equal to or greater than 10 PW.

In this case, the output roughly approximates the square wave input. If CR is made smaller than $\frac{1}{10}$ PW, then the output even more closely resembles the square wave input.

For the waveform (b) in Figure 2-7, CR is equal to the pulse width. In Section 2-1, it was shown that the capacitor is charged to 63.2% of the input voltage after time $t = CR$. However, the settled waveform has an amplitude which is less than 63.2% of E. Under these conditions, the waveform of capacitor voltage begins to approach a triangular shape.

When CR is made equal to ten times the pulse width, the waveform of Figure 2-7(c) results. In this case the CR circuit, as arranged, is referred to as an *integrator*. To understand how the circuit integrates, it is necessary to calculate the output voltage levels in relation to time. At $CR = 10 \times$ PW, use Equation 2-2 to obtain

$$e_{c2} = E - (E - E_o)\epsilon^{-\text{PW}/10\,\text{PW}}$$

for $E_o = 0$,
$$e_{c2} = E - E\epsilon^{-1/10}$$

$$\simeq E(1 - 0.9)$$

$$\simeq 0.1\,E$$

To calculate e_{c1} at $\frac{1}{2}$ PW, use

$$e_{c1} = E - E\epsilon^{(-1/2\text{PW})/10\text{PW}}$$
$$= E - E\epsilon^{-1/20}$$

$$\simeq E(1 - 0.95)$$

$$\simeq 0.05\,E$$

This result shows that after time t_1, $e_{c1} \approx 0.05\,E$; and after $t_2 = 2t_1$, $e_{c2} \approx 0.1\,E$. That is, $e_{c2} \simeq 2e_{c1}$ when $t_2 = 2t_1$. [See Figure 2-7(c).] Thus, the capacitor voltage is growing almost linearly. To further examine the output waveform when CR is 10 (or more) times the pulse width, consider Example 2-10.

EXAMPLE 2-10

The circuit shown in Figure 2-8 has the following pulse inputs applied: (a) $E = 10$ V, PW $= 1$ ms; (b) $E = 10$ V, PW $= 2$ ms; (c) $E = 20$ V, PW $= 1$ ms. Calculate the level of e_c at the end of each pulse. The initial charge on C is assumed to be zero.

Solution

Eq. 2-2,
$$e_c = E - (E - E_O)\epsilon^{-t/CR}$$

For input (a),

$$e_{c(a)} = 10\text{ V} - (10\text{ V} - 0\text{ V})\epsilon^{-1\,\text{ms}/(20\,\mu\text{F}\times 10\,\text{k}\Omega)}$$

$$\simeq 50\text{ mV [output (a) in Figure 2-8]}$$

For input (b),

$$e_{c(b)} = 10 \text{ V} - (10 \text{ V} - 0 \text{ V})\epsilon^{-2\,\text{ms}/(20\,\mu\text{F} \times 10\,\text{k}\Omega)}$$

$$\simeq 100 \text{ mV} \text{ [output (b) in Figure 2-8]}$$

For input (c),

$$e_{c(c)} = 20 \text{ V} - (20 \text{ V} - 0 \text{ V})\epsilon^{-1\,\text{ms}/(20\,\mu\text{F} \times 10\,\text{k}\Omega)}$$

$$\simeq 100 \text{ mV} \text{ [output (c) in Figure 2-8]}$$

Since the charging current remains substantially constant during the input PW, this problem can also be solved by using Eq. 2-8:

$$V = \frac{It}{C}$$

$$= \frac{E}{R} \times \frac{t}{C}$$

For input (a),

$$e_{c(a)} = \frac{10 \text{ V} \times 1 \text{ ms}}{10 \text{ k}\Omega \times 20 \text{ }\mu\text{F}} = 50 \text{ mV}$$

For input (b),

$$e_{c(b)} = \frac{10 \text{ V} \times 2 \text{ ms}}{10 \text{ k}\Omega \times 20 \text{ }\mu\text{F}} = 100 \text{ mV}$$

For input (c),

$$e_{c(c)} = \frac{20 \text{ V} \times 1 \text{ ms}}{10 \text{ k}\Omega \times 20 \text{ }\mu\text{F}} = 100 \text{ mV}$$

Example 2-10 shows that when the pulse width is doubled, the output voltage is doubled. Because the charging rate is very nearly linear for $CR \geqslant 10$ PW, the output amplitude is proportional to the pulse width. Also, when the pulse amplitude is doubled, the output voltage is doubled. In this case, the charging rate is increased in proportion to the input voltage. Thus, the output voltage is proportional to the *pulse area*, that is, to the product of the pulse width (PW) and the pulse amplitude (PA):

$$e_c \propto \text{PA} \times \text{PW}$$

Figure 2-8 and Example 2-10 show that the output voltage from an integrating circuit is proportional to the area of the pulse expressed as (volts × time). An *integrating circuit* is a *CR* circuit with the output taken across the capacitor and $CR \geqslant (10 \times \text{PW})$. In other integrator circuits discussed in Chapters 5 and 9, the capacitor charging current is held constant by the use of additional components.

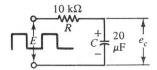

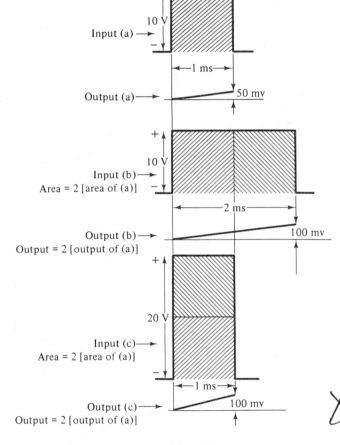

Figure 2-8   Integration of pulses with different amplitudes and widths. In all cases the final output (capacitor) voltage is directly proportional to the area under the pulse.

Integration of a Sine Wave

It will be demonstrated that the integrator output is zero when the sine wave input is at a peak level. Therefore, although a sine wave commences at zero and goes positive, the instant at which the peak level commences to go negative is taken as a starting point to determine the output waveform.

The sine wave input in Fig. 2-9 is divided into sections of equal widths. The height of each section corresponds approximately to the instantaneous sine wave amplitude. Thus, the sine wave is represented by a series of pulses of varying amplitudes.

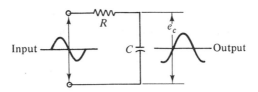

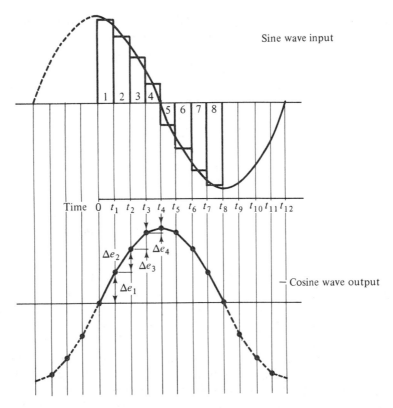

Figure 2-9   Sine wave divided into equal width pulses to show how it is integrated to give a negative cosine waveform.

The first pulse causes a linear increase in capacitor voltage from time 0 to t_1. This produces output voltage Δe_1. The second pulse, from t_1 to t_2, also produces a linear increase in the capacitor voltage. However, the pulse amplitude is now smaller, so the rate of increase in capacitor voltage is reduced. Thus, Δe_2 is less than Δe_1. Similarly, the third and fourth pulses produce linear voltage increases at decreasing rates. Since pulse 5 is negative, it causes the capacitor voltage to decrease by a small amount. Pulse 5 and pulse 4 are equal in amplitude, so the decrease in e_c due to pulse 5 is equal to the increase Δe_4 produced by pulse 4. Also, negative pulses 6, 7, and 8 linearly decrease the output (capacitor) voltage. Extending the waveforms (broken lines), we see that integration of the sine wave input produces a negative cosine wave output.

2-6 DIFFERENTIATING CIRCUITS

Differentiation of Pulse Waveforms

When the output from a CR circuit is taken across R, the output voltage is the differential of the input. As in the case of the integrating circuit, the relationship between CR and the puse width is important. Figure 2-10 shows the various output waveforms that can occur, depending upon PW and CR.

The voltage across R is the product of the charging current and the resistance, that is, $e_R = i_c \times R$. When the time constant CR is 10 times the pulse width (or greater), the capacitor charges very little during the pulse time, and the charging current falls only a small amount from its initial level. [See waveform (a) in Figure 2-10.] Thus, e_R remains almost constant during the PW. During the space width the capacitor is discharged, and i_c is a negative quantity. The resistor voltage is now negative and, again, remains nearly constant for the discharge time.

If CR is made equal to the pulse width, the capacitor is charged to approximately 60% of the input voltage during the pulse time. Consequently, the charging current falls by about 60% of its initial value, giving an output waveform with a very pronounced tilt. [See the waveform in Figure 2-10(b).]

When CR is less than one-tenth of the pulse width, the capacitor is charged very rapidly. Only a brief pulse of current is necessary to charge and discharge the capacitor at the beginning and end of the pulse. The resultant waveform of the resistor voltage is a series of positive and negative spikes at the pulse leading and lagging edges, respectively. [See Figure 2-10(c).] The *differential* of a quantity is a measure of the rate of change of the quantity. At the leading edge of the input pulse, the input voltage is changing rapidly in a positive direction. At the trailing edge of the pulse, the input voltage is changing rapidly in a negative direction. During both the pulse width and the space width, the input voltage does not change at all. Thus, it is seen that the positive and negative spikes with intervening spaces [Fig. 2-10(c)] do indeed represent a differentiated square wave.

When a ramp voltage is applied to the input of a differentiating circuit, the resultant output is a constant dc voltage level (Figure 2-11). While the input voltage continuously increases, the capacitor cannot become completely charged. Hence, the instantaneous capacitor voltage is always slightly less than the instantaneous input voltage. This small difference in E and e_c is developed across R, giving a constant level of charging current, and thus a constant level of e_R. While the ramp increases positively, the capacitor is charged with the polarity, positive on the left, negative on the right, as illustrated, and i_c produces a positive level of e_R. When the ramp goes negative, i_c is reversed, and consequently, e_R is negative.

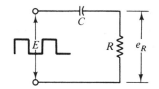

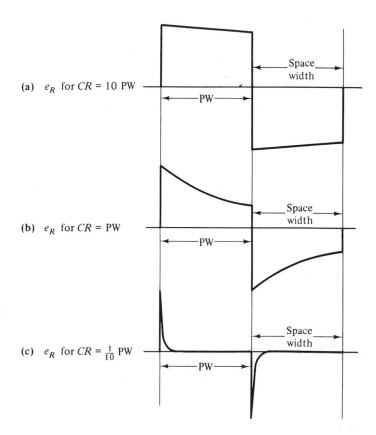

Figure 2-10   Differentiating circuit and output waveforms for various CR and pulse-width relationships. The circuit functions as a differentiator only when CR is equal to or less than PW/10.

EXAMPLE 2-11

For input waveforms (a) and (b) shown in Figure 2-11, calculate the levels of the outputs from the differentiating circuit.

Solution   At the end of the input ramp, $e_c \simeq E_{max}$. Since the charging current is constant, Eq. 2-8 may be applied.

(a) For the 10 V ramp,

$$I = \frac{CE_{max}}{t} = \frac{1 \ \mu F \times 10 \ V}{100 \ ms} = 0.1 \ mA$$

and $e_R = I \times R = 0.1 \ mA \times 1 \ k\Omega = 0.1 \ V$ [output (a) in Figure 2-11]

(b) For the 20 V ramp,

$$I = \frac{CE_{max}}{t} = \frac{1 \ \mu F \times 20 \ V}{100 \ ms} = 0.2 \ mA$$

and $e_R = I \times R = 0.2 \ mA \times 1 \ k\Omega = 0.2 \ V$ [output (b) in Figure 2-11]

In Example 2-11, the rate of change of the 10 V ramp is 10 V/100 ms, that is, 0.1 V/ms. For the 20 V ramp, the rate of change is 0.2 V/ms. Thus, the differentiated output doubles when the rate of change of input voltage is doubled.

From the example, an equation for the output from a differentiating circuit with a ramp input is

$$e_R = CR \times \frac{\Delta E}{t} \qquad\qquad (2\text{-}17)$$

That is, $e_R = CR \times$ (rate of change of input). This equation may be applied in the case of pulse waveforms with known rise and fall times, as illustrated in Figure 2-12 and Example 2-12.

EXAMPLE 2-12

Calculate the amplitude of the differentiated output waveform for the circuit and input pulse shown in Figure 2-12.

Solution
From Eq. 2-17

$$e_{R1} = CR \times \frac{\Delta E}{t_r} = 100 \ pF \times 1 \ k\Omega \times 8 \ V/\mu s$$

$$= 0.8 \ V$$

$$e_{R2} = CR \times \frac{\Delta E}{t_f} = 100 \ pF \times 1 \ k\Omega \times (-8 \ V/3 \ \mu s)$$

$$= -0.27 \ V$$

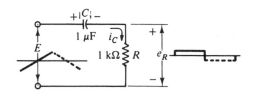

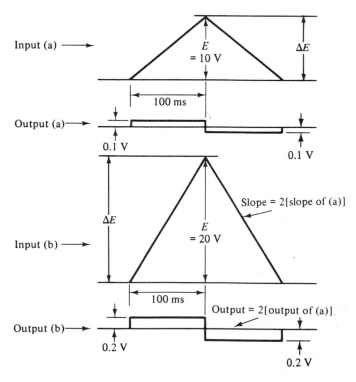

Figure 2-11   Differentiation of ramp voltages with different rates of change. In each case the circuit output (resistor) voltage is directly proportional to the rate of change of the ramp voltage.

Referring again to Eq. 2-17, it is seen that for the output voltage e_R to be a maximum, the rise time t_r must be equal to or smaller than CR. This gives $e_R = \Delta E$. Note that e_R cannot exceed the input voltage level. Sometimes a differentiator is used as a means of generating positive and negative spikes from a square wave input. In this case, the time constant is usually determined by just making CR much smaller than the input pulse width. The spike output waveform that results normally has a peak-to-peak amplitude equal to twice the peak-to-peak amplitude of the input. Consider,

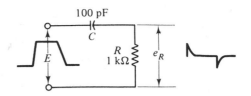

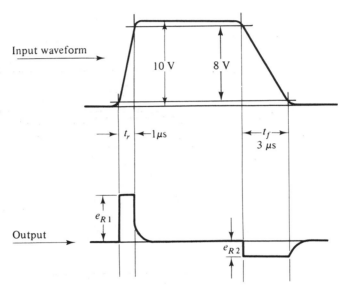

Figure 2-12 Differentiation of a pulse waveform with different rise and fall times. The circuit output is directly proportional to the rate of change of the input.

for example, Figure 2-13, which shows a differentiating circuit with a square wave input having an amplitude of $\pm E$. When the input is $+E$, the capacitor charges to E volts with the polarity shown. When the input goes to $-E$, the output spike amplitude is the sum of the input and the capacitor voltages, i.e., $-2\,E$. Similarly, while the input is negative, the capacitor charges with the opposite polarity. Consequently, when the input goes positive again, the output is $+2\,E$.

Differentiation of a Sine Wave

The process by which a sine wave is differentiated is illustrated in Figure 2-14. Although the input sine wave commences at zero volts and goes positive, the output waveform is investigated by starting at the positive peak of the input.

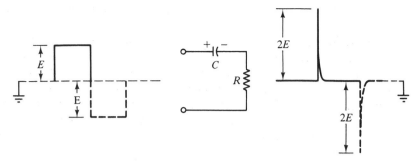

Figure 2-13 Differentiation of a symmetrical square wave with fast rise and fall times.

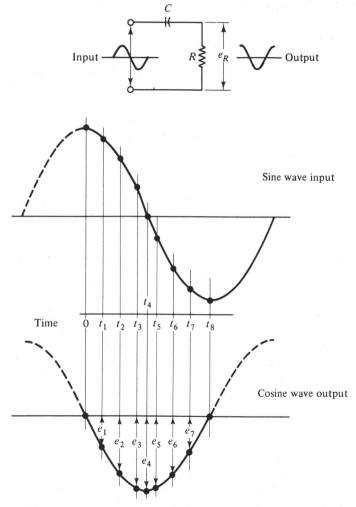

Figure 2-14 When a sine wave is differentiated, each instantaneous output level is directly proportional to the instantaneous rate of change of the input. This results in a cosine wave output.

At the peak of the sine wave ($t = 0$), the rate of change of the voltage is zero. Thus, the differentiated output voltage is zero. At time t_1 the sine wave amplitude is decreasing, producing a negative rate of change. Consequently, the differentiated output voltage is $-e_1$. At t_2 and t_3 the negative rate of change increases and finally becomes maximum at t_4. The differentiated output, therefore, increases negatively through $-e_2$ and $-e_3$ to $-e_4$. Beyond t_4, the negative rate of change decreases progressively to zero at t_8. Extending the waveform (broken lines) shows the differential of a sine wave to be a cosine wave.

2-7 SOURCE AND LOAD EFFECTS AND CAPACITOR POLARITY

Resistive and Capacitive Sources and Loads

For all circuitry considered so far, it was assumed that the signal source impedance and the circuit load impedance had no effect on the circuit performance. This is not always the case.

Consider the integrator and differentiator circuits in Figures 2-15(a) and (b), which include signal source resistance R_s and coupling capacitor C_c. In both circuits, where R_s is very much smaller than R, R_s may be neglected in circuit calculations. In general, if $R_s = 10\%$ of R, then there will be a 10% error in the output. Where $R_s = 1\%$ of R, a 1% output error may be expected.

Where a coupling capacitor C_c is employed, it should be very much larger than the circuit capacitor C, in both integrator and differentiator circuits. Another way of stating this is that the impedance of C_c (at any frequency) must be very much smaller than C if C_c is not to significantly affect circuit performance. As in the case of the source resistance and circuit resistance, the relationship between the capacitor impedances gives the output error that may be expected. Where X_{Cc} is 10% of X_c, a 10% error is likely.

Now refer to Figures 2-15(c) and (d), in which integrating and differentiating circuits with load resistances R_L and load capacitances C_L are illustrated. In the case of the integrator circuit, C_L in parallel with C simply adds to the circuit capacitance. (Total capacitance equals the sum of the two capacitances in parallel.) Therefore, for C_L to have a negligible effect on circuit performance, it must be very much smaller than C. In this case, if $C_L = 10\%$ of C, a 10% error may be expected in the integrator output.

Neglecting C_L, and assuming zero source impedance in Figure 2-15(c), the resistance seen from the terminals of capacitor C is R_L in parallel with R. Consequently, the presence of R_L has the effect of reducing the total circuit resistance. R_L and R also operate as a potential divider to attenuate the signal. To minimize these effects, R_L should be very much larger than R. With $R_L = 100\,R$, a 1% error may be expected in the output. When $R_L = 10\,R$, the error should be approximately 10%.

In the differentiating circuit in Figure 2-15(d), load resistance R_L is seen to be in parallel with R. Here again, the total circuit resistance is reduced, and R_L should be

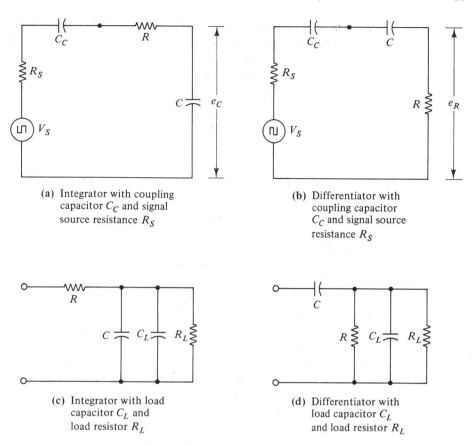

(a) Integrator with coupling
capacitor C_C and signal
source resistance R_S

(b) Differentiator with
coupling capacitor
C_C and signal source
resistance R_S

(c) Integrator with load
capacitor C_L and
load resistor R_L

(d) Differentiator with
load capacitor C_L
and load resistor R_L

Figure 2-15   Source resistance R_s and coupling capacitance C_c can affect the performance of integrator and differentiator circuits. Load resistance R_L and load capacitance C_L can also affect circuit performance.

very much larger than R in order to have minimum effect on the circuit. As before, a value of $R_L = 100 R$ may be expected to give a 1% error.

Looking from the (output) terminals of R in Figure 2-15(d), and again assuming zero source impedance, load capacitor C_L appears in parallel with C. Since parallel capacitances add together to give the total capacitance, C_L should again be very much smaller than C if the load capacitance is not to have a measurable effect on the circuit. This relationship can also be stated in terms of capacitor impedances as $X_{CL} \gg X_c$. Once again, when $C_L = 1\%$ of C, a 1% error is likely to result.

To summarize, series-connected additional components should have impedances very much smaller than the circuit component impedances; that is,

$$R_s \ll R \text{ and } X_{Cc} \ll X_c$$

Parallel-connected additional components should have impedances very much larger than the circuit component impedances; that is,

$$R_L \gg R \text{ and } X_{CL} \gg X_c$$

Oscilloscope Input C and R

Every oscilloscope has an input capacitance C_i and input resistance R_i, as illustrated in Figure 2-16(a). In addition, when an oscilloscope input is switched to ac, a coupling capacitor C_c is introduced. The coupling capacitor usually has a value of 1 μF, and because its impedance is very much smaller than R_i from low to very high frequencies, its only effect is to remove any dc component present in the input waveform. However, at very low frequencies, X_{Cc} is not very much larger than R_i. Consequently, as discussed in Section 2-3, C_c introduces tilt on pulse waveforms, as well as attenuation and phase shift on sinusoidal inputs.

A typical oscilloscope input capacitance and resistance is $C_i = 30$ pF and $R_i = 1$ MΩ. These quantities behave as loads on all circuits connected to the oscilloscope. Thus, they have the same effects as C_L and R_L discussed above. Additional parallel capacitance is introduced by the coaxial cable normally used on an oscilloscope probe, to connect to the circuit being tested. To minimize the input capacitance and resistance, an *oscilloscope 10:1 probe* is employed.

The circuit of an oscilloscope 10:1 probe is illustrated in Figure 2–16(b). A 9 MΩ resistor R_1 and an adjustable capacitor C_1 are connected as shown. The input signal is now attenuated by the resistive potential divider R_1 and R_i, and by the capacitive

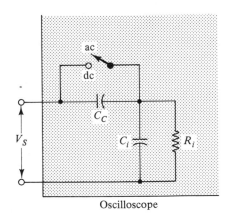

(a) Oscilloscope input capacitance, resistance, and coupling capacitor

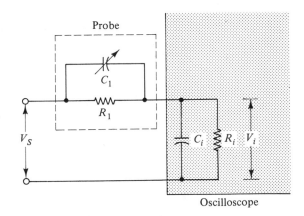

(b) Oscilloscope input with probe

Figure 2-16   An oscilloscope has input capacitance C_i and input resistance R_i; when switched to ac, a coupling capacitor C_c is also included. In an oscilloscope 10:1 probe, a resistor R_1 and adjustable capacitor C_1 compensate for the effects of C_i.

potential divider C_1 and C_i. Analysis of the circuit reveals that, when $X_{ci}/X_{c1} = R_i/R_1$, the resistive and capacitive potential divider effects are equal at all frequencies, and no distortion is introduced. The input resistance offered by the probe and oscilloscope is 10 MΩ, and the input capacitance is approximately $C_i/10$.

Transistor Circuits as Loads

One loading problem that occurs frequently is the case of a transistor which is to be switched *on* by the output of a differentiating circuit. Figure 2-17(a) illustrates the situation. When the transistor is not connected, the differentiated waveform developed across R_1 consists of positive-going and negative-going spikes. With Q_1 connected, the voltage across R_1 cannot exceed the V_{BE} of the transistor. (This is typically 0.7 V for a

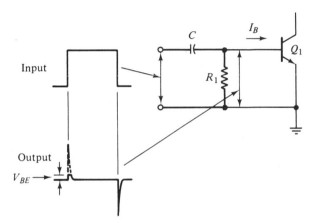

(a) Positive spike clipped by transistor

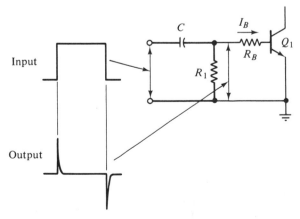

(b) $R_1 \ll R_B$, waveform not clipped

Figure 2-17 An *npn* transistor switching circuit may clip positive output spikes from a differentiating circuit, unless an additional resistor is included in series with its base.

silicon transistor, and 0.2 V for a germanium device.) Of course, this affects only the positive spikes when an *npn* transistor is used. During the negative spikes the device is *off* and the spike amplitude is unaltered. Whenever a *pnp* transistor is involved, the reverse is true: negative spikes are clipped, positive spikes unaltered.

If it is intended that the transistor should be triggered *on* at the edge of the input pulse, then the spike-clipping may not be important. However, when the spikes are not to be clipped, the circuit must be modified slightly. Figure 2-17(b) illustrates the necessary modification. Resistor R_B limits the transistor base current to any desired level and allows the voltage across R_1 to climb to the required amplitude.

R_B is calculated as explained in Chapter 4. The total input resistance offered by the transistor and R_B is now

$$R_i = R_B + \frac{V_{BE}}{I_B}$$

The differentiating circuit resistance (during positive pulses when an *npn* transistor is used) now becomes

$$R = R_1 \| R_i$$

During negative-going spikes Q_1 is biased *off*, I_B ceases to flow, and R becomes equal to R_1. If R_1 is made much smaller than R_i, R is always approximately equal to R_1, and the transistor has a negligible effect on the performance of the differentiating circuit.

The same kind of loading effect may present a problem with an integrating circuit. If the transistor is directly connected, the output waveform may be clipped. [See Figure 2-18(a).] Again, the solution is to employ a base resistor with the transistor—R_B in Figure 2-18(b). In this case, the capacitor current I must be made very much larger than I_B for the transistor to have a negligible effect on the performance of the integrating circuit.

Capacitor Polarity

Some capacitors (notably those that are electrolytic) are designed to operate with one terminal more positive than the other. These are termed *polarized capacitors*. In the capacitor circuit symbol, the curved line represents the negative terminal and the straight line represents the positive terminal. If a polarized capacitor is incorrectly connected, a large leakage current may flow, and this can seriously affect the bias conditions of a circuit. More important, an incorrectly connected polarized capacitor *may explode*.

The dc voltage levels in a circuit dictate capacitor polarity. Consider Figure 2-19(a). The voltage at the junction of R_1 and R_2 is clearly more positive than ground. Also, the left side of the capacitor is grounded via the signal generator. Therefore, the capacitor must be connected as illustrated.

In the circuit of Figure 2-19(b), the transistor collector terminal is usually more positive than the junction of R_2 and R_3, and the capacitor must then be connected as

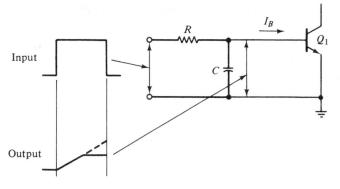

(a) Output wave clipped

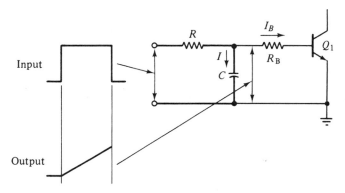

(b) $I \gg I_B$, waveform not clipped

Figure 2-18 An *npn* transistor switching circuit may clip off a portion of a positive ramp, unless measures are taken to minimize the loading effect when the transistor switches on.

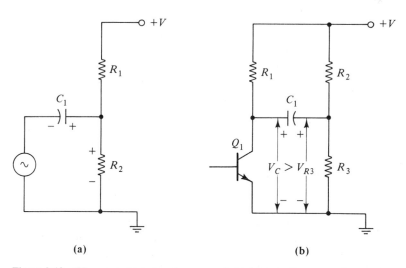

(a)

(b)

Figure 2-19 Where polarized capacitors are used, it is important to connect the capacitor with the correct terminal polarity. The positive terminal must not be allowed to go negative with respect to the negative terminal.

shown. However, this might not always be the case, so the actual voltage levels should be determined when the circuit is designed.

Now look at the circuits in Figure 2-17. The transistor base is grounded via R_1, and the input is a purely positive pulse. Clearly, then, the capacitor polarity should be positive to the signal source, negative to the grounded base. If the input waveform had a negative half cycle, then the capacitor terminal polarity would be reversed during the negative portion of the input. In this circumstance, and in all situations where the terminal polarity can be reversed, an unpolarized capacitor must be employed.

REVIEW QUESTIONS AND PROBLEMS

2-1. A capacitor C is charged from a voltage source E via a resistance R. The general formula for the capacitor voltage is

$$e_c = E - (E - E_o)\epsilon^{-t/CR}$$

(a) Derive an expression for the time required for the capacitor voltage to go from 10% to 90% of its maximum level. (b) Derive expressions for e_c when $t = CR$ and when $t = 5$ CR.

2-2. The CR circuit shown in Figure 2-8 has a 20 V dc voltage input. Plot the graphs of e_c and e_R *versus* time.

2-3. A 10 µF capacitor is charged via a 10 kΩ resistance from a 5 V source. If the capacitor has an initial charge of -2 V, calculate its charge after 50 ms.

2-4. For Problem 2-3, determine the level of charging current after 35 ms.

2-5. For Problem 2-3, calculate the time required for the capacitor to charge to 4.5 V.

2-6. A 4.7 µF capacitor is to be charged from a 1 mA constant-current current source. Calculate the capacitor voltage at 10 ms and at 17 ms from commencement of charge.

2-7. A 1000 µF capacitor is to be charged from zero to $e_c = 3$ V over a time period of 1 hour. Determine the required level of constant charging current.

2-8. Sketch a low pass CR circuit and its voltage/frequency response graph. Derive the equation relating upper cutoff frequency f_H and rise time t_r at the output when an input step is applied.

2-9. Sketch a high-pass CR circuit and its voltage/frequency response graph. Derive the equation relating lower cutoff frequency f_L and tilt on the output when an input pulse waveform is applied.

2-10. An amplifier has an input coupling capacitor of 30 µF and an input resistance of 20 kΩ. Determine the lowest square wave frequency that can be applied as an input if the output tilt is not to exceed 5%.

2-11. An oscilloscope displays a 5 Hz square waveform with a 6% tilt. The signal input has no tilt and is coupled to the oscilloscope via a 4.7 µF capacitor. Calculate the oscilloscope input resistance.

2-12. A pulse waveform displayed on an oscilloscope has a rise time of 100 ns. The waveform, as derived from a pulse generator, has a rise time very much less than 100 ns. If the pulse generator output resistance is 1 kΩ, determine the input capacitance of the oscilloscope.

2-13. A 10 V step is switched on to a 22 kΩ resistor in series with a 300 pF capacitor. Calculate the rise time of the capacitor voltage, the time for the capacitor to charge to 63.2% of its maximum voltage, and the time for the capacitor to become completely charged.

2-14. If the square wave input to the circuit shown in Figure 2-6(a) has a frequency of 250 Hz and an amplitude of 15 V, determine the capacitor voltage at $t = 7$ ms.

2-15. For Problem 2-14, determine the maximum and minimum levels of capacitor voltage when the waveform has settled. Sketch the waveform of e_c to show its relationship to the input square wave.

2-16. For Problem 2-14, determine the maximum and minimum levels of charging current when the waveform has settled. Sketch the waveform of i_c to show its relationship to the e_c waveform.

2-17. A 1000 pF capacitor is charged from a 5 V source via a 4.7 kΩ resistor. Using the normalized charge and discharge graphs in Figure 2-3, determine (a) the time required for the capacitor to charge to 3.5 V, (b) the capacitor voltage 15 μs after commencing to discharge from its fully charged level.

2-18. For the circuit described in Problem 2-13, use the normalized charge and discharge graphs in Figure 2-3 to determine (a) the capacitor voltage 10 μs after $e_c = 0$ V, (b) the time required for the capacitor to charge to $e_c = 9$ V.

2-19. For the circuit and input shown in Figure 2-20, (a) determine the levels of e_c and i_c at $t = 2.5$ ms. (b) Sketch the settled waveform of e_c.

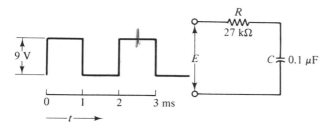

Figure 2-20 Circuit and input for Problem 2-19.

2-20. Sketch an integrating circuit with a square wave input. Show the output waveforms for (a) $CR \simeq 10 \times \text{PW}$, (b) $CR \simeq \frac{1}{10} \text{PW}$, (c) $CR \simeq \text{PW}$.

2-21. Explain why the output of an integrating circuit represents the integration of the input waveform.

2-22. Sketch the shape of the output waveform from an integrator when the input is a cosine wave.

2-23. A pulse having an amplitude of 5 V and a PW of 100 μs is applied to the CR circuit shown in Figure 2-20. Determine the amplitude of e_c at the end of the pulse. If the input PW goes to 150 μs and the amplitude goes to 7.5 V, calculate the new level of e_c at the end of the pulse time.

2-24. Sketch a differentiating circuit with a square wave input. Show the waveforms for (a) $CR \simeq \frac{1}{10} \text{PW}$, (b) $CR \simeq 10 \times \text{PW}$, (c) $CR \simeq \text{PW}$.

2-25. Explain why the output of a differentiating circuit represents the differential of the input waveform.

2-26. Sketch the shape of the output waveform from a differentiator when the input is a cosine wave.

2-27. A 100 Hz triangular wave with a peak-to-peak amplitude of 9 V is applied to a differentiating circuit. $R = 1$ MΩ and $C = 100$ pF. Calculate the output amplitude, and sketch the waveform of the output.

2-28. A pulse with a rise time $t_r = 500$ ns, a fall time $t_f = 1$ μs, PA = 12 V, and PW = 10 μs is applied to a differentiating circuit with $C = 200$ pF and $R = 470$ Ω. Determine the amplitude of the differentiated outputs, and sketch the output waveform.

2-29. Discuss the loading problems that occur with differentiating and integrating circuits, and explain how they may be overcome.

Chapter 3

Diode Switching

INTRODUCTION

Because it passes a large current when forward biased and an extremely small current when reverse biased, a semiconductor diode can be employed as a switch. The speed with which a diode can be switched is determined by the *reverse recovery time* of the device. Diodes are widely applied to *clip* unwanted portions from a waveform, or to *clamp* the peak of a waveform at a desired dc level. Zener diodes may be used as reference voltage sources in clipping and clamping circuits.

3-1 THE DIODE AS A SWITCH

Diode Characteristics

Typical characteristics for a low-current silicon diode are shown in Figure 3-1. It is seen that when the *forward bias voltage* V_F is approximately 0.7 V, the *forward current* I_F is approximately 10 mA. Above the 10 mA level, I_F increases substantially with very small increases in V_F. The reverse characteristics show an approximately constant *reverse leakage current* I_s of 0.05 μA for *reverse voltages* ranging up to *reverse breakdown* at 75 V. (Some diodes can survive reverse biases much greater than 75 V.) Since the typical reverse current is on the order of 1/200,000 of the forward current, the reverse leakage current can be neglected for many purposes. The diode is then thought of as a *one-way device*. It stimulates a switch, which is closed when the device is forward biased and open when it is reverse biased.

A switch is characterized by a zero voltage drop when it is closed and a zero current when it is open. This is also true of an ideal diode. Figure 3-2(a) illustrates the

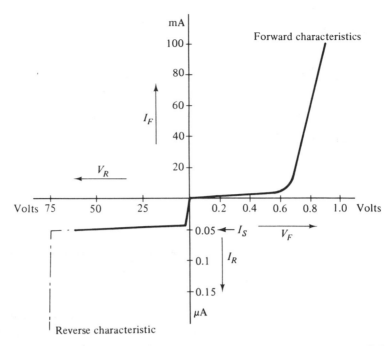

Figure 3-1 Forward and reverse characteristics for a typical low-current silicon diode. Note that the reverse leakage current is very much smaller than forward current levels beyond $V_F \approx 0.7$ V. Thus, the diode can be treated as a one-way device.

characteristics of a switch, and Figures 3-2(b) and (c) show similar approximate diode characteristics. The silicon device has a typical forward voltage drop of 0.7 V, while V_F for the germanium diode is approximately 0.3 V. For both silicon and germanium, the reverse current is normally extremely small.

In a great many applications, diode characteristics can be ignored. The device is assumed to have a constant forward voltage drop V_F when forward biased, and a constant (temperature-dependent) *reverse leakage current* I_S when reverse biased. To select a diode for a particular application, it is necessary to determine the forward current that must be passed, the power dissipation, the reverse voltage, and the maximum reverse leakage current that can be tolerated. Another item that must be considered is the required operating frequency of the diode.

Diode Switching

The effect of a sudden change from forward bias to reverse bias is illustrated in Figure 3-3(a). Instead of switching *off* sharply when the input becomes negative, the diode initially conducts in reverse. The reverse current I_R is at first equal to I_F; then it falls off to the reverse leakage current level I_S. At the instant of reverse bias there are charge carriers crossing the junction depletion region, and these must be removed. This removal of charge carriers constitutes the reverse current I_R. The *reverse recovery time* t_{rr} is the

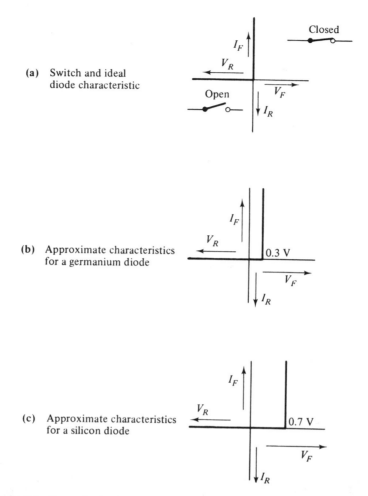

(a) Switch and ideal
     diode characteristic

(b) Approximate characteristics
     for a germanium diode

(c) Approximate characteristics
     for a silicon diode

Figure 3-2 Except for the device forward voltage drop, the approximate characteristics
of a diode are similar to the characteristics of a switch.

time required for the reverse current to fall to I_S. Typical values of t_{rr} for switching
diodes range from 4 ns to 50 ns.

   If the diode forward current is reduced to a very small level before the device is
reverse biased, then the initial level of reverse current will also be very small. Even
when there is a large forward current, the reverse current can be kept small if the
forward current is reduced slowly. This means that for minimum reverse current, the
fall time of forward current should be much longer than the diode reverse recovery
time [see Figure 3-3(b)].

   When a diode is switched *on* (from reverse bias to forward bias), there is a finite
turn-on time. However, the turn-on time is so much smaller than the reverse recovery
time, that it is usually neglected.

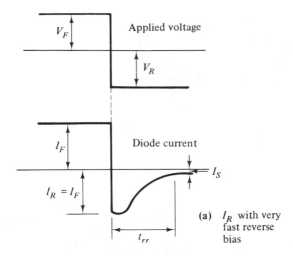

(a) I_R with very fast reverse bias

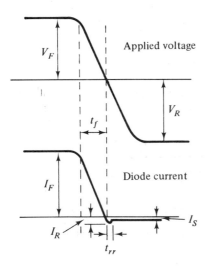

(b) I_R minimized by making $t_f \gg t_{rr}$

Figure 3-3 A diode switched from *on* to *off* conducts in reverse for a time known as the reverse recovery time t_{rr}. If the fall time of the forward current is made much larger than t_{rr}, the reverse current can be minimized.

3-2 THE ZENER DIODE

The characteristics of a *Zener diode* are shown in Figure 3-4. This device is a semiconductor diode designed to operate in the *reverse breakdown region* of its characteristics. If the reverse current is maintained within certain limits, the voltage drop across the diode remains at a reliable constant level. Thus the Zener diode (also known as an *avalanche diode* or *breakdown diode*) is useful as a voltage reference source.

From Figure 3-4, V_Z is the *Zener voltage* measured at *test current* I_{ZT}. The *knee current* I_{ZK} is the minimum current that should pass through the device to keep it in breakdown, and thus maintain a constant V_Z. The maximum Zener current that may be

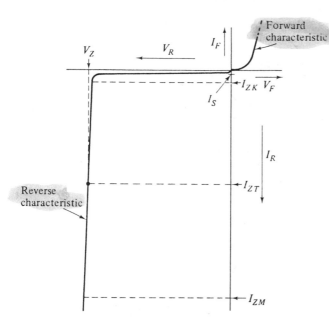

Figure 3-4 Zener diode characteristics. The reverse characteristics of a Zener diode show that the breakdown voltage V_Z is constant over a wide range of reverse current levels. The forward characteristics are similar to those of an ordinary low-current diode.

passed is designated as I_{ZM}. This maximum current level ensures that the device power dissipation does not exceed a maximum safe level.

For correct operation, the Zener diode voltage must be positive on the cathode and negative on the anode; that is, the diode must be reverse biased. When the reverse voltage is smaller than V_Z, only the normal diode *reverse leakage current I_S* flows. *When the Zener diode is forward biased, it behaves as an ordinary diode*; a large forward current flows, and the diode voltage is typically $V_F = 0.7$ V.

Zener diode specifications are listed on the data sheet in Appendix 1-3. Zener breakdown voltages range from 3.3 V for the 1N746 to 12 V for the 1N759. It should be noted that these voltages have a ±10% tolerance. The 1N758, for example, has a V_Z ranging from 9 V to 11 V. Devices with ±5% voltage tolerances are available.

3-3 DIODE SERIES CLIPPER CIRCUITS

Series Clipper

A *clipper* (or *limiter*) *circuit* is one that clips off a portion of an input waveform. Two clipping circuits are shown in Figure 3-5. These are essentially half-wave rectifier circuits. In the *negative series clipper* [Fig. 3-5(a)], the diode is forward biased when the input becomes positive. Thus, the output voltage at this time is the peak input voltage minus the diode voltage drop. When the input becomes negative, the diode is reverse biased and the reverse leakage current I_S flows through resistor R_1. The output then is a very small negative voltage $-(I_S \times R_1)$. The resultant output waveform from the circuit is essentially the input with the negative portion clipped off.

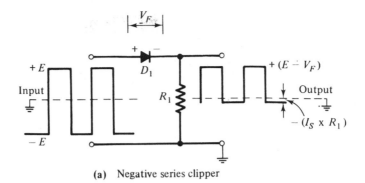

(a)   Negative series clipper

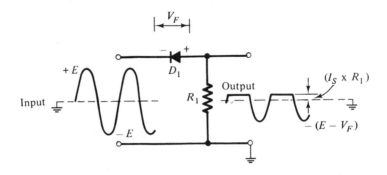

(b)   Positive series clipper

Figure 3-5   Negative and positive series clipping circuits. These are essentially half-wave rectifier circuits.

The *positive series clipper* operates in the same way as the negative clipper except that, in this case, the diode is turned around, and the positive portion of the input waveform is clipped off. The input waveform may be square, sinusoidal, or any other shape.

EXAMPLE 3-1

The negative series clipping circuit in Figure 3-5(a) is to have an input of $E = \pm 50$ V. The output current from the circuit is to be $I_L = 20$ mA, and the negative output voltage $-V_o$ is not to exceed 0.5 V. Calculate the value of R_1, and specify the diode in terms of forward current, power dissipation, and peak reverse voltage.

Solution   For a negative input,

$$I_S = 5 \ \mu\text{A (maximum likely; see Appendix 1-1)}$$

$$-V_o = I_S \times R_1$$

$$R_1 = \frac{V_o}{I_S} = \frac{0.5 \text{ V}}{5 \text{ μA}} = 100 \text{ k}\Omega$$

Use a 100 kΩ standard value (see Appendix 2).
Diode peak reverse voltage,

$$-E = -50 \text{ V}$$

For a positive input,

$$I_{R1} = \frac{E - V_F}{R_1} = \frac{50 \text{ V} - 0.7 \text{ V}}{100 \text{ k}\Omega}$$

$$\simeq 0.5 \text{ mA}$$

Power dissipation in R_1,

$$P_{R1} = \frac{E^2}{R} = \frac{(50 \text{ V})^2}{100 \text{ k}\Omega}$$

$$= 25 \text{ mW}$$

Diode forward current,

$$I_F = I_L + I_{R1}$$

$$= 20 \text{ mA} + 0.5 \text{ mA}$$

$$= 20.5 \text{ mA}$$

Diode power dissipation,

$$P_{D1} = V_F \times I_F$$

$$= 0.7 \text{ V} \times 20.5 \text{ mA (for a silicon diode)}$$

$$= 14.35 \text{ mW}$$

EXAMPLE 3-2

From the diode data sheets in Appendix 1, select a suitable device for the circuit designed in Example 3-1.

Solution   From Example 3-1,

Reverse leakage current $I_S \leqslant 5$ μA

Peak reverse voltage $E \geqslant 50$ V

Forward current $I_F \geqslant 20.5$ mA

The diode selected must have a reverse current I_R *not greater than* 5 μA, a maximum reverse voltage V_R *not less than* 50 V, and a maximum forward current I_F *not less than* 20.5 mA. The 1N914, 1N915, and 1N916 diodes fulfill all of the required conditions. The 1N917 can withstand a maximum reverse voltage of only 30 V; therefore it is not suitable.

Series Noise Clipper

Frequently a signal has unwanted voltage fluctuations (called *noise*) which can trigger sensitive circuits. To eliminate noise, a *series noise clipping circuit* may be employed. If the noise is considerably smaller than the normal forward voltage drop of a diode and the signal voltages are larger than V_F, then the diode noise clipper shown in Figure 3-6 may be employed. Since the peaks of noise voltage are not large enough to forward-bias either D_1 or D_2, the output during the time between signals is zero. The wanted signals readily forward-bias the diodes, and the output peak voltage is $(E - V_F)$. A *dead zone* of $\pm V_F$ exists around ground level at the output. This simply indicates that for signals to be passed to the output, they must exceed $\pm V_F$.

When the noise voltage is too large for a single diode, two or more diodes may be connected in series to give a larger dead zone. Thus, D_1 and D_2 in Figure 3-6 would each be replaced with two or more series-connected diodes.

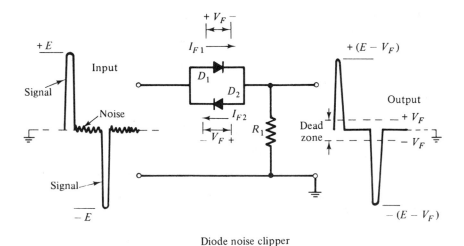

Diode noise clipper

Figure 3-6   A series noise clipping circuit uses the voltage drop across diodes to block unwanted noise voltages. The noise voltage amplitudes must not be large enough to forward-bias the diodes.

EXAMPLE 3-3

The diode noise clipper in Figure 3-6 has input signals of $E = \pm2$ V. The input noise has an amplitude of ±0.5 V. Calculate the resistance of R_1 and the amplitude of the output signals.

Solution   The output signal amplitude is

$$V_{R1} = V_o = \pm(E - V_F)$$

$$= \pm(2\text{ V} - 0.7\text{ V})$$

$$= \pm 1.3\text{ V}$$

In the absence of a load current, R_1 must pass enough current to keep the diode conducting when the signal is present. A suitable minimum level of I_F for a low-current diode is

$$I_F = 1\text{ mA}$$

Thus,
$$R_1 = \frac{V_{R1}}{I_{R_1}} = \frac{1.3\text{ V}}{1\text{ mA}}$$

$$= 1.3\text{ k}\Omega$$

Use a standard-value resistor ($R_1 = 1.2\text{ k}\Omega$).

3-4 DIODE SHUNT CLIPPER CIRCUITS

Shunt Clipper

Negative and positive shunt clipping circuits are shown in Figure 3-7. In each case the clipping circuit is applied to protect the base-emitter junction of a transistor from excessive reverse bias. Most transistors will not survive more than 5 V applied in reverse across the base-emitter junction. Consequently, when input signals are greater than 5 V, some sort of protective circuitry is needed.

In the circuit of Figure 3-7(a), the negative portion of the input signal is clipped off to protect the *npn* transistor. When the signal becomes positive, D_1 is reverse biased, and all of the current I_o flows through the transistor base. The voltage at the transistor base, i.e., the clipper output, is

$$\text{(Input voltage)} - \text{(Voltage drop across } R_1) = E - I_o R_1$$

When the input becomes negative, the transistor base-emitter junction is reverse biased and the diode is forward biased. Since the diode is in parallel with the transistor input, the maximum reverse base-emitter voltage is limited to the diode forward voltage drop V_F.

Diode D_1 in Figure 3-7(b) is forward biased when the input is positive. Thus, the transistor base-emitter voltage is limited to V_F, positive on the base and negative on the emitter, which is reverse bias for the *pnp* transistor. When the input becomes negative, D_1 is reverse biased and I_o flows through the transistor base.

As in the case of the series clippers, shunt clipping circuits can be employed with sine or other input waveforms, as well as with square waves.

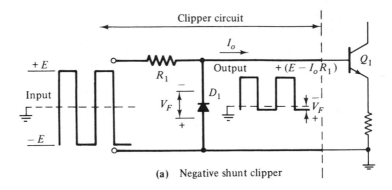

(a)   Negative shunt clipper

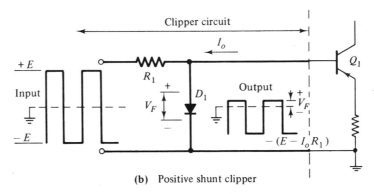

(b)   Positive shunt clipper

Figure 3-7   Negative and positive shunt clipping circuits. In each case the output voltage cannot exceed the forward voltage drop across the diode when the input forward-biases the diode.

EXAMPLE 3-4

The negative shunt clipper circuit in Figure 3-7(a) is to have an output voltage of 9 V and an output current of approximately 1 mA. If the input voltage is ± 10 V, calculate the value of R_1 and the diode forward current.

Solution   When the input is $+10$ V,

$$\text{Output} = 9 \text{ V} = E - I_o R_1$$

$$I_o R_1 = E - 9 \text{ V} = 10 \text{ V} - 9 \text{ V} = 1 \text{ V}$$

$$R_1 = \frac{1 \text{ V}}{I_o} = \frac{1 \text{ V}}{1 \text{ mA}} = 1 \text{ k}\Omega$$

When the input is -10 V, D_1 is forward biased and

$$V_F \approx 0.7 \text{ V}$$

$$V_F = E - I_F R_1$$

$$I_F = \frac{E - V_F}{R_1} = \frac{10 \text{ V} - 0.7 \text{ V}}{1 \text{ k}\Omega}$$

$$= 9.3 \text{ mA}$$

Biased Shunt Clipper

The shunt clipping circuits discussed so far clip off either the positive or the negative portion of an input waveform. The unwanted output is limited to a maximum of V_F above or below ground. In the circuit of Figure 3-8(a), diode D_1 has its cathode connected

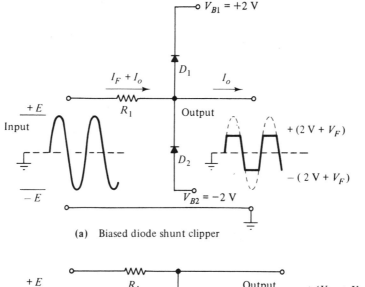

(a)   Biased diode shunt clipper

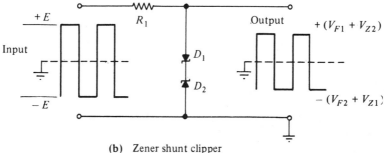

(b)   Zener shunt clipper

Figure 3-8   A biased shunt clipper circuit clips the output at $(V_B + V_F)$, the sum of the bias voltage and diode forward voltage. A Zener diode shunt clipper performs the same function, with V_z acting as a bias voltage.

to a bias of $+2$ V, and D_2 has its anode connected to -2 V. In this case, D_1 will not be forward biased while the output of the clipping circuit is below 2 V. The presence of D_1 and V_{B_1} limits the positive output to a maximum of $(2$ V $+ V_F)$. Similarly, D_2 will be reverse biased until the output is more negative than -2 V. This limits the negative output to a maximum of $-(2$ V $+ V_F)$.

The biased shunt clipper is normally used to protect a device or circuit which has both positive and negative input signals. The bias voltage is selected to prevent the input (either positive or negative) from exceeding a maximum safe level.

EXAMPLE 3-5

A biased shunt clipper circuit as shown in Figure 3-8(a) is to be designed to protect a circuit which cannot accept voltages exceeding $V_o = \pm 2.7$ V. The input to the clipper is a ± 8 V square wave, and the output current is to be a maximum of 1 mA. Calculate the value of R_1 and specify the diodes to be used.

Solution   Take the minimum forward current I_F as 1 mA.

$$V_o = \text{bias voltage } V_B + \text{diode voltage drop } V_F$$

$$V_B = V_o - V_F$$

$$= 2.7 \text{ V} - 0.7 \text{ V (using a silicon diode)}$$

$$= 2 \text{ V}$$

Voltage across $R_1 = (I_F + I_o) \times R_1$

$$(I_F + I_o) \times R_1 = E - V_B - V_F$$

$$R_1 = \frac{E - V_B - V_F}{I_F + I_o}$$

$$= \frac{8 \text{ V} - 2 \text{ V} - 0.7 \text{ V}}{1 \text{ mA} + 1 \text{ mA}}$$

$$= 2.65 \text{ k}\Omega \text{ [use 2.2 k}\Omega \text{ standard value]}$$

The diodes selected should be low-current devices with $V_F \approx 0.7$ V at $I_F = 1$ mA, and should have a peak reverse voltage greater than 10 V.

Zener Diode Shunt Clipper

The Zener diode clipper shown in Figure 3-8(b) performs a function similar to that of the circuit of Figure 3-8(a). In this case, however, no separate bias voltage supplies are necessary. When the input signal becomes sufficiently positive, D_1 operates like an ordinary forward-biased diode while D_2 goes into Zener breakdown. The output voltage at this time is $(V_{F1} + V_{Z2})$. When the input is negative, D_1 is in Zener breakdown and D_2 is forward biased. The output voltage now is $-(V_{F2} + V_{Z1})$.

EXAMPLE 3-6

Design a Zener diode shunt clipper circuit as shown in Figure 3-8(b) to be connected between a ±25 V square wave signal and a circuit which cannot accept inputs greater than ±11 V. The output current is to be a maximum of 1 mA.

Solution   The clipper circuit output is

$$V_o = \pm11 \text{ V}$$

$$V_o = V_F + V_Z$$

$$V_Z = V_o - V_F$$

$$= 11 \text{ V} - 0.7 \text{ V} = 10.3 \text{ V}$$

From the breakdown diode data sheet in Appendix 1-3,

$$V_Z = 9.1 \text{ V for the 1N757 device (use a 1N757)}$$

and $$V_{R1} = E - V_o$$

$$= E - (V_F + V_Z)$$

$$= 25 \text{ V} - 9.1 \text{ V} - 0.7 \text{ V} = 15.2 \text{ V}$$

To ensure that $I_Z > I_{ZK}$, make

$$I_Z \approx I_{ZT}/4$$

$$= 20 \text{ mA}/4 = 5 \text{ mA}$$

$$I_{R1} = I_z + I_o$$

$$= 5 \text{ mA} + 1 \text{ mA} = 6 \text{ mA}$$

$$R_1 = \frac{V_{R1}}{I_{R1}} = \frac{15.2 \text{ V}}{6 \text{ mA}} = 2.5 \text{ k}\Omega \quad \begin{array}{l}\text{[use a 2.2 k}\Omega \text{ standard value} \\ \text{resistor]}\end{array}$$

Note, from Appendix 1-3, that V_z for the 1N757 can be as low as 8.19 V or as high as 10.01 V. This means that the output could be a maximum of $\pm(10.01$ V $+ 0.7$ V$) = \pm10.71$ V, or a minimum of $\pm(8.19$ V $+ 0.7$ V$) = \pm8.89$ V.

3-5 DIODE CLAMPER CIRCUITS

Negative and Positive Clamping Circuits

The *clamper circuit* (also known as a *dc restorer circuit*) changes the dc level but does not affect the shape of a waveform. When the input is positive in the *negative voltage clamper* circuit of Figure 3-9(a), diode D_1 is forward biased and capacitor C_1 charges with the polarity shown. During the positive input peak, the output cannot exceed the

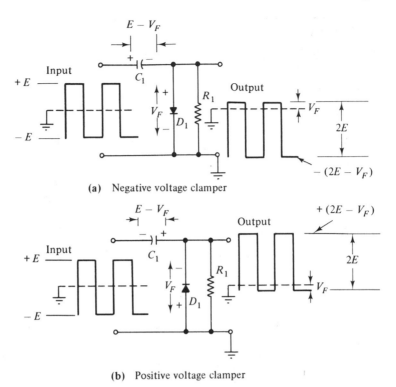

Figure 3-9 Negative and positive clamping circuits. In each case the output upper or lower level is clamped to V_F above or below ground level. The peak-to-peak amplitude of the waveform is unaffected.

diode forward-bias voltage V_F. At this time, therefore, the voltage on the right side of the capacitor is V_F, while on the left side of the capacitor the voltage is $+E$. Thus, the capacitor is charged to $E - V_F$, positive on the left and negative on the right, as illustrated.

When the input switches to negative, the diode is reverse biased, and it has no further effect on the capacitor voltage. Also, R_1 has a high resistance value and cannot discharge the capacitor significantly during the negative (or positive) portion of the input waveform. While the input is negative, the output voltage is the sum of the input voltage and the capacitor voltage. Since the polarity of the capacitor voltage is the same as the (negative) input, the result is a negative output larger than the input voltage. Thus,

$$\text{Negative output} = -E - (E - V_F)$$
$$= -(2E - V_F)$$

The peak-to-peak (*p-to-p*) output is the difference between the negative and positive peak voltages:

$$p\text{-to-}p \text{ output} = (\text{positive peak}) - (\text{negative peak})$$

$$= V_F - [-(2E - V_F)]$$

$$= 2E$$

It is seen that the amplitude of the output waveform from the *negative voltage clamper* is exactly the same as that of the input. Instead of being symmetrical above and below ground, however, the output positive peak is clamped to a level V_F above ground. The difference between *clipping* and *clamping* circuits is that while the clipper *clips off* an unwanted portion of the input waveform, the clamper simply *clamps* the maximum positive or negative peak to a desired dc level.

The function of R_1 is to discharge C_1 over several cycles of the input waveform. This would not be necessary if the input signal never changed. However, if the input is made smaller, C_1 must be partially discharged for the positive output peak to rise to V_F once again.

The *positive voltage clamping circuit* [Figure 3-9(b)] functions in exactly the same way as the negative voltage clamper. The diode, connected as shown, clamps the negative output peak at $-V_F$. Capacitor C_1 charges to $E - V_F$ positive on the right and negative on the left. (Note the capacitor terminal polarity.) The positive output then becomes $2E - V_F$.

To design a clamping circuit, C_1 should be selected so that it becomes completely charged during about five cycles of the input waveform. Since a capacitor takes approximately five time constants to become fully charged, and charging occurs only during the pulse width,

$$5\,CR = 5 \times \text{PW},$$

or

$$CR = \text{PW}$$

where PW is the width of the pulse which forward-biases the diode. In this case, R is the total resistance in series with the capacitor when it is being charged. This is the sum of the source resistance R_S and the diode forward resistance R_F. [See Figure 3-10(a).]

$$C(R_s + R_F) = \text{PW}$$

Usually $R_S \gg R_F$.

So

$$CR_S = \text{PW} \qquad\qquad (3\text{-}1)$$

When the capacitor partially discharges during the negative input peak (for a *negative* voltage clamper), some tilt appears on the output, as illustrated in Figure 3-10(b). The acceptable amount of tilt determines the value of the discharge resistor R_1. The voltage across R_1 during this time is approximately $2E$. Therefore, the discharge current is

$$I \approx \frac{2E}{R_1}$$

The diode reverse resistance is also involved [see Figure 3-10(b)], but since it is parallel with R_1 and is very much larger than R_1, it may be neglected. The current I remains

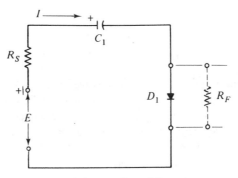

(a) Charge of C_1 via R_S and R_F

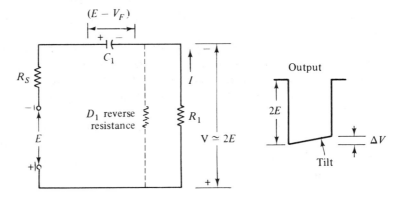

(b) Discharge of C_1

Figure 3-10 Capacitor charge and discharge must be considered when designing a clamping circuit. The relationship between C_1 and R_1 determines the amount of tilt on the output.

approximately constant during the discharge time, so Eq. 2-8 may be applied. In this case, a change in capacitor voltage ΔV is involved. Replacing I with $(2E)/R_1$ in the equation gives

$$\Delta V = \frac{2E}{R_1} \times \frac{t}{C}$$

or
$$R_1 = \frac{2Et}{\Delta V\, C} \tag{3-2}$$

EXAMPLE 3-7

A negative voltage clamper has a 1 kHz square wave input with an amplitude of ± 10 V. The signal source resistance R_S is 500 Ω, and the tilt on the output waveform is not to exceed 1%. Design a suitable circuit.

Solution For the input,

$$T = \frac{1}{f} = \frac{1}{1 \text{ kHz}} = 1 \text{ ms}$$

and
$$\text{PW} = T/2 = 500 \text{ } \mu\text{s}$$

From Eq. 3-1,

$$C = \frac{\text{PW}}{R_S}$$

$$= \frac{500 \text{ } \mu\text{s}}{500 \text{ } \Omega} = -1 \text{ } \mu\text{F}$$

For 1% tilt on the output,

$$\Delta V = 0.01(2E)$$

From Eq. 3-2,
$$R_1 = \frac{2 \, Et}{0.01(2E) \times C}$$

$$= \frac{t}{0.01 \, C}$$

and for $t = \text{PW} = 500$ μs,

$$R_1 = \frac{500 \text{ } \mu\text{s}}{0.01 \times 1 \text{ } \mu\text{F}} = 50 \text{ k}\Omega \quad \text{(use 56 k}\Omega \text{ standard value to give a} \\ \text{tilt slightly less than 1\%)}$$

Biased Clamping Circuits

In the biased clamping circuit shown in Figure 3-11(a), the cathode of diode D_1 is connected to a 2 V bias level. When the input becomes positive, the output level is clamped to the bias level plus the diode voltage drop, that is, 2 V + V_F. At this time the voltage on the right of C_1 is 2 V + V_F, and that on the left is E. Therefore, C_1 charges to $E - (2 \text{ V} + V_F)$, positive on the left and negative on the right. When the input goes to $-E$, the output becomes $-E$ plus the capacitor voltage. That is,

$$\text{Negative output} = -[E + (E - 2 \text{ V} - V_F)]$$

$$= -(2E - 2 \text{ V} - V_F)$$

and the peak-to-peak output is equal to (positive peak) − (negative peak), or

$$p\text{-to-}p \text{ output} = (2 \text{ V} + V_F) - [-(2E - 2 \text{ V} - V_F)]$$

$$= 2E$$

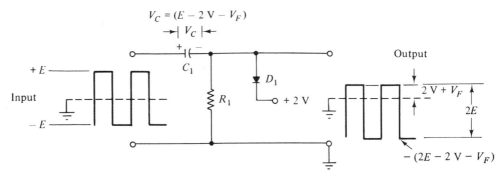

(a) Circuit to clamp output at approximately + 2 V maximum

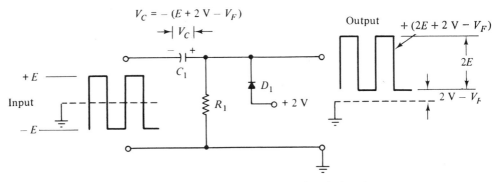

(b) Circuit to clamp output at approximately + 2 V minimum

Figure 3-11 A biased clamping circuit clamps the output waveform at $(V_B + V_F)$ or $(V_B - V_F)$, depending on the diode polarity.

It is seen that, although the output is clamped to a maximum dc level of 2 V + V_F, the wave shape and amplitude are unchanged. As before, the function of R_1 is to partially discharge C_1 over several cycles of the input. The clamper circuit on Figure 3-11(b) is similar to that in Figure 3-11(a) except that the diode is inverted. Since the capacitor charges with a different polarity, C_1 also is inverted from its condition in Figure 3-11(a). The anode of the diode is always at the bias voltage level, which is 2 V in this case. When the diode is forward biased, its cathode voltage is 2 V − V_F, and the cathode cannot go below this level. Therefore, the lowest level of output voltage is 2 V − V_F.

The diode is forward biased during the time that the input voltage is −E. Capacitor C_1 charges, during this time, to −E on the left, and to 2 V − V_F on the right. Therefore,

$$V_c = \text{Capacitor voltage} = (-E) - (2 \text{ V} - V_F)$$
$$= -(E + 2 \text{ V} - V_F)$$

This is positive on the right and negative on the left. When the input becomes +E, the capacitor voltage and the input have the same polarities and add together to give

$$\text{Output} = E + E + 2\,\text{V} - V_F$$
$$= 2E + 2\,\text{V} - V_F$$

The peak-to-peak value of the output waveform is again $2E$, and its lower level is clamped to $2\,\text{V} - V_F$.

Zener Diode Clamper

The Zener diode clamping circuits in Figure 3-12 perform the same function as biased clamping circuits. In Figure 3-12(a), the Zener diode behaves like a bias source with a

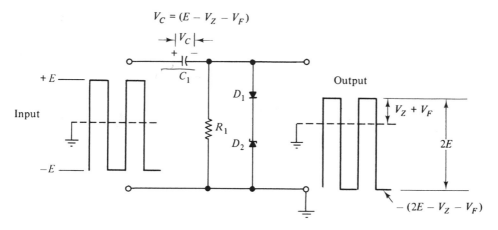

(a)   Circuit to clamp output at approximately $+ V_Z$ maximum

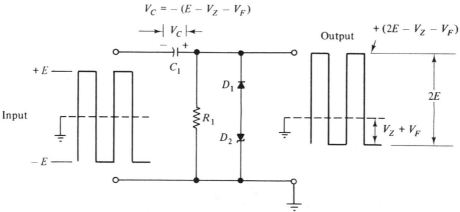

(b)   Circuit to clamp output at approximately $- V_Z$ minimum

Figure 3-12   Zener diode clamping circuits perform the same function as biased clamping circuits. V_Z acts as the bias voltage.

voltage of V_Z. When it is thought of in this way, its operation is seen to be exactly the same as the biased clamper in Figure 3-11(a). The Zener diode circuit in Figure 3-12(b) clamps the negative output at $-(V_Z + V_F)$. The capacitor charge then causes the positive output to be $2E - V_Z - V_F$. As always, the peak-to-peak output voltage is $2E$.

EXAMPLE 3-8

Design the biased positive voltage clamper circuit shown in Figure 3-11(b). The input waveform is ±20 V with a frequency of 2 kHz, and the tilt on the output is not to exceed 2%. The signal source resistance is 600 Ω.

Solution   For the input waveform, $T = 1/f = 1/2$ kHz $= 0.5$ ms,

and $PW = T/2 = 250 \ \mu s$

   From Equation 3-1,

$$C = \frac{PW}{R_S} = \frac{250 \ \mu s}{600 \ \Omega} \approx 0.42 \ \mu F \ \text{[use a 0.5 } \mu\text{F standard-value capacitor}$$
$$\text{(see Appendix 2)]}$$

For 2% tilt, $\Delta V = 0.02 \times 2E$

   From Equation 3-2,

$$R_1 = \frac{2E \times PW}{0.02 \times 2E \times C}$$

$$= \frac{2E \times 250 \ \mu s}{0.02 \times 2E \times 0.5 \ \mu F}$$

$$= 25 \ \text{k}\Omega \ \text{(use 27 k}\Omega \text{ standard value)}$$

The bias voltage should be $+2$ V as shown in the figure. The capacitor voltage should be rated at $V_i + V_B$; i.e., for the circuit designed, the capacitor selected should survive at least 22 V. The diode has to survive a reverse voltage of $2E$, i.e., 40 V for the above circuit.

EXAMPLE 3-9

A square wave having an amplitude of ±15 V and a source resistance R_s of 1 kΩ is to be clamped to a maximum positive level of approximately 9 V. The square wave frequency ranges from 500 Hz to 5 kHz, and the output tilt is not to exceed 1%. Design a Zener diode clamping circuit, as in Figure 3-12(a).

Solution   Maximum tilt occurs when the PW is longest, i.e., when f is a minimum:

$$\text{Maximum } T = \frac{1}{f_{min}} = \frac{1}{500 \text{ Hz}} = 2 \text{ ms}$$

$$\text{PW} = T/2 = 1 \text{ ms}$$

From Equation 3-1,

$$C = \frac{\text{PW}}{R_s} = \frac{1 \text{ ms}}{1 \text{ k}\Omega} = 1 \text{ }\mu\text{F } [\text{standard value (see Appendix 2)}]$$

For 1% tilt, $V = 0.01 \times 2E$

From Equation 3-2,

$$R_1 = \frac{2E \times \text{PW}}{0.01 \times 2E \times C}$$

$$= \frac{2E \times 1 \text{ ms}}{0.01 \times 2E \times 1 \text{ }\mu\text{F}} = 100 \text{ k}\Omega \text{ [standard value]}$$

$$V_o = V_Z + V_F = 9 \text{ V}$$

$$V_Z = V_o - V_F$$

$$= 9 \text{ V} - 0.7 \text{ V} = 8.3 \text{ V}$$

From the Zener diode data sheet (Appendix 1-3), the 1N756 has $V_Z = 8.2$ V; therefore, use a 1N756 Zener diode and a low-current diode such as a 1N914. Note that the diode should survive a reverse voltage of $2E = 30$ V. The capacitor voltage rating should be at least $V_i + V_Z$; i.e., minimum capacitor voltage is 23.2 V for this circuit.

3-6 VOLTAGE-MULTIPLYING CIRCUITS

The *voltage-doubling circuit* is simply a two-stage clamper circuit connected as illustrated in Figure 3-13(a). During the positive half-cycle of the input, capacitor C_1 charges to approximately the peak level (E volts) with the polarity illustrated. When the input goes negative, a voltage with a peak value of approximately $-2E$ volts appears at the output of the first stage (i.e., across D_1), as already explained in Section 3-5. During the time that $-2E$ is present across D_1, diode D_2 is forward biased, and capacitor C_2 is charged to $2E$ with the polarity illustrated. The dc (output) voltage from the terminals of C_2 is now double the peak value of the input voltage to the circuit.

If further diode-capacitor clamper stages are cascaded, as illustrated in Figure 3-13(b), each capacitor is charged to $2E$ volts. The voltage across C_2 and C_4 is now $4E$ volts. When a large number of stages are employed, high dc voltage levels can be obtained from very low-level supplies.

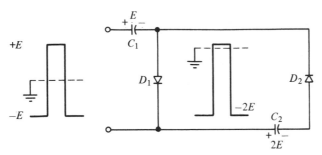

(a) Voltage doubling circuit

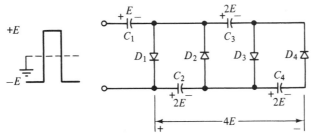

(b) Voltage multiplier circuit

Figure 3-13 Voltage-multiplying circuits consist of cascaded clamping circuit stages. Each capacitor charges to approximately 2 E volts. The total capacitor voltage can then add up to many times the input amplitude.

REVIEW QUESTIONS AND PROBLEMS

3-1. Sketch typical characteristics for a low-current silicon diode. Briefly explain why the diode can be thought of as a one-way device.

3-2. Sketch ideal diode characteristics, and approximate characteristics for silicon and germanium diodes. Briefly discuss the parameters that should be considered when selecting a diode.

3-3. Explain the origin of *reverse recovery time* for a semiconductor diode. By means of sketches, explain why a large reverse current flows when a very fast reverse bias is applied to a diode. Also, show how the reverse current can be minimized.

3-4. Sketch typical characteristics for a Zener diode. Indicate all important quantities related to the characteristics, and define each quantity.

3-5. Sketch the circuit of a positive series clipper, showing the input and output waveforms. Briefly explain its operation.

3-6. Repeat Problem 3-5 for a negative series clipper.

3-7. A negative series clipping circuit employs a diode with $V_F = 0.3$ V and $I_S = 10$ µA. The input voltage is ± 9 V, and the output current is to be a maximum of 10 mA. Calculate the value of the resistance R_1. Specify the diode in terms of forward current, power dissipation, and peak reverse voltage. The negative output voltage is to be maximum at 0.2 V.

3-8. Design a circuit to clip the positive peaks off a ±20 V square wave. A silicon diode is available with a maximum reverse leakage current of 10 μA. The positive output voltage is not to exceed 0.5 V. Calculate the amplitude of the negative output peak.

3-9. From the diode data sheets in Appendix 1, select a suitable device for the circuit designed in Problem 3-8.

3-10. Sketch the circuit of a diode noise clipper, showing typical input and output waveforms. Briefly explain how the circuit operates.

3-11. A *pnp* transistor which can take a maximum of 5 V in reverse at its base-emitter junction is to be protected from excessive input signal amplitude. Identify the required circuit and sketch the input and output waveforms. Briefly explain the operation of the circuit.

3-12. Repeat Problem 3-11 for an *npn* transistor.

3-13. A negative shunt clipper circuit has a square wave input of ±15 V. The output voltage is to be 13 V and −0.7 V, and the output current is to be 250 μA. Calculate the required resistance value and the diode forward current.

3-14. Sketch the circuit of a biased diode shunt clipper that has an output limited to a maximum of approximately ±4 V. Explain the operation of the circuit.

3-15. The input to the circuit of Problem 3-14 is ±16 V, and the output current is to be 500 μA. Determine the required resistance value, allowing the diode forward currents to be 10 mA.

3-16. Sketch a Zener diode shunt clipper circuit, and select suitable diodes which will clip off input peaks greater than approximately 6 V. Explain the operation of the circuit.

3-17. A ±14 V square wave is applied to the circuit of Problem 3-16. The output current is to be 2 mA maximum. Design a suitable circuit.

3-18. Define the difference between clipping and clamping circuits. A ±10 V square wave is applied to the input terminals of a negative voltage clamping circuit and to the input of a negative shunt clipper. Sketch the output waveform that will result in each case.

3-19. Sketch a negative voltage clamping circuit, showing input and output waveforms. Briefly explain the operation of the circuit.

3-20. Repeat Problems 3-19 for a positive voltage clamper.

3-21. A negative voltage clamper has a 5 kHz square wave input with an amplitude of ±6 V. The signal source resistance is 1 kΩ, and the tilt on the output waveform is not to exceed 1%. Design a suitable circuit.

3-22. Sketch the output waveforms you would expect from each of the circuits shown in Figure 3-14. Assume that the input to each circuit is a ±12 V square wave.

3-23. Sketch the output waveforms you would expect from each of the circuits shown in Figure 3-15. Assume that the input to each circuit is a ±9 V square wave.

3-24. Design a biased clamper circuit to clamp a ±12 V square wave to a minimum level of +3 V. The input waveform has a frequency which ranges from 1 kHz to 10 kHz, and the signal source resistance is 500 Ω. The tilt on the output is not to exceed 1%.

3-25. A square wave having an amplitude of ±18 V and a source resistance of 700 Ω is to be clamped to a maximum positive level of approximately 10 V. The square wave frequency is 800 Hz, and the output tilt is not to exceed 0.5%. Design a suitable Zener diode clamping circuit.

3-26. Sketch a diode-capacitor *voltage-multiplier* circuit, and explain how it operates.

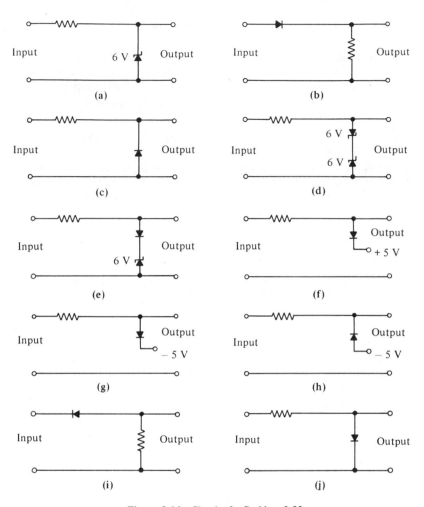

Figure 3-14 Circuits for Problem 3-22.

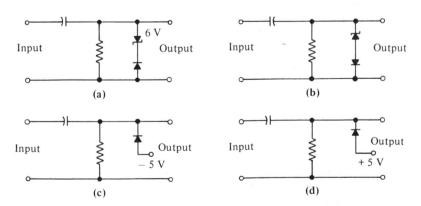

Figure 3-15 Circuits for Problem 3-23.

Chapter 4

Transistor Switching

INTRODUCTION

A bipolar transistor can be made to approximate an ideal switch. When the transistor is off, a small collector-base *leakage current* flows through the load. When it is on, there is a small collector-emitter *saturation voltage* across the device. A transistor will not switch *on* or *off* instantaneously. *Turn-on* and *turn-off times* depend upon the device and the circuit conditions. Input signals to transistor switching circuits may be direct coupled or capacitor coupled. Field effect transistors used as switches have some advantages over bipolar devices.

4-1 IDEAL TRANSISTOR SWITCH

Figure 4-1(a) shows a common emitter transistor circuit arranged to function as a switch. A collector resistance R_C is connected from the transistor collector to the supply voltage V_{CC}. The emitter terminal of the device is grounded. For the transistor to simulate a switch, the terminals of the switch are the transistor collector and emitter. The input voltage, or controlling voltage, for the transistor switch is the base-emitter voltage V_{BE}. The collector-emitter voltage V_{CE} is equal to the supply voltage minus the voltage drop across R_C:

$$V_{CE} = V_{CC} - I_C R_C \tag{4-1}$$

When the transistor base-emitter voltage is zero, or reverse biased, as in Figure 4-1(b), the base current I_B is zero, and the collector current I_C is also zero. The transistor switch is now in its *off* condition. Since there is no collector current, there can be no voltage drop across the load resistor. When $I_C = 0$, Eq. 4-1 gives

$$V_{CE} = V_{CC} - (0 \times R_C)$$
$$= V_{CC}$$

Thus, when an ideal transistor switch is *off*, its collector-emitter voltage equals the supply voltage.

When the transistor base is made positive with respect to the emitter [Figure 4-1(c)], a base current I_B flows. The collector current I_C is equal to I_B multiplied by the transistor *common emitter dc current gain* h_{FE}; that is, $I_C = h_{FE} \times I_B$. If I_B is made

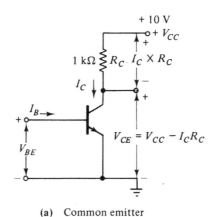

(a)   Common emitter
       circuit

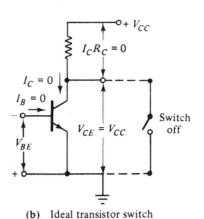 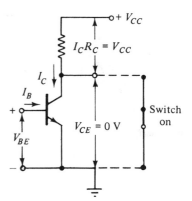

(b)   Ideal transistor switch   (c) Ideal transistor switch
       in *off* condition   in *on* (saturated)
                                              condition

Figure 4-1   Common emitter transistor circuit as a switch. An ideal transistor switching circuit would behave as an open switch when *off*—the collector current would be zero. When *on*, an ideal transistor switch would simulate a closed switch—the collector-emitter voltage would be zero.

large enough, $I_C \times R_C$ can become equal to the supply voltage V_{CC}. Then, by Equation 4-1,

$$V_{CE} = V_{CC} - V_{CC}$$

$$V_{CE} = 0$$

When an ideal transistor switch is *on*, its collector-emitter voltage equals zero. The transistor can also simulate a switch in that, *ideally*, it dissipates zero power when *on* or *off*. The only time power is dissipated in the device is when it is switching between *on* and *off*. Transistor power dissipation is given by,

$$P_D = I_C \times V_{CE}$$

When *off*,

$$I_C = 0, \text{ so}$$

$$P_D = 0 \times V_{CE} = 0$$

When *on*,

$$V_{CE} = 0, \text{ so}$$

$$P_D = I_C \times 0 = 0$$

As described above, the transistor can be operated as a switch which is *off* when V_{BE} is zero or negative, and which is *on* when V_{BE} is sufficiently positive. Ideally, $V_{CE} = V_{CC}$ when the transistor is *off*, and $V_{CE} = 0$ V when the device is *on*. With a practical transistor, these conditions are not precisely achieved; however, they can be approximated.

4-2 PRACTICAL TRANSISTOR SWITCH

To understand how a practical transistor switch differs from the ideal case, it is necessary to consider the common emitter characteristics. In Figure 4-2, the dc load line for the circuit of Figure 4-1(a) is drawn upon the transistor common emitter characteristics. Using Eq. 4-1, the procedure for drawing the load line is as follows:

When $I_C = 0$,

$$V_{CE} = V_{CC} - 0$$

$$= 10 \text{ V (For the circuit shown)}$$

Plot point A on the characteristics at $I_C = 0$ and $V_{CE} = 10$ V.
        When $V_{CE} = 0$,

$$0 = V_{CC} - I_C R_C$$

$$I_C = \frac{V_{CC}}{R_C}$$

$$= \frac{10 \text{ V}}{1 \text{ k}\Omega} = 10 \text{ mA} \quad \text{[for the circuit of Figure 4-1(a)]}$$

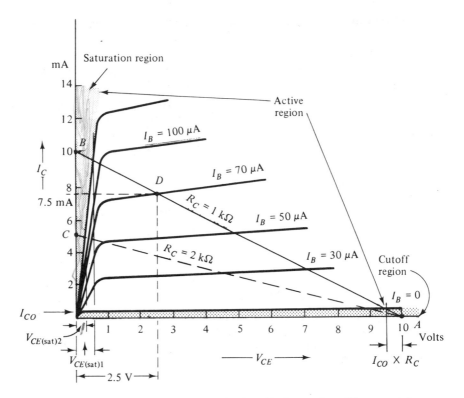

Figure 4-2 Characteristics and dc load line for a bipolar transistor. When operated as a switch, the transistor is normally biased into either the saturation or cutoff region of the characteristics.

Plot point B on the characteristics at $V_{CE} = 0$, $I_C = 10$ mA. Draw the dc load line (for $R_C = 1$ kΩ) through points A and B.

The dc load line defines all corresponding current and voltage conditions that can exist in the circuit. For any given level of I_C, a particular V_{CE} is dictated by Equation 4-1 and is illustrated by the load line.

The common emitter characteristics are divided into three regions, as shown in Figure 4-2. The *active region* of the characteristics usually is employed only in amplifier circuits. Here, a linear change in the base current produces a nearly linear collector-emitter voltage change. When the collector current is so large that V_{CE} is less than approximately 0.7 V, the device is said to be operating in the *saturation region* of the characteristics. The cutoff region exists below the level of $I_B = 0$. On the characteristics shown in Figure 4-2, the cutoff region is exaggerated. Normally, the cutoff current is so small that it could not be shown on the current scale employed in Figure 4-2.

Again, with reference to the load line, it is seen that when $I_B = 0$, I_C is *not* zero. Instead, a small current I_{CO} flows. This is the collector-base *reverse leakage current*, or *collector cutoff current*, sometimes designated I_{CBO}. For most recent silicon transistors,

I_{CO} at 25° C is in the nanoampere range; however, at high temperatures this may increase into the microamps range.

Refer to the data sheet for 2N3903 and 2N3904 transistors in Appendix 1-4. The collector cutoff current is designated I_{CEX}. This is the collector-base leakage current measured under particular conditions specified by the manufacturer. I_{CEX} can be regarded as essentially equal to I_{CO}. From the data sheet, the maximum collector cutoff current for 2N3903 and 2N3904 transistors is 50 nA. The presence of I_{CO} makes V_{CE} slightly less than V_{CC} when the transistor is *off*. [See Figure 4-3(a).]

$$V_{CE} = V_{CC} - I_{CO} R_C \qquad (4\text{-}2)$$

For $V_{CC} = 10$ V, $R_C = 1$ kΩ, and $I_{CO} = 1$ μA,

$$V_{CE} = 10 \text{ V} - (1 \text{ } \mu\text{A} \times 1 \text{ k}\Omega)$$

$$= 9.999 \text{ V}$$

$$\approx V_{CC}$$

When the transistor is in saturation, a small collector-emitter saturation voltage $V_{CE(sat)}$ exists. [See Figure 4-2.] Typically about 0.2 V, $V_{CE(sat)}$ largely depends upon I_C and the resistance of the semiconductor material that forms the transistor collector. The load line for $R_C = 2$ kΩ (broken line in Figure 4-2) reveals that when saturation occurs with smaller levels of I_C, $V_{CE(sat)}$ is reduced. The 2N3903 and 2N3904 data sheet in Appendix 1-4 specifies maximum levels of $V_{CE(sat)}$ as 0.3 V at $I_C = 50$ mA, and 0.2 V at $I_C = 10$ mA.

The saturated transistor circuit in Figure 4-3(b) has typical voltages of $V_{BE} = 0.7$ V and $V_{CE(sat)} = 0.2$ V. Thus, the base terminal is 0.5 V positive with respect to the collector terminal, and the normally reverse-biased collector-base junction is actually forward biased. As will be seen later, this forward bias at the collector-base junction limits the switching speed of the transistor.

The forward bias at the collector-base junction when a transistor is saturated reduces the dc current gain h_{FE}. This happens because, to draw the maximum number of charge carriers from emitter to collector, the collector-base junction must be reverse biased. For saturation to occur, the transistor current gain must have a minimum value $h_{FE(min)}$, depending upon the circuit conditions. Suppose that a transistor has a base current of $I_B = 50$ μA and requires a collector current I_C of 1 mA for saturation. Then $h_{FE(min)} = I_C/I_B = 1$ mA/50 μA $= 20$. If h_{FE} is less than 20 in this case, I_C will be less than 1 mA, and saturation will *not* occur. If h_{FE} is greater than 20, I_C will tend to be greater than the required 1 mA, and saturation will occur.

EXAMPLE 4-1

For the circuit of Figure 4-1(a), $I_B = 0.2$ mA. (a) Determine the value of $h_{FE(min)}$ for saturation to occur. (b) If R_C in Figure 4-1(a) is changed to 220 Ω and a 2N3904 transistor is employed, will the transistor be saturated?

Solution (a)   For saturation,

$$I_C \approx \frac{V_{CC}}{R_C}$$

$$= \frac{10 \text{ V}}{1 \text{ k}\Omega} = 10 \text{ mA}$$

$$h_{FE(min)} = \frac{I_C}{I_B}$$

$$= \frac{10 \text{ mA}}{0.2 \text{ mA}} = 50$$

Solution (b)   For saturation,

$$I_C \approx \frac{V_{CC}}{R_C}$$

$$= \frac{10 \text{ V}}{220 \text{ }\Omega} \approx 45 \text{ mA}$$

$$h_{FE(min)} = \frac{I_C}{I_B}$$

$$= \frac{45 \text{ mA}}{0.2 \text{ mA}}$$

$$= 225$$

From the 2N3904 data sheet in Appendix 1-4, at $I_C = 50$ mA, $h_{FE(min)} =$ 60. Therefore, the transistor will *not* be saturated.

Suppose a 2N3904 transistor is employed in the case of Example 4-1(a). From the 2N3904 data sheet in Appendix 1-4, at $I_C = 10$ mA, $h_{FE(min)} = 100$ and $h_{FE(max)} = 300$. This suggests that for $I_B = 0.2$ mA, I_C could be any value between 100×0.2 mA and 300×0.2 mA, that is, from 20 mA to 60 mA. In fact, the maximum collector current cannot exceed $I_C = V_{CC}/R_C = 10$ mA, as calculated in the example. Thus, with an h_{FE} value greater than 50, more base current flows than is needed to drive the transistor into saturation. The extra base current flows out through the emitter terminal, and in this situation the transistor is said to be *overdriven*.

Although transistors in switching circuits are usually switched from cutoff to saturation, and *vice versa*, they can also be switched between cutoff and the active region. For example, if the base current is limited to 70 µA for the load line shown in Figure 4-2 (point *D*), then V_{CE} is 2.5 V. In this case the transistor circuit is referred to as a *nonsaturating switch* [Figure 4-3(c)].

The power dissipation is very small with a practical transistor in saturation or

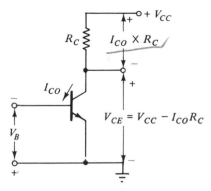

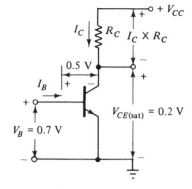

(a) Transistor in cutoff

(b) Transistor in saturation

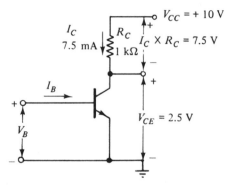

(c) Transistor in active region of characteristics

Figure 4-3 Transistors operating in (a) the cutoff region of the characteristics ($I_C = I_{CO}$), (b) the saturation region ($V_{CE} = V_{CE(sat)}$), and (c) the active region ($V_{CE} = V_{CC} - I_C R_C$).

cutoff. For a nonsaturating transistor switch, the power dissipation is much larger than for either the cutoff or saturated cases.

EXAMPLE 4-2

If the circuit of Figure 4-1(a) employs a 2N3904 transistor, calculate the transistor power dissipation (a) at cutoff, (b) at saturation, and (c) when $V_{CE} = 2$ V.

Solution (a)   For cutoff:
From the 2N3904 data sheet, $I_C \approx I_{CEX} = 50$ nA,

so that
$$P_D \approx I_C \times V_{CC}$$
$$= 50 \text{ nA} \times 10 \text{ V}$$
$$= 0.5 \text{ } \mu\text{W}$$

Solution (b)   For saturation,

$$I_C \approx \frac{V_{CC}}{R_C}$$

$$= \frac{10 \text{ V}}{1 \text{ k}\Omega} = 10 \text{ mA}$$

From the 2N3904 data sheet, at $I_C = 10$ mA, $V_{CE(\text{sat})} = 0.2$ V.

Thus, $P_D = I_C \times V_{CE(\text{sat})}$

$$= 10 \text{ mA} \times 0.2 \text{ V}$$

$$= 2 \text{ mW}$$

Solution (c)   At $V_{CE} = 2$ V:
From Equation 4-1,

$$I_C = \frac{V_{CC} - V_{CE}}{R_C} = \frac{10 \text{ V} - 2 \text{ V}}{1 \text{ k}\Omega}$$

$$= 8 \text{ mA}$$

$$P_D = I_C \times V_{CE}$$

$$= 8 \text{ mA} \times 2 \text{ V}$$

$$= 16 \text{ mW}$$

4-3 TRANSISTOR SWITCHING TIMES

One most important characteristic of a switching transistor is the speed with which it can be switched *on* and *off*. Consider Figure 4-4, where the time relationship between collector current and base current is shown. When the input current I_B is applied, the transistor does not switch *on* immediately. The time between application of base current and commencement of collector current is termed the *delay time* t_d. (See Figure 4-4.) The delay time is defined as the time required for I_C to reach 10% of its final level, after I_B has commenced. Even when the transistor begins to switch *on*, a finite time elapses before I_C reaches its maximum level. The *rise time* t_r is defined as the time it takes for I_C to go from 10% to 90% of its maximum level. The *turn-on time* t_{on} for the transistor is the sum of t_d and t_r. (See Figure 4-4.)

Similarly, a transistor cannot be switched *off* instantaneously. The *turn-off time* t_{off} is composed of a *storage time* t_s and a *fall time* t_f (Figure 4-4). The storage time results from the fact that the collector-base junction is forward biased when the transistor is in saturation. Charge carriers crossing a forward-biased junction are trapped (or stored)

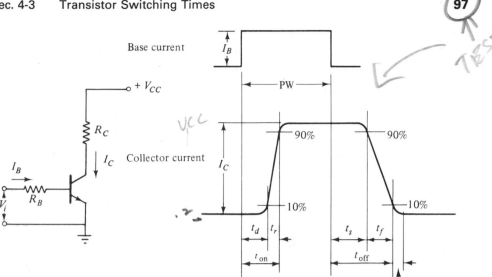

Figure 4-4   Time relationship between base current and collector current in a transistor switching circuit. The turn-on time t_{on} is made up of the delay time t_d and the rise time t_r. The turn-off time t_{off} consists of the storage time t_s and the fall time t_f.

in the depletion region when the junction is reversed. These charge carriers must be withdrawn or made to recombine with charge carriers of an opposite type before the collector current begins to fall. The storage time t_s is defined as the time between I_B switch-*off* and I_C falling to 90% of its maximum level. The fall time t_f is the time required for I_C to fall from 90% to 10% of its maximum. A further quantity, *the decay time*, is sometimes included in the turn-off time. This is the time required for I_C to go from its 10% level to I_{CO}. Usually this is not an important quantity, since the transistor is regarded as being *off* when I_C falls to the 10% level.

   Refer to the data sheet for the 2N3904 general-purpose transistor in Appendix 1-4. The turn-on and turn-off times given are

$$\text{Turn-on time} = t_d + t_r = 35 \text{ ns} + 35 \text{ ns} = 70 \text{ ns}$$

$$\text{Turn-off time} = t_s + t_f = 200 \text{ ns} + 50 \text{ ns} = 250 \text{ ns}$$

The 2N4418 high-speed switching transistor specified in Appendix 1-7 has total turn-on and turn-off times listed as 20 ns and 22 ns, respectively.

   In the case of a nonsaturating transistor switch (Section 4-2), the collector-base junction is reverse biased when the transistor is *on*. Therefore, no storage time is involved, and the turn-off time is not much larger than the fall time. This faster turn-off time is the major advantage of the nonsaturating switch.

   Figure 4-5 shows the time relationship of the input and output voltages as well

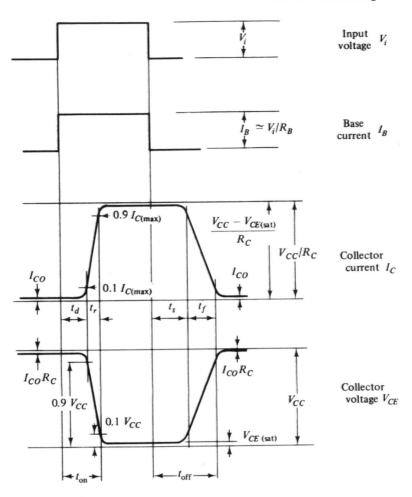

Figure 4-5 Time relationship between voltages and currents in a transistor switching circuit. When I_C increases from zero to 0.1 $I_{C(max)}$, V_C falls from approximately V_{CC} to 0.9 V_{CC}. When I_C becomes 0.9 $I_{C(max)}$, V_C is 0.1 V_{CC}.

as the I_B and I_C waveforms for the circuit in Figure 4-4. I_B commences almost immediately when V_i is applied. The approximate level of I_B is input voltage V_i divided by base resistor R_B, that is, V_i/R_B (ignoring V_{BE}). The output voltage at any instant depends upon the instantaneous level of I_C. Thus, V_{CE} is initially ($V_{CC} - I_{CO}R_C$) and falls to 90% of V_{CC} when I_C becomes 10% of $I_{C(max)}$ after t_d. (See Figure 4-5.) When I_C is 90% of $I_{C(max)}$, V_{CE} is 10% of V_{CC}, and finally falls to $V_{CE(sat)}$ when I_C reaches its maximum level. When I_B goes to zero, the storage time elapses before I_C commences to fall. Then V_{CE} again becomes approximately 0.1 V_{CC} when I_C is 90% of its maximum level, and V_{CE} becomes 0.9 V_{CC} when I_C is 10% of maximum. Finally, V_{CE} returns to $V_{CC} - I_{CO}R_C$ when I_C falls to the level of the reverse leakage current.

EXAMPLE 4-3

The circuit in Figure 4-4 has $V_{CC} = 12$ V and $R_C = 3.3$ kΩ. The transistor employed is a 2N3904, and the input voltage has a PW of 5 μs. Calculate the level of V_{CE} (a) before the input pulse is applied, (b) at the end of the delay time, and (c) at the end of the turn-on time. Also, determine the time from commencement of the input pulse until the transistor switches *off*.

Solution   (a) In the 2N3904 data sheet (Appendix 1-4) the collector cutoff current is defined as $I_{CEX} = 50$ nA. Before the transistor switches *on*,

$$V_{CE} = V_{CC} - I_{CEX}R_C$$
$$= 12 \text{ V} - (50 \text{ nA} \times 3.3 \text{ kΩ})$$
$$= 11.9998 \text{ V}$$

(b) At the end of the delay time,

$$V_{CE} = V_{CC} - (0.1\ I_{C(\max)}\ R_C)$$
$$= V_{CC} - \left(0.1 \times \frac{V_{CC}}{R_C} \times R_C\right)$$
$$= 12 \text{ V} - (0.1 \times 12 \text{ V})$$
$$= 10.8 \text{ V}$$

(c) At the end of the turn-on time,

$$V_{CE} = V_{CC} - \left(0.9 \times \frac{V_{CC}}{R_C} \times R_C\right)$$
$$= 12 \text{ V} - (0.9 \times 12 \text{ V})$$
$$= 1.2 \text{ V}$$

For the 2N3904, $t_{off} = 250$ ns. The time from commencement of input to the transistor switching *off* is PW + t_{off}.

Thus, $$\text{PW} + t_{off} = 5 \text{ μs} + 250 \text{ ns}$$
$$= 5.25 \text{ μs}$$

4-4 IMPROVING THE SWITCHING TIMES

Speed-up or Commutating Capacitor

If the base-emitter junction of the transistor is reverse biased before switch-on, the delay time is longer than in the case when V_{BE} is initially zero. This is because the transistor input capacitance is charged to the reverse-bias voltage, and must be discharged before V_{BE} can become positive. Therefore, to minimize the turn-on time, V_{BE} should be zero or have a very small reverse bias before switch-on.

Both the delay time and the rise time can be reduced if the transistor is *overdriven*, i.e., if I_B is made larger than the minimum required for saturation. With a larger I_B, the junction capacitances are charged faster, thus reducing the turn-on time. A major disadvantage of overdriving is that the storage time is extended by the larger current flow across the forward-biased collector-base junction when the transistor is in saturation. Therefore, although an overdriven transistor will turn on faster, it has a longer turn-off time than a transistor which has just enough base current for saturation.

One way to shorten the turn-off time is to provide a large negative input voltage at switch-off. This produces a reverse base current flow which causes the junction capacitance to discharge rapidly. In this case, the disadvantage is that the turn-on time is increased because of the initial large reverse bias of the base-emitter junction.

Ideally, for fast switching, V_{BE} should start at zero volts, and I_B should be large at switch-on, but should rapidly settle down to the minimum required for saturation. Also, switch-off should be accomplished by a large reverse bias voltage which quickly returns to zero. Exactly these conditions are achieved when a capacitor is connected in parallel with R_B, as shown in Figure 4-6. This capacitor, termed a *speed-up capacitor* (or a *commutating capacitor*), is initially uncharged before the input voltage pulse is applied. When the input voltage rises, the capacitor commences to charge to ($V_i - V_{BE}$) [Figure 4-6(a)]. The capacitor charging current flows into the transistor base terminal. Thus, I_B is initially large, but quickly settles down to its dc level as the capacitor becomes charged. [See the waveforms in Figure 4-6(a).] At switch-off [Figure 4-6(b)], the capacitor discharge produces a reverse base current which rapidly returns to zero.

The speed-up capacitor tends to reduce t_d and t_s as well as t_r and t_f. However, if C_1 is so small that it becomes completely charged within the delay time, then it will not have a significant effect upon the rise time. Similarly, if C_1 is completely discharged during the storage time, it will not produce a marked improvement in the fall time.

Consider the circuit of Figure 4-7. The settled base current level (i.e., the level after the capacitor is completely charged) can be calculated by using V_i, R_B, and R_s:

$$I_B = \frac{V_i - V_{BE}}{R_s + R_B}$$

$$= \frac{5\ \text{V} - 0.7\ \text{V}}{1\ \text{k}\Omega + 8.2\ \text{k}\Omega} \approx 0.5\ \text{mA}$$

The initial level of capacitor charging current is approximately the signal voltage divided by the signal source resistance (because $V_{C1} = 0$):

$$I_1 \approx \frac{V_i - V_{BE}}{R_s}$$

$$= \frac{5\ \text{V} - 0.7\ \text{V}}{1\ \text{k}\Omega} = 4.3\ \text{mA}$$

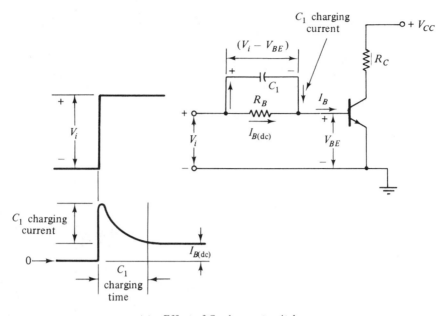

(a)   Effect of C_1 charge at switch-on

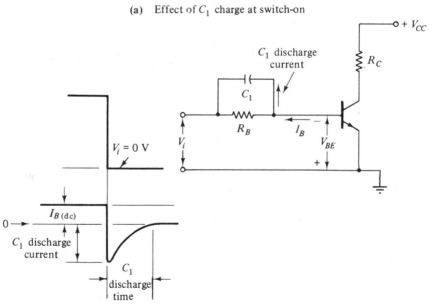

(b)   Effect of C_1 discharge at switch-off

Figure 4-6   Effect of speed-up capacitor C_1. At switch-on, the capacitor charging current provides an increased base current to turn the transistor *on* rapidly. At switch-off, the capacitor discharge current assists in device turn-off.

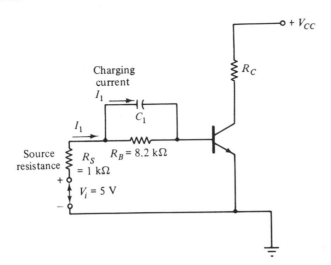

Figure 4-7 The value of the speed-up capacitor is calculated from a knowledge of the signal source resistance and the transistor turn-on time.

This is considerably greater than the dc level of I_B. Therefore, an improvement in switching speed may be expected. For the best possible improvement in switching speed, a speed-up capacitor should be selected that is large enough to maintain the charging current (i.e., base current) nearly constant at its maximum level during the transistor turn-on time. The charging current will drop by only 10% from its maximum level if the capacitor is allowed to charge by 10% during the turn-on time. From Eq. 2-11, C_1 charges by 10% during a time of $0.1\ C_1 R_s$. Therefore,

$$t_{on} = 0.1\ C_1 R_s$$

and

$$C_1 = \frac{t_{on}}{0.1 R_s} \tag{4-3}$$

For $t_{on} = 300$ ns, and $R_s = 1$ kΩ,

$$C_1 = \frac{300\ \text{ns}}{0.1 \times 1\ \text{k}\Omega} = 3000\ \text{pF}$$

A larger capacitor than this is not likely to offer any greater improvement in switching time. Also, if a ten times improvement in switching time is achieved, then a capacitor of 300 pF might be almost as effective as one with a value of 3000 pF. This is because t_{on} in the above calculation would be reduced from 300 ns to 30 ns, and, consequently, C_1 would be calculated as 300 pF. To achieve such an improvement in switching time, however, the transistor must initially operate well below its maximum switching speed. Also, the input pulse must have a rise time very much less than the minimum switching time sought.

The upper limit to the value of C_1 that may be used depends upon the maximum signal frequency. When the transistor is switched *off*, C_1 discharges through R_B. For correct switching, C_1 should be at least 90% discharged during the time interval between transistor *switch-off* and *switch-on*. The time required for the capacitor to return to its discharged condition is variously referred to as the *settling time*, the *resolving time*, or the *recovery time* t_{re} of the circuit. In this case, the transistor is *off* and the capacitor is discharged through R_B. From Eq. 2-10, C_1 discharges by 90% in a time $t = 2.3C_1R_B$.

Thus,
$$t_{re} = 2.3\ C_1 R_B$$

or
$$\text{maximum } C_1 = \frac{t_{re}}{2.3R_B} \qquad (4\text{-}4)$$

EXAMPLE 4-4

An inverter circuit similar to the one in Figure 4-7 has $R_S = 600\ \Omega$, $R_B = 5.6$ kΩ, and a 2N3904 transistor. (a) Determine the size of speed-up capacitor to give maximum improvement in transistor turn-on time. (b) Calculate the maximum square wave input frequency that may be used with the circuit.

Solution   (a) From Appendix 1-4, $t_{on} = 70$ ns.

Eq. 4-3,
$$C_1 = \frac{t_{on}}{0.1R_s} = \frac{70 \text{ ns}}{0.1 \times 600\ \Omega}$$
$$= 1167 \text{ pF}$$

(b) From Eq. 4-4,
$$t_{re} = 2.3C_1R_B$$
$$= 2.3 \times 1167 \text{ pF} \times 5.6 \text{ k}\Omega$$
$$= 15\ \mu\text{s}$$
$$f = \frac{1}{2T} = \frac{1}{2t_{re}} = \frac{1}{2 \times 15\ \mu\text{s}}$$
$$\approx 33.3 \text{ kHz}$$

Nonsaturating Switch

Operation of a transistor switch in a nonsaturating condition is briefly discussed in Section 4-2. Instead of driving the transistor into saturation when it is switched *on*, the base current is limited to a level that prevents the device from saturating. The level of V_{CE} may be very low, but the collector-base junction does not become forward biased. This effectively eliminates the storage time (see Figure 4-4) when the transistor is to be switched *off*.

Another method of avoiding transistor saturation is illustrated in Figure 4-8. Diode D_1, connected between the base and collector terminals of the transistor, prevents the

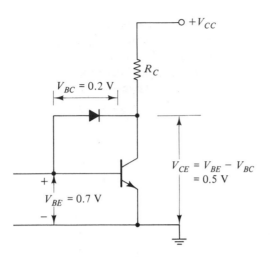

Figure 4-8   A diode connected between the base and collector terminals of a transistor limits the forward voltage at the collector-base junction, thus minimizing the storage time when the transistor is to be turned *off*.

collector-base junction from becoming forward biased by more than the diode voltage drop V_F. The diode is selected to have a forward voltage less than the normal base-collector voltage of the transistor at saturation. In this way, the transistor is prevented from becoming completely saturated.

In the case of a silicon transistor, $V_{BC} \approx 0.5$ V when the device is saturated. If the diode has $V_F = 0.2$ V, then V_{BC} cannot exceed 0.2 V. This produces a considerable reduction in storage time from the case when $V_{BC} \approx 0.5$ V, consequently, a faster transistor turn-off time is achieved.

Of course, the forward-biased diode must be switched *off* before transistor turn-off can commence. To achieve an improved performance, the diode must be selected to have a small reverse recovery time, as well as a low forward voltage drop. The *Schottky diode* is just such a device. Constructed as a junction of metal and *n*-type semiconductor material, the Schottky diode has $V_F \approx 0.2$ V and a reverse recovery time on the order of 4 ns.

4-5 DIRECT-COUPLED INVERTER

Inverter Operation

A transistor *inverter circuit* is essentially an overdriven common emitter circuit. The input may be a square wave, a pulse waveform, or even a sine wave, provided that the input amplitude is sufficient to drive the transistor alternately into saturation and cutoff. Figure 4-9(a) shows an inverter circuit with a positive pulse wave input. When the input voltage is zero, there is no collector current, and the output is approximately V_{CC}. When the input becomes positive, the transistor switches into saturation, and the output becomes $V_{CE(sat)}$. Thus, a positive-going input produces a negative-going output,

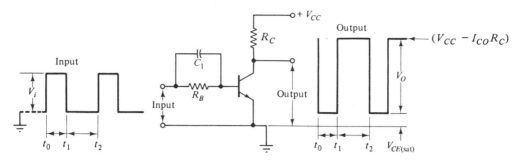

(a) Inverter with a positive pulse input waveform

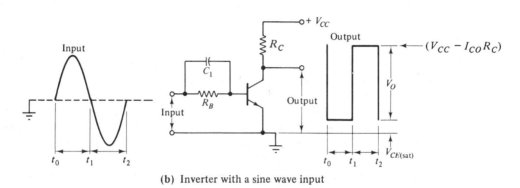

(b) Inverter with a sine wave input

Figure 4-9 Transistor inverter circuits produce a pulse-type output waveform which is phase inverted in relation to the input. Any type of input waveform may be employed, provided its amplitude is large enough to drive the transistor into saturation and cutoff.

and *vice versa*. The output waveform is then the inverse of the input (although the output amplitude may differ from the input), hence the name *inverter*.

Figure 4-9(b) illustrates the effect of a sine wave input to an inverter. If the input amplitude is large enough to alternately switch the transistor rapidly between saturation and cutoff, then an inverted square wave output is the result.

When the input waveform to an inverter has a large amplitude, the base-emitter junction of the transistor may be destroyed by an excessive reverse voltage. Most transistors can take only about 5 V in reverse at the base-emitter junction. To protect the transistor, a diode connected as a clipper may be employed. (See Figure 3-7.) Alternatively, a diode may be connected in series with the transistor emitter terminal, as shown in Figure 4-10. Since diodes can normally survive reverse bias voltages on the order of at least 50 V, the combined base-emitter junction and diode will withstand a large reverse bias.

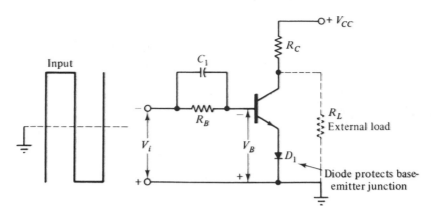

Figure 4-10   A diode may be connected in series with the emitter terminal of a transistor to protect it against excessive reverse base-emitter voltage.

Inverter Design

The design of a transistor inverter circuit should begin with selection of the collector resistance R_C, unless it is already specified. In general, R_C should be much smaller than the external load to be connected to the inverter output (Figure 4-10). This is to ensure that the external load has no significant effect on the circuit performance. However, R_C should also be made as large as possible in order to keep I_C to a minimum. If this principle is followed in all circuit design, the current demand from power supplies is kept to a minimum. Also, with current maintained as small as possible, power dissipation in all components is minimized. The lower limit for I_C is dependent upon the particular transistor used. The data sheet in Appendix 1-4 shows that for the 2N3904 transistor, $h_{FE(min)}$ is only 40 at $I_C = 0.1$ mA, and $h_{FE(min)} = 100$ at $I_C = 10$ mA. For the 2N930 transistor (App. 1-6), $h_{FE(min)} = 100$ at an I_C of only 10 μA, and $h_{FE(min)}$ is 150 at $I_C = 500$ μA. Obviously, the 2N3904 should not be operated with a collector current much below 100 μA, while the 2N930 can easily be operated with I_C as low as 10 μA. One disadvantage of operating a transistor at very low current levels is that the resultant large resistance values make the circuit more susceptible to picking up unwanted (noise) signals.

If V_{CC} and I_C are known, R_C is calculated simply as the voltage across R_C divided by the current through R_C. That is,

$$R_C = \frac{V_{CC} - V_{CE(sat)}}{I_C} \qquad (4\text{-}5)$$

Usually, the calculated value of R_C is not exactly equal to an available standard resistance value. In this case, the next higher standard value should be selected. The voltage drop across R_C must at least equal $V_{CC} - V_{CE(sat)}$ for transistor saturation. A value of R_C that is larger than calculated gives saturation with a lower I_C level. If a value of R_C smaller than calculated is used, I_C must be increased to ensure saturation. When the design commences with R_C being chosen much smaller than the external load, I_C is calculated from Eq. 4-5.

The minimum base current for saturation is calculated as

$$I_{B(min)} = \frac{I_C}{h_{FE(min)}} \qquad (4\text{-}6)$$

Here, the use of $h_{FE(min)}$ is necessary to ensure transistor saturation. If the particular transistor used has a larger than minimum value of h_{FE}, then, for a given I_B, I_C will tend to be larger than necessary, and saturation will be achieved.

The value of R_B is determined by dividing the voltage across R_B by the current through it (see Fig. 4-11):

$$R_B = \frac{V_i - V_{BE}}{I_{B(min)}}$$

Again, an available standard resistance must be selected for R_B, but this time the next *lower* standard value should be selected. This is because the voltage across R_B is a fixed quantity $(V_i - V_{BE})$, and

$$I_B = \frac{V_i - V_{BE}}{R_B}$$

If an R_B larger than the calculated value is selected, then I_B will be less than the value of $I_{B(min)}$ required to saturate the transistor. If R_B is smaller than calculated, I_B is greater than $I_{B(min)}$, and transistor saturation will occur.

Speed-up capacitor C_1 is calculated for maximum signal frequency, as explained in Sec. 4-4.

EXAMPLE 4-5

Design an inverter circuit as in Fig. 4-11 using a 2N3904 transistor. The value of V_{CC} is 12 V, and the input is a ±3 V square wave. Use $I_C = 1$ mA.

Solution   At saturation,

$$I_C R_C = V_{CC} - V_{CE(sat)}$$

$$R_C = \frac{V_{CC} - V_{CE(sat)}}{I_C} = \frac{12 \text{ V} - 0.2 \text{ V}}{1 \text{ mA}}$$

$$= 11.8 \text{ k}\Omega \text{ (use 12 k}\Omega \text{ standard value)}$$

$$I_{B(min)} = \frac{I_C}{h_{FE(min)}}$$

From the 2N3904 data sheet, $h_{FE(min)} = 70$ at $I_C = 1$ mA.

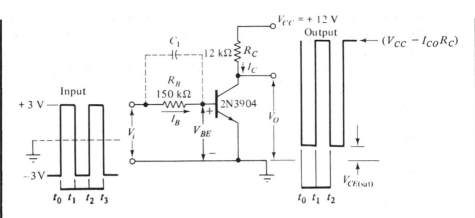

Figure 4-11   Direct-coupled inverter circuit designed in Example 4-5.

Thus,
$$I_B = \frac{1\ \text{mA}}{70} \approx 14.3\ \mu\text{A}$$

$$R_{B(\text{max})} = \frac{V_i - V_{BE}}{I_B} = \frac{3\ \text{V} - 0.7\ \text{V}}{14.3\ \mu\text{A}}$$

$$\approx 160\ \text{k}\Omega\ (\text{use }150\ \text{k}\Omega\ \text{standard value})$$

EXAMPLE 4-6

The square wave input to the circuit designed in Example 4-5 has a maximum frequency of 45 kHz. Determine the maximum value of the speed-up capacitor C_1.

Solution C_1 must discharge via R_B by about 90% during the negative (or *off*) portion of the square wave input.

$$\text{Resolving time } t_{re} = \frac{T}{2} = \frac{1}{2f}$$

$$= \frac{1}{2 \times 45\ \text{kHz}} \approx 11\ \mu\text{s}$$

Eq. 4-4,

$$C = \frac{t_{re}}{2.3\ R_B} = \frac{11\ \mu\text{s}}{2.3 \times 150\ \text{k}\Omega}$$

$$\approx 32\ \text{pF}\ (\text{use }30\ \text{pF}\ \text{standard value})$$

EXAMPLE 4-7

Design a transistor inverter circuit to handle a square wave input of ± 10 V. V_{CC} is 15 V, and the external load has a resistance of 100 kΩ. Use a 2N3903 transistor, and determine the amplitude of the output waveform.

Solution   Since the input can be -10 V, diode D_1 (Figure 4-10) is necessary to protect the transistor.

$$R_C \ll R_L$$

Make

$$R_C \approx \frac{1}{10} R_L$$

$$= \frac{1}{10} \times 100 \text{ k}\Omega = 10 \text{ k}\Omega \text{ (standard value)}$$

At saturation,

$$I_C \approx \frac{V_{CC} - V_{CE(\text{sat})} - V_{D1}}{R_C}$$

$$= \frac{15 \text{ V} - 0.2 \text{ V} - 0.7 \text{ V}}{10 \text{ k}\Omega}$$

$$= 1.41 \text{ mA}$$

From the 2N3903 data sheet in Appendix 1-4, $h_{FE(\text{min})} = 35$ at $I_C = 1$ mA.

$$I_{B(\text{min})} = \frac{I_C}{h_{FE(\text{min})}}$$

$$= \frac{1.41 \text{ mA}}{35}$$

$$\approx 40 \text{ }\mu\text{A}$$

and

$$R_{B(\text{max})} = \frac{V_i - V_{BE} - V_{D1}}{I_{B(\text{min})}}$$

$$= \frac{10 \text{ V} - 0.7 \text{ V} - 0.7 \text{ V}}{40 \text{ }\mu\text{A}} = 215 \text{ k}\Omega \text{ (use 180 k}\Omega \text{ standard value)}$$

D_1 should be a low-current diode with reverse breakdown voltage greater than 10 V. The 1N914 is a suitable device; see the 1N914 data sheet in Appendix 1-1. When the transistor is saturated,

$$V_{o(\text{min})} = V_{D1} + V_{CE(\text{sat})}$$

$$= 0.7 \text{ V} + 0.2 \text{ V} = 0.9 \text{ V}$$

At cutoff, R_C and R_L act as a potential divider, and so

$$V_{o(\text{max})} = \frac{V_{CC} \times R_L}{R_L + R_C} = \frac{15 \text{ V} \times 100 \text{ k}\Omega}{100 \text{ k}\Omega + 10 \text{ k}\Omega}$$

$$= 13.6 \text{ V}$$

Peak-to-peak output,

$$V_o = V_{o(\text{max})} - V_{o(\text{min})}$$

$$= 13.6 \text{ V} - 0.9 \text{ V}$$

$$= 12.7 \text{ V}$$

4-6 CAPACITOR-COUPLED INVERTER CIRCUITS

Normally-on Inverter

Sometimes a transistor is required to remain biased in the *on* condition, until an input signal is applied to switch it *off*. In this case the signal may be capacitor coupled. Such a circuit is shown in Figure 4-12. Bias resistor R_B provides base current from the supply, to keep the device in saturation while no signal is present. Capacitor C_c couples the negative-going pulse input to the transistor base. When the signal pulls the base below the emitter voltage level, the transistor switches *off*. The value of C_c is calculated from a knowledge of the input voltage amplitude and pulse width. For reasons of economy, physical size, and shortest capacitor recharge time, it is best to choose the lowest possible value of coupling capacitor.

Consider the voltage waveforms shown in Figure 4-12. The circuit input terminal is normally at ground level (i.e., before the signal pulse is applied). The other terminal of C_c (the transistor base terminal) is at V_{BE}. Therefore, the capacitor charge is normally V_{BE}, positive on the right as illustrated in the diagram. When the input pulse with an amplitude of $-V_i$ is applied, the transistor base is pulled down to $-(V_i - V_{BE})$. The transistor switches *off* and the capacitor immediately commences to charge via R_B, so that the negative pulse appearing at the base has tilt, as illustrated. The tilt must not be so great that V_B rises above ground; otherwise the transistor may switch *on* before the signal pulse has ended. The capacitor charging current is approximately constant and can be calculated by dividing the voltage across R_B by R_B:

$$I \approx \frac{V_{CC} - V_i}{R_B}$$

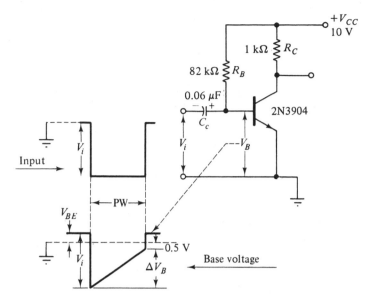

Figure 4-12 In a normally-on capacitor-coupled inverter circuit, the base bias resistor R_B provides sufficient base current to keep the transistor in saturation while the input voltage is zero.

If V_B is allowed to rise to -0.5 V, then the voltage change on the capacitor is $\Delta V_B = (V_i - V_{BE}) - 0.5$ V. When the capacitor charging time t is equal to the pulse width, the simple constant-current capacitor equation 2-8 may be employed:

$$C = \frac{It}{\Delta V}$$

The design of a normally-on inverter circuit is demonstrated in Example 4-8.

EXAMPLE 4-8

The capacitor-coupled inverter circuit of Figure 4-12 has a signal pulse input with an amplitude of -4V, and PW $= 1$ ms. V_{CC} is 10 V, and I_C is to be 10 mA. Using a 2N3904 transistor, design a suitable circuit.

Solution

$$R_C = \frac{V_{CC} - V_{CE(\text{sat})}}{I_C} = \frac{10 \text{ V} - 0.2 \text{ V}}{10 \text{ mA}}$$

$$= 980 \ \Omega \text{ (use a 1 k}\Omega \text{ standard value)}$$

From the 2N3904 data sheet, $h_{FE(\text{min})} = 100$ at $I_C = 10$ mA;

$$I_{B(min)} = \frac{I_C}{h_{FE(min)}} = \frac{10 \text{ mA}}{100}$$

$$= 100 \text{ } \mu A$$

$$R_B = \frac{V_{CC} - V_{BE}}{I_{B(min)}} = \frac{10 \text{ V} - 0.7 \text{ V}}{100 \text{ } \mu A}$$

$$= 93 \text{ k}\Omega \text{ (use 82 k}\Omega \text{ standard value)}$$

$$\Delta V_B = V_i - V_{BE} - 0.5 \text{ V}$$

$$= 4 \text{ V} - 0.7 \text{ V} - 0.5 \text{ V}$$

$$= 2.8 \text{ V}$$

The capacitor charging current is

$$I \approx \frac{V_{CC} - V_i}{R_B}$$

$$= \frac{10 \text{ V} - (-4 \text{ V})}{82 \text{ k}\Omega}$$

$$= 0.17 \text{ mA}$$

$$t = PW = 1 \text{ ms}$$

$$C_c = \frac{I \times t}{\Delta V_B} = \frac{0.17 \text{ mA} \times 1 \text{ ms}}{2.8 \text{ V}} = 0.06 \text{ } \mu F \text{ (standard value)}$$

Although the calculation of C_c in Example 4-8 assumes that the input pulse is negative with respect to ground, actually the pulse could be negative with respect to any other initial voltage level. For example, the pulse could go from +8 V to +4 V, and have exactly the same effect on the inverter circuit as a completely negative pulse. The calculated value of C_c would be the same as in the example, but the capacitor polarity would have to be reversed.

The circuit in Fig. 4-13 is similar to that in Fig. 4-12, except that the transistor is a *pnp* device and all voltage polarities are inverted. For the *pnp* transistor to switch *off*, a positive-going input pulse must be applied. Also, the voltage at the transistor base should be maintained positive until the end of the pulse. The design procedure for this circuit is exactly the same as for the *npn* inverter.

Normally-off Inverter

The converse of the *normally-on* inverter is the *normally-off* circuit, shown in Fig. 4-14. Here R_B connects the base and emitter terminals, thus keeping V_{BE} equal to zero until a positive-going input pulse is applied. While no input pulse is present, the only

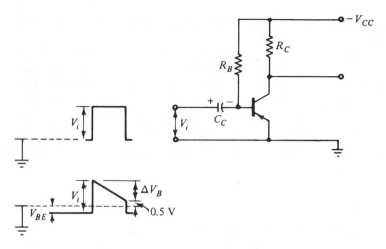

Figure 4-13 Normally-on capacitor-coupled inverter using a *pnp* transistor. This is similar to the *npn* circuit in Figure 4-12, except that all voltage polarities are reversed.

current through R_B is the reverse leakage current I_{CO}. The voltage drop across R_B must not be large enough to partially forward-bias the base-emitter junction; otherwise, the collector current may become larger than I_{CO}. With a maximum I_{CO} of 10 μA (at the highest ambient temperature) and maximum $V_{RB} = 0.1$ V, a typical resistance for R_B is 0.1 V/10 μA = 10 kΩ. Any resistance below 22 kΩ is usually suitable for R_B in this

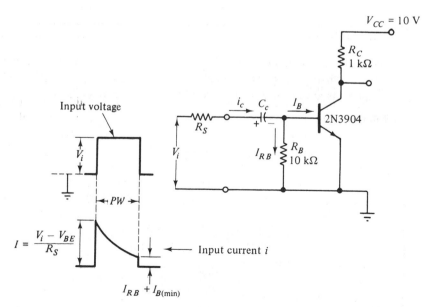

Figure 4-14 In a normally-off capacitor-coupled inverter circuit, resistor R_B keeps $V_{BE} = 0$, thus maintaining the device in an *off* condition while no input is present.

type of circuit. To avoid overloading the signal source, a resistance less than 1 kΩ should normally not be used for R_B.

In the circuit of Fig. 4-14, C_c starts at zero volts and charges via the signal source resistance while the input pulse is present. The capacitor charging current i splits up, to flow through R_B and the transistor base terminal. This current begins at a maximum of $I = (V_i - V_{BE})/R_S$, and then falls off as C_c becomes charged. At the end of the input pulse, the minimum capacitor current i_c must still be large enough to provide current through R_B and sufficient base current to saturate the transistor. Using the equation for capacitor charging current (Eq. 2-4), an expression for the capacitance of the coupling capacitor can be determined.

$$i_c = I\epsilon^{-t/CR}$$

giving

$$\frac{t}{CR} = \ln\frac{I}{i_c}$$

or

$$C_c = \frac{t}{R\ln(I/i_c)} \tag{4-7}$$

In this expression, R is the signal source resistance R_S, t is the input pulse width, I is the initial charging current, and i_c is the charging current at the end of the signal pulse. As in the circuit of Example 4-8, the input pulse does not have to start at ground level.

EXAMPLE 4-9

The inverter circuit in Figure 4-14 has an input pulse of 4 V amplitude and a pulse width of 1 ms. The signal source resistance R_S is 1 kΩ, and the other components are as illustrated. Determine a suitable value for C_c.

Solution

$$I_C = \frac{V_{CC} - V_{CE(\text{sat})}}{R_C}$$

$$= \frac{10\text{ V} - 0.2\text{ V}}{1\text{ k}\Omega} = 9.8\text{ mA}$$

$$I_{B(\text{min})} = \frac{I_C}{h_{FE(\text{min})}} = \frac{9.8\text{ mA}}{100} = 98\ \mu\text{A}$$

$$I_{RB} = \frac{V_{BE}}{R_B} = \frac{0.7\text{ V}}{10\text{ k}\Omega} = 70\ \mu\text{A}$$

At the end of the pulse,

$$i_c = I_{B(min)} + I_{RB}$$

$$= 98\ \mu A + 70\ \mu A$$

$$= 168\ \mu A$$

The initial charging current is,

$$I = \frac{V_i - V_{BE}}{R_S}$$

$$= \frac{4\ V - 0.7\ V}{1\ k\Omega} = 3.3\ mA$$

$$t = PW = 1\ ms$$

Eq. 4-7, $$C_c = \frac{t}{R_S \ln(I/i_c)} = \frac{1\ ms}{1\ k\Omega\ \ln(3.3\ mA/168\ \mu A)}$$

$$= 0.33\ \mu F\ \text{(standard value)}$$

Another normally-*off* capacitor-coupled inverter circuit is shown in Figure 4-15. In this case the transistor base is biased to a negative voltage $-V_{BB}$, to improve the turn-off time. The base-emitter junction is reverse biased, so that capacitor C_c has an

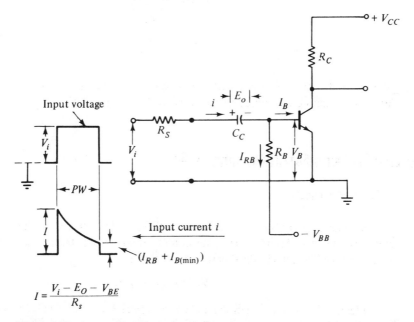

$$I = \frac{V_i - E_O - V_{BE}}{R_s}$$

Figure 4-15   A normally-off capacitor-coupled inverter circuit may employ a negative bias voltage $-V_{BB}$ to improve the turn-off time of the transistor.

initial charge of $E_o = V_{BB}$, with the polarity shown. When a positive input voltage V_i is applied, the transistor base voltage is

$$V_B = V_i - iR_s - E_o$$

Thus, the transistor base will not rise above ground, and the device will not switch *on*, unless V_i is greater than $(iR_s + E_o)$.

When the transistor switches *on*, its base voltage becomes $+V_{BE}$ above ground level. The current through R_B now is

$$I_{RB} = \frac{V_{BE} - V_{BB}}{R_B}$$

Also, the transistor minimum base current remains

$$I_{B(\text{min})} = \frac{I_C}{h_{FE(\text{min})}}$$

For the transistor to remain conducting during the input PW, the input (capacitor) current at the end of the PW must be at least

$$i_c = I_{RB} + I_{B(\text{min})}$$

The initial capacitor charging current for this circuit is

$$I = \frac{\text{initial voltage across } R_s}{R_s}$$

$$= \frac{V_i - E_o - V_{BE}}{R_s}$$

When i_c and I are determined as explained above, the value of C for the circuit in Figure 4-15 can be calculated in the same way as in Example 4-9.

4-7 JUNCTION FET SWITCHING CIRCUITS

JFET as a Switch

An *n*-channel *junction field effect transistor* (JFET), connected in common source configuration, is shown in Figure 4-16. The output voltage from the circuit equals the supply voltage minus the voltage drop across R_D. That is,

$$V_o = V_{DD} - I_D R_D \tag{4-8}$$

Ideally, $V_o = V_{DD}$ when the device is *off*, and $V_o = 0$ when the FET is *on*. The input signal biases the transistor *off* when V_i is negative, so that the *drain current I_D* is zero. When the input signal is at ground level, the *gate* potential equals that at the source terminal, and the drain current flows through the channel. When the drain current

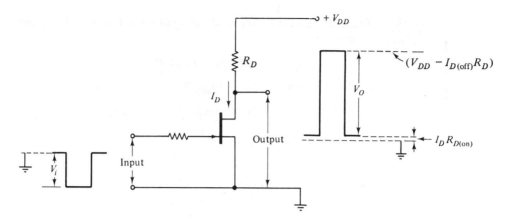

Figure 4-16 Common source *n*-channel JFET connected to function as a direct-coupled inverter circuit. The FET is normally in the *on* condition when the input voltage is zero. To turn the device *off*, a negative input voltage exceeding $V_{GS(off)}$ must be applied.

is small, there will be only a small voltage drop along the channel, due to the small channel resistance. or *drain-source on resistance* $R_{D(on)}$.

When the JFET is biased *on* [Figure 4-17(a)], the output voltage is

$$V_{DS(on)} = I_D \times R_{D(on)} \qquad (4\text{-}9)$$

Typically, $R_{D(on)}$ is 30 Ω or less. With a drain current of 100 μA, the typical level of $V_{DS(on)}$ can be quite small:

$$V_{DS(on)} = 100 \ \mu A \times 30 \ \Omega$$

$$= 3 \ mV$$

A comparison of $V_{DS(on)}$ with the typical $V_{CE(sat)}$ of 0.2 V for a bipolar transistor shows that $V_{DS(on)} \ll V_{CE(sat)}$, and this is a major advantage of the JFET switch.

The *off*-biased JFET has a small *drain-gate leakage current* $I_{D(off)}$ flowing across the reverse-biased gate-channel junctions. As illustrated in Figure 4-17(b), $I_{D(off)}$ causes a very small voltage drop across R_D.

Another advantage of the JFET switch over a bipolar transistor switch is that the JFET has a much higher input resistance than a bipolar transistor. Thus, the JFET can be easily switched by signals that have high source resistances.

EXAMPLE 4-10

The circuit in Figure 4-16 uses a 2N4857 JFET and has $V_{DD} = 15$ V and $R_D = 6.8$ kΩ. Determine the level of the output voltage (a) when the device is cut *off*, and (b) when the transistor is switched *on*.

Solution From the 2N4857 data sheet in Appendix 1-8, the maximum drain current at cutoff is

$$I_{D(off)} = 0.25 \text{ nA (i.e., at 25° C)}$$

$$V_o = V_{DD} - I_{D(off)}R_D$$

$$= 15 \text{ V} - (0.25 \text{ nA} \times 6.8 \text{ k}\Omega)$$

$$= 15 \text{ V} - 1.7 \text{ }\mu\text{V}$$

$$\approx 15 \text{ V}$$

In the data sheet, $R_{D(on)}$ is identified as

$$r_{ds(on)} = 40 \text{ }\Omega$$

and when the FET is *on*,

$$I_D \approx \frac{V_{DD}}{R_D}$$

$$= \frac{15 \text{ V}}{6.8 \text{ k}\Omega} = 2.2 \text{ mA}$$

$$V_o = I_D r_{ds(on)}$$

$$= 2.2 \text{ mA} \times 40 \text{ }\Omega$$

$$= 88 \text{ mV}$$

JFET Inverter Circuits

The direct-coupled inverter illustrated in Figure 4-18(a) is similar to the circuit in Figure 4-16, except that it uses a *p*-channel FET instead of an *n*-channel device. The supply voltage for the circuit in Figure 4-18(a) is a negative quantity, and the input pulse has to be positive to turn the FET *off*. When the input is at ground level, the device is *on*. Drain current I_D flows, producing a voltage drop of almost V_{DD} across drain resistor R_D. To ensure that the FET switches *off*, the input voltage amplitude must exceed the gate source cutoff voltage $V_{GS(off)}$ for the device. A typical value of $V_{GS(off)}$ is 8 V, and this would require an input of approximately 9 V. The only function served by resistor R_G in Figure 4-18(a) is to limit the gate current in the event that the gate-source junction becomes forward biased. Typically, R_G is selected as 1 MΩ.

Figure 4-18(b) shows a capacitor-coupled *n*-channel JFET inverter circuit. Again, R_G is typically 1 MΩ, this time to present a high input impedance to signals. The circuit shown has the gate biased to the same voltage as the source terminal; therefore, the FET is normally *on*. To switch the device *off*, the negative-going input signal amplitude must exceed $V_{GS(off)}$.

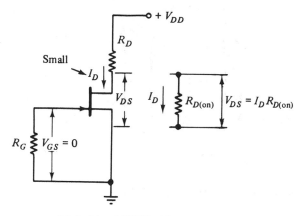

(a) *On*-biased JFET with small I_D

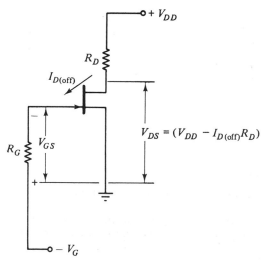

(b) *Off*-biased JFET

Figure 4-17 A field effect transistor operated as a switch is usually biased into
either the *channel ohmic region* or the
cutoff region of the characteristics. When
on, $V_{DS} = I_D R_{D(on)}$. When *off*, there is a
small drain-gate leakage current $I_{D(off)}$.

Appendix 1-10 shows typical turn-*on* and turn-*off* times of 15 ns and 20 ns, respectively, for a junction field effect transistor designed for switching applications.

Design of FET inverter circuits essentially involves calculation of R_D from the
supply voltage and desired drain current level. Coupling capacitor values are determined
in the same way as for bipolar inverter circuits. While the device is to remain *off*, the
gate-source reverse voltage must not be allowed to fall below $V_{GS(off)}$. Because the
circuit input resistance is very high, the charging current to the coupling capacitor is
extremely small, so C_C can be quite a small-value capacitor.

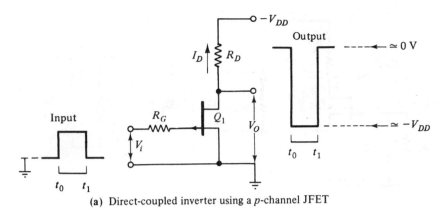

(a) Direct-coupled inverter using a *p*-channel JFET

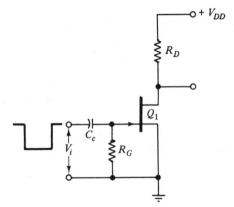

(b) Normally-*on* capacitive-coupled inverter using an *n*-channel JFET

Figure 4-18 JFET inverter circuits may be direct-coupled, normally-*on* capacitor-coupled, or normally-*off* capacitor-coupled. In all cases, gate resistor R_G is usually selected as 1 MΩ.

4-8 MOSFET SWITCHING CIRCUITS

MOSFET Switches

N-channel and *p*-channel *metal oxide semiconductor field effect transistor (MOSFET)* switching circuits are shown in Figure 4-19 together with input and output waveforms. With the *enhancement* devices illustrated, no channel exists while the gate is at the same potential as the source. Therefore, unlike JFETs, MOSFETs require no external bias voltage to switch them *off*; that is, they can be operated from a single polarity supply. For the *n*-channel MOSFET, a positive input pulse is necessary for switch-on. When the input signal becomes positive, I_D flows, and the output voltage drops from

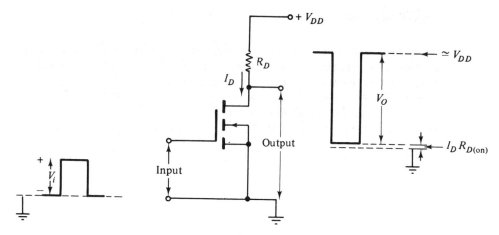

(a) n-channel enhancement MOSFET switch

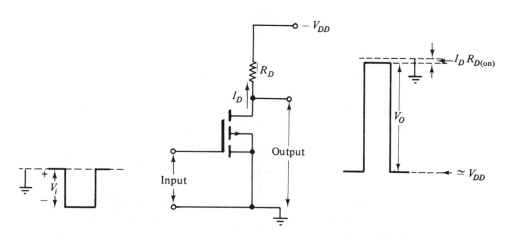

(b) p-channel enhancement MOSFET switch

Figure 4-19   An n-channel enhancement MOSFET used as a switch is *off* until a positive input voltage is applied to the gate. A similar p-channel device requires a negative gate voltage for switch-on.

V_{DD} to $I_D R_{D(on)}$. In the case of the p-channel device, the output is $-V_{DD}$ while no drain current is flowing. A negative signal on the gate terminal switches the transistor *on*, causing the output level to change to $-I_D R_{D(on)}$.

Since the gate terminal in a MOSFET is insulated from the channel, there is no drain-gate leakage current. This results in an input resistance even higher than that of a JFET circuit. There is a small drain-source leakage current, which causes some voltage drop across R_D when the MOSFET is *off*.

CMOS Inverter

When two devices are identical in every way except for their supply and input voltage polarities, they are termed *complementary devices*. For example, if two bipolar transistors have identical parameters, but one is an *npn* device and the other is *pnp*, they are complementary. Similarly, two MOSFETS, one of which is *p*-channel and the other *n*-channel, can be complementary. When complementary *p*-channel and *n*-channel MOSFETS are combined, the resulting circuitry is termed *complementary MOS*, or CMOS.

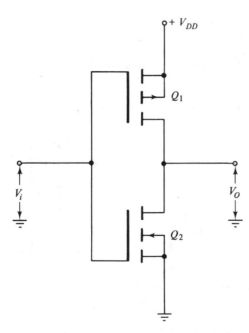

Figure 4-20   A complementary MOSFET (CMOS) inverter uses *n*-channel and *p*-channel devices. When the common gate input voltage is zero, Q_1 is *on* and Q_2 is *off*. A positive input turns Q_1 *off* and Q_2 *on*.

Figure 4-20 shows the circuit of a CMOS switch. Both devices are enhancement MOSFETS, so that no channel exists until one of them is switched *on*. When the input voltage is zero at the common gate terminal, the *p*-channel device Q_1 is biased *on* and the *n*-channel device Q_2 is *off*. In this condition there is only a very small voltage drop from the drain to the source for the *p*-channel device, and the output voltage is very close to V_{DD}. When the input voltage becomes positive, the *n*-channel transistor is biased *on* and the *p*-channel device is *off*. There is now only a very small voltage drop along the *n*-channel device, and consequently, the output voltage is very close to ground.

The major advantage of CMOS circuits is that the power dissipation is extremely small compared to other circuits. The small power dissipation, together with the small voltage drop across the *on* transistor and the high input impedance, makes the CMOS inverter approach the ideal switch. CMOS is discussed further in Chapter 12.

REVIEW QUESTIONS AND PROBLEMS

4-1. Sketch a circuit to show a bipolar transistor employed as a switch. Compare the transistor to an ideal switch.

4-2. Define the following terms: saturating switch, nonsaturating switch, saturation voltage, collector-base leakage current, and $h_{FE(min)}$. Discuss the importance of the latter three items in relation to a transistor switch.

4-3. Sketch typical transistor common emitter characteristics. Identify the various regions of the characteristics, and show how $V_{CE(sat)}$ differs with different transistor load resistances.

4-4. (a) A common emitter transistor circuit has $V_{CC} = 20$ V, $R_C = 2.2$ kΩ, and $I_B = 0.3$ mA. Determine $h_{FE(min)}$ if the transistor is to be saturated. (b) If a 2N3903 transistor is used in the above circuit, calculate the minimum I_B level at which the transistor will become saturated.

4-5. A common emitter circuit, using a 2N3904 transistor, has $V_{CC} = 25$ V. The collector resistance can be 22 kΩ or 2.2 kΩ. Calculate the minimum level of base current needed to achieve saturation in each case.

4-6. For the circuit described in Problem 4-5, calculate the transistor power dissipation for each collector resistance, at both saturation and cutoff. Also calculate the transistor power dissipation for each load resistance when the collector-emitter voltage is 3 V.

4-7. For a transistor switch, sketch the waveforms of the input voltage, base current, collector current, and collector voltage. Show the various components of transistor turn-on time and turn-off time, and discuss their origins.

4-8. A switching circuit using a 2N4418 transistor (data sheet in Appendix 1-7) has $V_{CC} = 15$ V and $R_C = 2.7$ kΩ. The input pulse width is 2 μs. Calculate the level of V_{CE} (a) before the pulse is applied, (b) at the end of the delay time, (c) at the end of the storage time. Also determine the times from commencement of the input pulse until the transistor switches *on* and until the transistor switches *off*.

4-9. Show how a *speed-up capacitor* may be employed to improve the turn-on and turn-off time of a transistor. Sketch the waveforms of the base current with and without the speed-up capacitor. Explain how the capacitor improves switching speed.

4-10. The circuit described in Problem 4-5 has a base resistance of 27 kΩ. (a) If the circuit is to be switched at the maximum frequency of 100 kHz, calculate the maximum value of the speed-up capacitor that should be used. (b) Determine the maximum switching signal frequency when a 100 pF speed-up capacitor is employed.

4-11. Explain how a junction field effect transistor can be employed as a switch. Sketch a circuit which uses a JFET switch. Sketch the input and output waveforms, and show how the JFET operates when *on* and when *off*. Compare the JFET switch to a bipolar transistor switch.

4-12. A 2N4856 JFET (data sheet in Appendix 1-8) is connected as a switch, with $V_{DD} = 20$ V and $R_D = 4.7$ kΩ. Determine the level of the output (drain-source) voltage (a) when the device is cut off and (b) when it is switched *on*.

4-13. Sketch *p*-channel and *n*-channel MOSFET switching circuits. Show input and output waveforms in each case, and discuss the advantages of MOSFET switches.

4-14. Define CMOS. Sketch the basic CMOS inverter circuit and explain how it operates. Describe the advantages and disadvantages of CMOS.

4-15. Sketch the complete circuit of a direct-coupled bipolar transistor inverter, and explain the function of each component. Show the output waveform when the following inputs are applied: (a) pulse wave, (b) square wave, (c) sine wave.

4-16. Show two methods of protecting a transistor base-emitter junction against excessive reverse input voltages.

4-17. Design a direct-coupled transistor inverter circuit using a 2N3903 transistor. For this circuit, V_{CC} is 9 V, the input is a ± 5 V square wave, and $I_C = 10$ mA.

4-18. If, in the circuit of Problem 4-17, a 200 pF speed-up capacitor is used, determine the maximum input signal frequency that may be employed.

4-19. A direct-coupled transistor inverter using a 2N3904 transistor has an input square wave of ± 9 V and $V_{CC} = 20$ V. An external load of 220 kΩ is connected to the inverter output terminals. Design a suitable circuit and determine the amplitude of the output voltage.

4-20. Sketch the circuit of a normally-*on* capacitor-coupled inverter using (a) an *npn* transistor and (b) a *pnp* transistor. In each case show the input voltage and current waveforms, and explain the operation of the circuit.

4-21. Repeat Problem 4-20 for a normally-*off* capacitor-coupled inverter.

4-22. A normally-*on* transistor inverter has a capacitor-coupled pulse input signal with PA = -3 V and PW = 600 μs; $V_{CC} = 12$ V and I_C is to be 1 mA. Design a suitable circuit.

4-23. The conditions specified in Problem 4-22 can be applied to a normally-*off* inverter if the pulse amplitude is $+3$ V. Design the circuit, taking the signal source resistance as 1 kΩ.

4-24. A normally-*off* inverter circuit using a 2N4418 transistor has $V_{CC} = 9$ V, and $V_{BB} = -3$ V. I_C is to be approximately 10 mA, and the input pulse has PA = 6 V, PW = 500 μs, and $R_s = 600$ Ω. Design the circuit.

4-25. Sketch the circuits of direct-coupled and capacitor-coupled JFET inverter circuits. Explain the operation of the circuits, and discuss their advantages and disadvantages compared to bipolar inverters.

Chapter 5

IC Operational Amplifiers in Switching Circuits

INTRODUCTION

IC operational amplifiers have two input terminals and one output. They have very high voltage gain, high input impedance, and low output impedance. They are extensively used in switching circuits, both as linear amplifiers and as switching devices. In these situations, some operational amplifiers require frequency compensation to suppress a tendency to oscillate at high frequency. Operational amplifiers can be used to improve the performance of such circuits as clippers, differentiators, and integrators.

5-1 IC OPERATIONAL AMPLIFIERS

General Description of an Operational Amplifier

An *operational amplifier* (op-amp) is a very high-gain dc amplifier with two input terminals and one output. One input terminal is known as the *inverting input*, because a positive-going signal at this input produces a negative-going output voltage, and *vice versa*. The other input terminal is designated the *noninverting* input. A positive-going signal at the noninverting input produces a positive-going output. The input impedance of the operational amplifier is extremely high, and the output impedance is very low.

The basic circuit symbol for an operational amplifier is shown in Figure 5-1(a). The noninverting and inverting input terminals are identified as + and −, respectively. The output terminal is at the point of the triangle opposite the inputs. Power supply voltages V_{CC} and $-V_{EE}$ normally are symmetrical with respect to ground. A typical supply voltage is ±15 V. Frequently, additional terminals are shown for connection of external components. The input terminals are usually biased close to ground level.

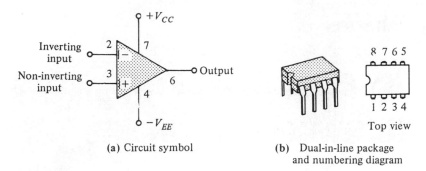

(a) Circuit symbol

(b) Dual-in-line package
and numbering diagram

Figure 5-1 An operational amplifier has one output terminal and two (inverting and noninverting) inputs. Plus and minus ($+V_{CC}$ and $-V_{EE}$) supply voltages are usually required.

Op-amp Performance

Consider the data sheet in Appendix 1-11 for the 741 IC operational amplifier. Note that the device parameters are specified for a supply voltage of $V_S = \pm 15$ V, although the supply may be a maximum of ± 22 V. The voltage gain is 50 000 minimum or 200 000 typical. This is the *open-loop voltage gain*, i.e., the gain without any negative feedback. In this case, the output voltage is the voltage at the output terminal measured with respect to ground, while the input voltage is the voltage difference between the two input terminals. Thus, an output of 10 V, for example, would require a typical input voltage of

$$V_i = \frac{10 \text{ V}}{200\ 000} = 50 \ \mu\text{V}$$

This might be achieved by one input terminal being at ground level, and the other input being at 50 μV above ground.

Other important quantities specified for the 741 are:

Output resistance = 75 Ω typical

Input resistance = 2 MΩ typical

Input bias current = 80 nA typical

The *input bias current* is the current flowing into each of the two input terminals when they are biased to the same voltage level. The *input offset current* is the difference between the two input bias currents. The *input offset voltage* is the voltage difference that may have to be applied between the two input terminals in order to adjust the output level to exactly zero. For the 741, this quantity is typically 1 mV.

Op-amp Frequency Compensation*

Operational amplifiers are widely applied as linear amplifiers, and in such applications resistors are employed to provide negative feedback from output to input. Figure 5-2(a) shows an op-amp connected to function as a noninverting amplifier. Resistors R_2 and R_3 are the feedback components. The negative feedback reduces the amplifier voltage gain (the *closed-loop gain*) to any desired value. Operational amplifiers are never used directly as amplifiers without negative feedback, because the device voltage gain is so high that the output would simply saturate at a level close to one of the supply voltages.

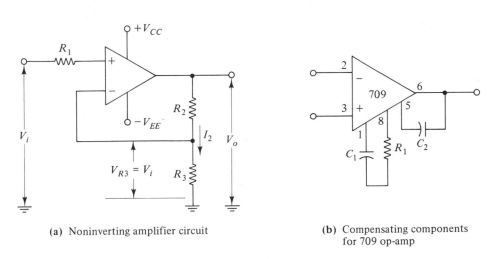

(a) Noninverting amplifier circuit

(b) Compensating components for 709 op-amp

Figure 5-2 In a noninverting amplifier, the signal voltage is applied to the noninverting input terminal of the op-amp. When the input goes in a positive direction, the output also goes positive. External compensating components are required with some operational amplifiers to prevent the circuit from oscillating.

In linear applications, many op-amps have a tendency to oscillate at a high frequency. This is because phase shifts within the amplifier convert the negative feedback into positive feedback at certain frequencies. Measures taken to prevent unwanted oscillations are termed *frequency compensation*. Frequency compensation involves connecting additional components external to the device. [See Figure 5-2(b).] Some operational amplifiers, notably the 741, have frequency-compensating components connected internally. This makes the amplifier very easy to use, but limits its frequency response. The 741 has an upper frequency limit of $f_H \approx 1$ MHz at a voltage gain of 1, while the 709 goes up to $f_H \approx 10$ MHz. With higher voltage gains, the upper cutoff frequencies

* David A. Bell, *Electronic Devices and Circuits*, 3rd Ed. (Englewood Cliffs, N.J.: Prentice-Hall, Inc., 1986), Chapter 14.

are reduced. The 709 requires external compensating components, the values of which depend upon the closed-loop voltage gain of the circuit.

Wherever possible, internally compensated op-amps should be used for convenience. Externally compensated devices must be employed in situations where high-frequency performance is required.

5-2 OP-AMPS AS LINEAR AMPLIFIERS

In many nonlinear, or pulse, applications, operational amplifiers are employed as linear amplifiers. It is important that such circuits be recognized as amplifiers, and that design calculations be understood. It is also important for frequency compensating-components to be correctly selected, where appropriate.

Noninverting Amplifier

In the *noninverting amplifier* shown in Figure 5-2, the output voltage always goes in a positive direction when the input goes positive, and in a negative direction when the input goes negative. Hence, the name noninverting amplifier.

In all linear applications of op-amps, both input terminals are maintained at the same dc potential. If they were not at the same potential, the difference would be amplified, to produce a large dc output voltage. This would result in a feedback voltage, which would reduce the input terminal voltage difference to near zero once again.

The voltage at the inverting input terminal of the circuit in Figure 5-2(a) always follows the voltage at the noninverting terminal. Suppose that the input is $V_i = +100$ mV with respect to ground. Then both the noninverting and inverting input terminals will be $+100$ mV above ground. Since the inverting terminal is connected to the junction of R_2 and R_3, this point will also be $+100$ mV above ground. The bottom of resistor R_3 is grounded. Therefore, input voltage V_i appears across R_3. Also, the output voltage V_o is developed across $(R_2 + R_3)$. Thus,

$$V_o = I_2(R_2 + R_3)$$

and
$$V_i = I_2 R_3$$

The voltage gain of the circuit is

$$A_v = \frac{V_o}{V_i} = \frac{I_2(R_2 + R_3)}{I_2 R_3}$$

or
$$A_v = \frac{R_2 + R_3}{R_3} \tag{5-1}$$

Design of the noninverting amplifier begins with the selection of I_2 very much larger than the maximum level of input bias current to the op-amp input terminals. Then the resistors are calculated as $R_3 = V_i/I_2$ and $(R_2 + R_3) = V_o/I_2$. Resistor R_1 is determined as $R_2 \| R_3$, so that each input terminal of the op-amp has approximately the same dc source resistance. If an op-amp that requires frequency compensation is used, the compensating components are determined from the manufacturer's data sheet. (See Appendix 1-12.) To avoid output saturation, op-amp supply voltages should be at least 3 V greater than the amplifier peak output voltage. The supply voltages should not be less than the minimum specified on the op-amp data sheet.

EXAMPLE 5-1

Design a noninverting amplifier using an (internally compensated) 741 operational amplifier. The amplifier is to have a voltage gain of 28, and the input is $V_i = 50$ mV.

Solution   From the 741 data sheet in Appendix 1-11, the input bias current $I_{B(max)} = 500$ nA.

Let
$$I_2 = 100\, I_{B(max)} = 100 \times 500 \text{ nA}$$
$$= 50\ \mu A$$

Then
$$R_3 = \frac{V_i}{I_2} = \frac{50\text{ mV}}{50\ \mu A}$$
$$= 1\ k\Omega \text{ (standard value)}$$
$$V_o = A_v \times V_i = 28 \times 50\text{ mV}$$
$$= 1.4 \text{ V}$$
$$R_2 + R_3 = \frac{V_o}{I_2} = \frac{1.4\text{ V}}{50\ \mu A}$$
$$= 28\ k\Omega$$
$$R_2 = (R_2 + R_3) - R_3 = 28\ k\Omega - 1\ k\Omega$$
$$= 27\ k\Omega \text{ (standard value)}$$
$$R_1 = R_2 \| R_3 \approx 1\ k\Omega$$

EXAMPLE 5-2

If the noninverting amplifier designed in Example 5-1 uses a 709 op-amp, select suitable compensating components.

Solution

$$A_v = 28 \approx 29 \text{ dB}$$

Refer to the 709 data sheet in Appendix 1-12. The graph of *frequency response for various closed-loop gains* shows compensating components for $A_V = 40$ dB and for $A_V = 20$ dB. Select the components for $A_V = 20$ dB. This will provide *overcompensation* and make the circuit quite stable:

$$C_1 = 500 \text{ pF}, R_1 = 1.5 \text{ k}\Omega, C_2 = 20 \text{ pF}$$

Inverting Amplifier

The circuit shown in Figure 5-3 is termed an *inverting amplifier*, because the output voltage always has a polarity opposite to that of the input. That is, a positive-going input produces a negative-going output, and *vice versa*.

As in all linear applications of op-amps, both input terminals are maintained at the same dc potential. Since the noninverting input terminal is grounded via resistor R_3, the voltage at both terminals always remains at ground level. Because the inverting input terminal remains at ground level, but is not grounded, it is sometimes termed a *virtual ground* or *virtual earth*. With V_i applied to the left side of R_1, and the right side of R_1 at ground level, V_i is developed across R_1. Also, the left side of R_2 is always at ground level, while its right side is at V_o. Consequently, V_o appears across R_2.

$$V_o = -I_1 R_2 \text{ and } V_i = I_1 R_1$$

The voltage gain of the circuit is

$$A_v = \frac{V_o}{V_i} = \frac{-I_1 R_2}{I_1 R_1}$$

or

$$A_v = -\frac{R_2}{R_1} \tag{5-2}$$

The design procedure for an inverting amplifier is very similar to that for a non-inverting circuit. The current I_1 is selected very much larger than the maximum level of the op-amp input bias current. The resistors are calculated as $R_1 = V_i/I_1$, and $R_2 = V_o/I_1$. Resistor R_3 is determined as $R_1\|R_2$.

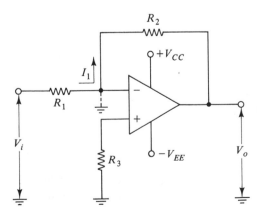

Figure 5-3 In an inverting amplifier, the noninverting input terminal of the op-amp is grounded via a resistor, and the signal voltage is applied to the inverting input terminal via another resistor. A positive-going signal produces a negative-going output, and *vice versa*.

Voltage Follower

An operational amplifier voltage follower circuit, as illustrated in Figure 5-4, is essentially a noninverting amplifier with a voltage gain of 1. The input is applied directly to the noninverting input terminal of the op-amp. The output is not potentially divided (as it is for the noninverting circuit in Figure 5-2), but is applied directly to the inverting input terminal. Thus, the voltage at the inverting input terminal and at the output follow every move of the input voltage.

Like its transistor counterpart, the emitter follower, the voltage follower is employed as a *buffer amplifier* to drive a low-impedance load from a high-impedance signal source. Unlike the emitter follower, there is virtually no difference between input and output voltages with a voltage follower. Also, the voltage follower normally has much higher input impedance and lower output impedance than an emitter follower.

No design calculations are required with a voltage follower, but the signal voltage amplitude must be considered when selecting an operational amplifier. The maximum input voltage that may be employed is defined on the device data sheet as the *input voltage range*. Appendix 1-11 reveals that the input voltage range for the 741 is ± 12 V when a ± 15 V supply is used. With the same supply voltages, a 709 op-amp has an input voltage range of only ± 8 V. (See Appendix 1-12.) If the signal amplitude exceeds the specified input voltage range of the op-amp, internal transistors can go into saturation, and then the output does not follow the input.

As for all linear operational amplifier applications, voltage followers using op-amps without internal compensation must have external compensating components. Because the circuit voltage gain is 1, the compensating capacitors required are larger than for other amplifier circuits.

5-3 OP-AMPS IN SWITCHING CIRCUITS

Frequency Compensation

Operational amplifiers employed in nonlinear applications, where they are switching from one output extreme to another, normally do not require frequency compensation. While the output is in a steady-state condition, there is no ac feedback. In switching

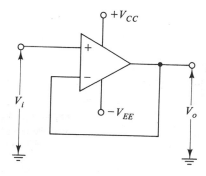

Figure 5-4   A voltage follower is a noninverting amplifier with a voltage gain of 1. Because the voltage follower has a high input impedance and a low output impedance, it is applied as a buffer amplifier.

applications, the feedback is usually a positive dc quantity, rather than negative ac. So the ac negative feedback considerations, important in linear applications, do not normally apply to op-amp switching circuits.

Output Voltage Swing

In switching applications, the output voltage of an op-amp is normally switched between the extreme positive and negative voltage levels. Usually, the output is approximately equal to $+(V_{CC} - 1 \text{ V})$, or $-(V_{EE} - 1 \text{ V})$. The 1 V difference is only a rough approximation; it might easily be 2 V or 0.5 V. The extremes of the output voltage are sometimes referred to as the *output saturation voltage*.

An input voltage of approximately $V_i = V_o/(\text{open-loop voltage gain})$ is required to drive the output to its extreme levels. For a ± 15 V supply,

$$V_i \approx \frac{\pm 15 \text{ V}}{200\,000} \approx \pm 75 \ \mu\text{V}$$

This is the minimum voltage difference that is necessary between the two input terminals. The actual voltage can be much higher than this minimum.

Slew Rate

As discussed in Chapters 2 and 3, the circuits with the highest frequency response tend to give the fastest output rise time when a step input is applied. Therefore, for fast switching applications, op-amps with the highest frequency response give the best

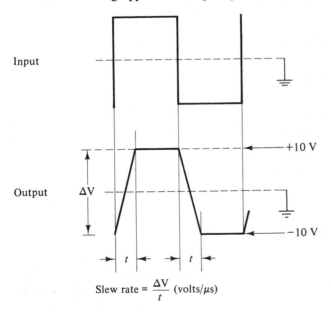

Slew rate = $\dfrac{\Delta V}{t}$ (volts/μs)

Figure 5-5 The *slew rate* of an operational amplifier defines how fast the op-amp output can respond to a step input.

performance. The response of an op-amp to an input step is usually defined in terms of the *slew rate*.

The slew rate is the rate of change of output voltage, or the speed with which the output changes in response to a step input. For the 741, the slew rate is specified as typically 0.5 V/μs. Thus, a time of 1 μs is required for the output to change by 0.5 V. Consider the input and output waveforms in Figure 5-5. The output voltage moves from -10 V to $+10$ V in a time (for a 741) of

$$t = \frac{\Delta V}{\text{(slew rate)}} = \frac{20 \text{ V}}{(0.5 \text{ V/μs})}$$

$$= 40 \text{ μs}$$

For many applications, the slew rate of the 741 is too slow, and another op-amp must be selected. The 715, for example, has a slew rate of 100 V/μs.

5-4 OP-AMP INVERTERS

Direct-Coupled Inverter

An IC operational amplifier employed as a direct-coupled inverter is shown in Figure 5-6(a). The noninverting terminal is connected directly to ground, and the input pulse is applied to the inverting input terminal. When V_i is more than about 75 μV below ground level, the output voltage is saturated at approximately $V_{CC} - 1$ V. At $V_i > 75$ μV above ground, the output becomes approximately $-(V_{EE} - 1 \text{ V})$. Thus, an IC operational amplifier can be directly employed as an inverter without using additional components. This circuit is sometimes termed a *zero-crossing detector*, because the output changes state each time the input voltage crosses the zero level.

Capacitor-Coupled Inverter

Sometimes an op-amp inverter must have a bias voltage provided at its inverting input terminal, in order to hold the output at its positive or negative extreme when no input signals are present. Such a circuit is shown in Figure 5-6(b). If bias voltage V_B is positive, the output is normally negative; if V_B is negative, the output is normally positive. V_B should usually be approximately ±0.5 V, so that a small input signal can easily drive the inverter output from one maximum level to the other.

To design the capacitor-coupled inverter, R_2 should be selected so that the bias current I_2 is very much larger than the input current of the device. Then, R_1 is determined as $R_1 \approx (V_{CC} - V_B)/I_2$. C_c should be selected for an acceptable level of tilt on the signal at the inverting input terminal.

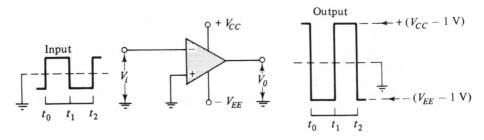

(a) Direct-coupled op-amp inverter

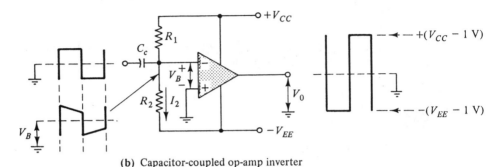

(b) Capacitor-coupled op-amp inverter

Figure 5-6 Operational amplifiers may be employed as direct-coupled inverters and as capacitor-coupled inverters. The output of the capacitor-coupled inverter may be *normally-high* or *normally-low*.

EXAMPLE 5-3

Using a 741 IC operational amplifier, design an inverter as in Fig. 5-6(b) to provide an output of $V_o \approx \pm 11$ V. The output normally should be negative when no input is present. The input voltage is a ± 6 V square wave with $f = 1$ kHz.

Solution For $V_o \approx \pm 11$ V, the supply should be ± 12 V.
To maintain a negative dc output, bias voltage V_B must be positive. Take $V_B \approx +0.5$ V.

$$V_{R_2} = V_B - (-V_{EE})$$

$$= 0.5 \text{ V} - (-12 \text{ V})$$

$$= 12.5 \text{ V}$$

From the 741 data sheet in Appendix 1-11, $I_{B(\text{max})} = 500$ nA.

Let $I_2 = 100 \times 500$ nA

        $= 50 \ \mu\text{A}$

Then
$$R_2 = \frac{V_{R_2}}{I_2} = \frac{12.5\ V}{50\ \mu A}$$

$$= 250\ k\Omega \ (\text{use } 220\ k\Omega \text{ standard value})$$

I_2 becomes
$$\frac{V_{R_2}}{R_2} = \frac{12.5\ V}{220\ k\Omega} = 56.8\ \mu A$$

$$R_1 = \frac{V_{CC} - V_B}{I_2}$$

$$= \frac{12\ V - 0.5\ V}{56.8\ \mu A}$$

$$= 202\ k\Omega \ (\text{use } 180\ k\Omega \text{ standard value})$$

Note that use of a *smaller-than-calculated* value of R_1 ensures that V_B is larger than required for the output to be negative.

During the pulse time, C_c charges via the circuit input resistance, so the input waveform has *tilt* as it appears at the inverting input terminal. For a ±6 V input, a tilt of two or three volts will have no effect on the inverter operation.

Let $\Delta V_C = 2\ V$.

$$\text{Input resistance } R_i = R_1 \| R_2$$

$$= 180\ k\Omega \| 220\ k\Omega$$

$$= 99\ k\Omega$$

For a ±6 V input, the approximate input current is

$$I \approx \frac{V_i}{R_i} = \frac{6\ V}{99\ k\Omega}$$

$$\approx 60.6\ \mu A$$

$$PW = \frac{1}{2f}$$

$$PW = \frac{1}{2 \times 1\ kHz} = 500\ \mu s$$

For I assumed constant,

Eq. 2-8,
$$C = \frac{It}{\Delta V} = \frac{60.6\ \mu A \times 500\ \mu s}{2\ V}$$

$$\approx 0.015\ \mu F \ (\text{standard capacitance value})$$

5-5 OP-AMP CLIPPER CIRCUITS

Precision Rectifiers

The circuit of an op-amp *negative voltage clipper* (see Section 3-3), or *precision rectifier*, is illustrated in Figure 5-7(a). The circuit is simply a voltage follower with a diode inserted between the op-amp output terminal and the circuit output point. When the input signal is positive, the diode is forward biased, and the output voltage follows the input. Note that the feedback causes the voltage at the inverting input terminal to follow that at the noninverting terminal. Since the output of the circuit and the inverting input terminal are common, the output follows the input (within about 75 µV). The diode forward voltage drop is not involved.

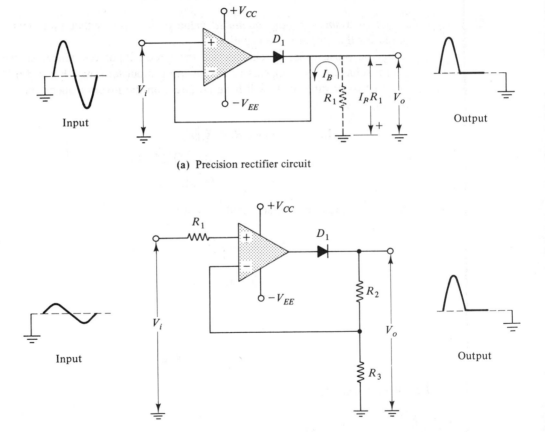

(a) Precision rectifier circuit

(b) Precision rectifier with amplification

Figure 5-7 Operational amplifier clipper circuits, or precision rectifiers. The diodes block the unwanted half-cycles of the input waveform. The circuits have low output impedance, and the diode forward voltage drop does not affect the output voltage.

While the input voltage is in its negative half-cycle, tne op-amp output is negative, and the diode is reverse biased. Consequently, the feedback path is interrupted, the inverting input terminal remains at ground level, and the op-amp output is saturated in a negative direction. The negative half-cycle of the input does not pass to the output—i.e., it is clipped off.

If there is no load resistor at the output of the precision rectifier, resistor R_1 [shown broken in Figure 5-7(a)] should be included. This is to ensure that the output and the op-amp inverting input terminal remain at ground level during the time that the diode is reverse biased. During this time, the input bias current to the op-amp produces a voltage drop $I_B \times R_1$ across resistor R_1, as illustrated. This gives a small unwanted negative output voltage. Obviously, R_1 should be selected so that this voltage is negligible. The diode should be selected as a low-current switching device, with a reverse breakdown voltage greater than V_{EE}.

The advantages of the op-amp clipper circuit over a simple diode clipper are (1) no diode voltage drop between input and output, and (2) low output impedance. If the diode polarity is reversed in Figure 5-7(a), the negative half-cycle of the input waveform will be passed to the output, and the positive half-cycle will be clipped off.

Figure 5-7(b) shows the circuit of a precision rectifier with voltage gain. Clearly, this is a noninverting amplifier with the diode included. The circuit is designed exactly as discussed in Section 5-2. Because of the presence of R_2 and R_3, no additional external resistor is required, as is the case in Figure 5-7(a).

Peak Clipper

Back-to-back series-connected Zener diodes are used in the circuit of Figure 5-8 to clip off the peaks of the output voltage. Without the diodes, the circuit is an op-amp

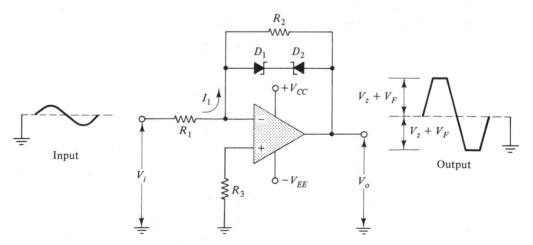

Figure 5-8 An op-amp peak clipper circuit is simply an inverting amplifier with back-to-back series-connected Zener diodes across the feedback resistor. The diodes limit the output voltage to $V_z + V_F$.

inverting amplifier. As discussed in Section 5-2, the output voltage appears across feedback resistor R_2. When this voltage equals the sum of the forward voltage drop of one Zener diode and the reverse breakdown voltage of the other, i.e., $V_F + V_z$, the output is limited, or clipped. When V_{z1} equals V_{z2}, the output voltage will be clipped symmetrically, as illustrated. If V_{z1} and V_{z2} are unequal, the positive and negative output amplitudes will be unequal.

To design a peak clipper circuit, the Zener diodes are first selected to clip the output at the desired levels. The resistors are then calculated exactly as for an inverting amplifier, except that current I_1 must be selected larger than the knee current I_{zk} for the Zener diodes. (See Figure 3-4.)

5-6 OP-AMP DIFFERENTIATOR

The CR differentiating circuits discussed in Section 2-6 can be improved by the use of operational amplifiers. The op-amp circuit may be designed to give an output amplitude larger than the input signal, which is not the case with the simple CR circuit. A further advantage is that the op-amp differentiator has a very low output impedance, whereas the CR circuit can be overloaded.

Consider the op-amp differentiating circuit in Figure 5-9. For dc considerations, C_1 is an open circuit. The noninverting input terminal is grounded via R_3, and the inverting input is connected to the output via R_2. Thus, the circuit behaves as a voltage follower with the (noninverting) input grounded. While no signal is present, the output of the circuit remains at ground level.

When the input signal is a positive-going voltage, a current I_1 flows into C_1, as illustrated. If I_1 is much larger than the maximum input bias current to the op-amp, then effectively all of I_1 flows through R_2. The left side of R_2, at the op-amp inverting input, remains close to ground, and the output voltage is

$$V_o = -I_1 R_2$$

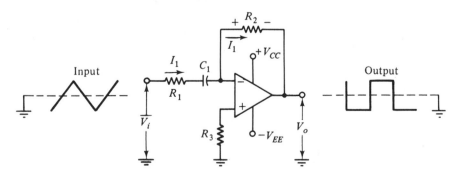

Figure 5-9 An operational amplifier differentiating circuit gives an inverted output with an amplitude that can be larger than the signal amplitude. It also has a low output impedance.

Since I_1 is the capacitor current, the voltage developed across R_2 is the differential of the input waveform, as in the case of the simple CR differentiator. Note that the differentiated output of this circuit is inverted: when the input is positive-going, the output is negative, and *vice versa*.

Resistor R_1 in series with capacitor C_1 is included to prevent the circuit from oscillating at high frequencies. To fully understand this, it is necessary to study the typical open-loop frequency and phase responses of an operational amplifier, and frequency compensation techniques. Suffice to say that, in the absence of R_1, the feedback network consisting of R_2 and C_1 can add a 90° phase lag between the output terminal and the inverting input terminal. At high frequencies, there is additional phase lag within the op-amp, which, when added to the feedback network phase lag, can make the circuit oscillate. If R_1 is much larger than X_{C_1} at the frequency at which the circuit might oscillate, then the feedback network phase lag is reduced to near zero, and the circuit is stable.

If the frequency at which the circuit might oscillate is known, R_1 can be readily calculated. But a suitable value of R_1 can usually be determined by selecting it to have no significant effect on the normal operation of the differentiator. A simple rule of thumb is to make $R_1 = R_2/100$. If the signal source resistance R_s is equal to or larger than the calculated value of R_1, then R_1 can be omitted. Note that the voltage gain for determining the op-amp compensating components is $A_v = -R_2/(R_1 + R_s)$.

Design of an op-amp differentiator commences with selection of I_1 very much larger than the $I_{B(max)}$ of the op-amp. Resistor R_2 is calculated as V_o/I_1, and Eq. 2-17 is employed to calculate C_1. Resistor R_3 is made equal to R_2, so that each input terminal of the op-amp has the same bias resistance. Resistor R_1 is determined as explained above. The operational amplifier selected must have a slew rate at least several times greater than the fastest rate of change of the input signal to be differentiated.

EXAMPLE 5-4

Design a differentiating circuit to give $V_o = 5$ V when the input changes by 1 V over a time period of 10 μs.

Solution

$$I_1 \gg I_{B(max)}$$

So let

$$I_1 = 500 \ \mu A$$

Then

$$R_2 = \frac{V_o}{I_1} = \frac{5 \text{ V}}{500 \ \mu A}$$

$$= 10 \text{ k}\Omega \text{ (standard value)}$$

From Eq. 2-17,

$$C_1 = \frac{e_r \times t}{R \times E_{(max)}} = \frac{V_o \times t_r}{R_2 \times V_i}$$

$$= \frac{5\,V \times 10\,\mu s}{10\,k\Omega \times 1\,V}$$

$$= 5000\,pF \text{ (standard value)}$$

$$R_1 = \frac{R_2}{100} = \frac{10\,k\Omega}{100}$$

$$= 100\,\Omega$$

$$R_3 = R_2 = 10\,k\Omega$$

5-7 MILLER INTEGRATOR CIRCUIT

Miller Effect

Consider the circuit of Figure 5-10 in which an operational amplifier is connected as an *inverting amplifier* (without resistors). Let the amplifier voltage gain be $-M$. Then

$$V_o = -MV_i$$

Note that because the input is applied to the op-amp inverting input terminal, the voltage at the left terminal of C_1 increases by V_i, while that at the right terminal of the capacitor decreases by MV_i, when V_i is positive. This results in a total capacitor voltage change of

$$\Delta V_1 = V_i + MV_i$$

$$= V_i(1 + M)$$

From Eq. 2-8, the charge supplied to the capacitor is

$$Q = C_1 \times \Delta V_1$$

$$= C_1 \times V_i(1 + M)$$

or
$$Q = (1 + M)C_1 \times V_i$$

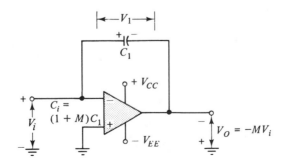

Figure 5-10 The Miller effect, or capacitance amplification, occurs at the input of an inverting amplifier. The amplifier input capacitance is $(1 + M)C_1$.

Thus, it appears that the input has supplied a charge to a capacitor with a value of $(1 + M)C_1$, instead of just C_1. Capacitance C_1 is said to have been *amplified* by a factor of $(1 + M)$. This is known as the *Miller effect*.

Miller Integrator

The *Miller integrator* utilizes the Miller effect to generate a linear ramp. In the circuit of Figure 5-11(a), a square wave input supplies charging current, alternately positive and negative, to C_1. The noninverting input terminal of the op-amp is grounded by a resistance R_2 (equal to resistance R_1). Because the noninverting terminal is grounded,

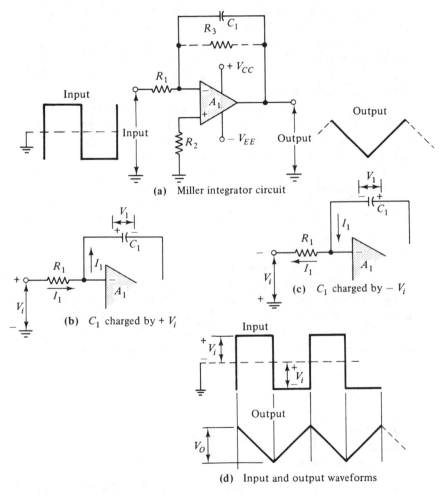

(a) Miller integrator circuit

(b) C_1 charged by $+ V_i$

(c) C_1 charged by $- V_i$

(d) Input and output waveforms

Figure 5-11 Miller integrator circuit. A symmetrical square wave input charges the capacitor linearly, first with one polarity, and then with reversed polarity. This produces a triangular wave output.

the inverting input terminal is always very close to ground level. Thus, the input voltage appears across R_1 and the input current is simply V_i/R_1, which remains constant.

If the input current I_1 is much greater than the input bias current of the amplifier, then I_1 will not flow into the amplifier. Instead, effectively all of I_1 flows through capacitor C_1. For a positive input voltage, I_1 flows into C_1, charging it positively on the left side and negatively on the right side [Figure 5-11(b)]. In this case the output voltage becomes negative, because the positive terminal, that is, the left terminal, of the capacitor is held at the virtual ground level of the inverting input terminal. A negative input voltage produces a flow of current out of C_1 [Figure 5-11(c)]. Thus, the capacitor is charged negatively on the left side and positively on the right side. Now the output becomes positive, because the negative terminal of the capacitor is held at virtual ground.

Since I_1 is a constant (+ or −) quantity, and since effectively all of I_1 flows through the capacitor, C_1 is charged linearly. Thus, the output voltage changes linearly, providing either a positive or negative ramp. When the input voltage is positive, the output is a negative-going ramp. When the input is negative, a positive-going output ramp is generated. When the input is a square wave, the output waveform is triangular. This is illustrated in Figure 5-11(d).

Consider the Miller circuit of Figure 5-11(a). If the op-amp inverting input is supposed to be at ground but is, say, 20 μV away from ground level, then the output voltage could be $(M \times 20 \text{ μV}) = \pm(200\ 000 \times 20 \text{ μV}) = \pm 4$ V. In this case the output is said to have *drifted* from its zero level. Even when the input terminal is maintained exactly at ground level, there could be a slight difference in the voltage at the amplifier inputs, due to small differences in the resistances of R_1 and R_2, for example. Thus, because of the very high gain of the operational amplifier, its output voltage is very likely to drift from the zero level. The output voltage drift produces a charge on capacitor C_1; this charge gives the output an *offset* so that it is not symmetrical above and below ground. (See Figure 5-12.)

To minimize the output voltage drift, a large resistance R_3 is connected between the output and the inverting input terminals. The effect of this resistance is to *cut down* the dc gain of the amplifier. When $R_3/R_1 = 10$, for example, the output drift will be only 10 times the input voltage difference. A ratio of 10:1 is typical for R_3/R_1.

The presence of R_3 has the disadvantage that it adversely affects the performance of the integrator at low frequencies. If the input frequency is so low that the impedance

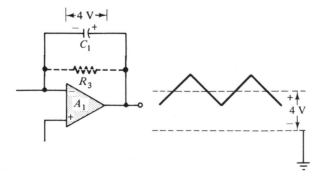

Figure 5-12 In the absence of dc feedback, voltage drift occurs at the output of an integrator. This can be corrected by including feedback resistor R_3.

of C_1 is very much larger than R_3, then the capacitor has a negligible effect and the circuit will not function as an integrator. Therefore, C_1 should be selected so that $X_{C_1} \ll R_3$ at the lowest operating frequency of the integrator. As a lower limit, let $X_{C_1} = R_3/10$, so that

$$\frac{1}{2\pi f C_1} = \frac{R_3}{10}$$

Or, the lowest operating frequency of the integrator is

$$f = \frac{10}{2\pi C_1 R_3} \qquad (5\text{-}3)$$

The design of a Miller integrator circuit begins with selection of the input current I_1 very much larger than the amplifier bias current. Then, R_1 is calculated as V_i/I_1. From Eq. 2-8, C_1 is determined using the desired output voltage, the time period, and the input current.

EXAMPLE 5-5

Design a Miller integrator circuit to produce a triangular waveform output with a peak-to-peak amplitude of 4 V. The input is a ± 10 V square wave with a frequency of 250 Hz. Use a 741 operational amplifier with a supply of ± 15 V. Calculate the lowest operating frequency for the integrator.

Solution   The circuit is shown in Figure 5-11(a).
From Appendix 1-11,

$$I_{B(\text{max})} = 500 \text{ nA}$$
$$I_1 \gg I_{B(\text{max})}$$

For convenient calculations, let $I_1 = 1$ mA.

$$R_1 = \frac{V_i}{I_1} = \frac{10}{1 \text{ mA}}$$

$$= 10 \text{ k}\Omega$$

Let $R_3 = 10 R_1 = 100 \text{ k}\Omega$

$$R_2 = R_3 \| R_1 \approx 10 \text{ k}\Omega$$

The ramp length is equal to one-half of the time period of the input, which is $1/(2f)$, or

$$t = \frac{1}{2 \times 250 \text{ Hz}} = 2 \text{ ms}$$

The ramp amplitude is equal to the peak-to-peak voltage output, which is 4 V. Therefore,

$$C_1 = \frac{It}{\Delta V} = \frac{1 \text{ mA} \times 2 \text{ ms}}{4 \text{ V}} = 0.5 \text{ }\mu\text{F}$$

From Equation 5-3, the lowest operating frequency is

$$f = \frac{10}{2\pi \times 0.5 \text{ }\mu\text{F} \times 100 \text{ k}\Omega}$$

$$= 32 \text{ Hz}$$

REVIEW QUESTIONS AND PROBLEMS

5-1. Sketch the circuit symbol for an operational amplifier, identifying the most important terminals. Then briefly describe an operational amplifier, and state typical values of open-loop voltage gain, input impedance, output impedance, and input bias current.

5-2. Briefly explain operational amplifier frequency compensation. State whether frequency compensation may be required for each of the following circuits: noninverting amplifier, inverting amplifier, voltage follower, direct-coupled inverter, capacitor-coupled inverter, precision rectifier, peak clipper, differentiator, Miller integrator.

5-3. Sketch noninverting amplifier, inverting amplifier, and voltage follower operational amplifier circuits. Briefly explain the operation and characteristics of each circuit.

5-4. Design an op-amp noninverting amplifier to have a voltage gain of 75 and an input of ± 100 mV.

5-5. A noninverting amplifier is to have an input of ± 0.3 V and an output of ± 12 V. Design a suitable op-amp circuit.

5-6. Assuming that the circuits designed for Problems 5-4 and 5-5 use 709 operational amplifiers, select suitable compensating components in each case.

5-7. Design an op-amp inverting amplifier to have a voltage gain of 100 and an input voltage of ± 50 mV.

5-8. An op-amp inverting amplifier is to have an input of ± 0.5 V, and an output of ± 10 V. Design a suitable circuit.

5-9. Select suitable compensating components for each of the circuits designed in Problems 5-7 and 5-8. Assume that 709 operational amplifiers are used.

5-10. Sketch the circuits of direct-coupled and capacitor-coupled operational amplifier inverter circuits. Show input and output waveforms, and briefly explain the operation of each circuit.

5-11. Using a 709 operational amplifier, design a capacitor-coupled inverter to give an output of ± 14 V. The output should normally be negative when no input is present. The input is a ± 4 V square wave with a frequency of 500 Hz.

5-12. A capacitor-coupled op-amp inverter is to have a normally-high output. The input is a ± 5 V square wave with a frequency of 1.5 kHz, and the output is to be ± 9 V. Design a suitable circuit, using a 741 op-amp. Calculate the approximate rise time of the output voltage.

5-13. An operational amplifier with a maximum input bias current of 750 nA and a slew rate of

10 V/µs is to be used as an inverter. The supply voltage is $V_{CC} = \pm 12$ V, and the inverter output is to be normally positive. The square wave input has a minimum amplitude of ± 3 V and a frequency of 200 Hz. Design a suitable circuit.

5-14. Sketch the circuit of an op-amp precision rectifier (a) with a voltage gain of 1, and (b) with a voltage gain greater than 1. Sketch input and output waveforms for each circuit, and briefly explain.

5-15. Design an op-amp precision rectifier circuit with a voltage gain of 5. The input voltage is ± 1 V.

5-16. Sketch an op-amp peak clipper circuit using Zener diodes. Show input and output waveforms, and briefly explain the circuit operation.

5-17. Redesign the circuit in Problem 5-8 to have the output waveform clipped at approximately ± 9 V.

5-18. Sketch an op-amp differentiating circuit and explain how it operates. Compare this circuit to a simple CR differentiating circuit.

5-19. Using a 741 op-amp, design a differentiating circuit to give an output of 2 V when the input changes by 1 V/µs.

5-20. Calculate the output voltages when the input waveform in Figure 2-12 is differentiated by the circuit designed in Problem 5-19.

5-21. Sketch the circuit of an operational amplifier Miller integrator. Show input and output waveforms, and briefly explain the operation of the circuit. Explain the Miller effect.

5-22. Design a Miller integrator circuit to produce a triangular output waveform with a peak-to-peak amplitude of 3 V. The input is a ± 8 V square wave with a frequency of 750 Hz. Use a 741 operational amplifier with a supply of ± 12 V. Calculate the lowest operating frequency for the integrator.

5-23. A Miller integrator as in Figure 5-11 has the following component values: $R_1 = 3.3$ kΩ, $R_2 = 3.3$ kΩ, $R_3 = 330$ kΩ, $C_1 = 0.025$ µF. Determine the amplitude of the input square wave that will give a ± 5 V triangular output with a frequency of 2 kHz.

Chapter 6

Schmitt Trigger Circuits and Voltage Comparators

INTRODUCTION

Essentially, a **Schmitt trigger circuit** is a fast-operating voltage level detector. When the input voltage arrives at the upper or lower triggering levels, the output voltage rapidly changes. The circuit operates from almost any input waveform and always gives a pulse-type output. Transistor Schmitt trigger circuits can be designed to trigger at specified upper and lower levels of input voltage. An IC operational amplifier circuit can also be employed as a Schmitt trigger circuit, and several Schmitt trigger circuits are available in a single IC package. **Voltage comparator circuits** produce an output pulse when the input voltage level becomes exactly equal to a reference voltage.

6-1 TRANSISTOR SCHMITT TRIGGER CIRCUIT

A transistor Schmitt trigger circuit is shown in Figure 6-1(a). The figure shows that transistors Q_1 and Q_2 have a common emitter resistor R_E. The Q_2 base voltage, V_{B2}, is derived via the potential divider (R_1 and R_2) from the collector of Q_1. Transistors Q_1 and Q_2 have collector resistances R_{C1} and R_{C2}, respectively. The arrangement is such that when transistor Q_1 is *on*, transistor Q_2 is *off*; then, for Q_2 to switch *on*, Q_1 must switch *off*.

To understand the operation of this type of circuit, first consider the dc conditions when Q_1 is *off*. When it is *off*, Q_1 can be regarded as an open circuit; therefore, it can be left out of the circuit, as shown in Figure 6-1(b). The Q_2 base voltage is now derived from V_{CC} via a potential divider consisting of R_1, R_2, and R_{C1}. Thus, Q_2 is *on*, and a collector current I_{C2} flows, producing a voltage drop across R_{C2}. The output voltage is $(V_{CC} - I_{C2}R_{C2})$.

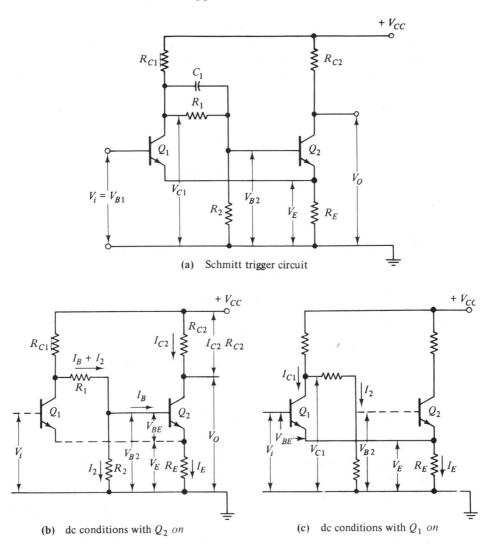

(a)   Schmitt trigger circuit

(b)   dc conditions with Q_2 *on*   (c)   dc conditions with Q_1 *on*

Figure 6-1   A transistor Schmitt trigger circuit switches between (Q_1 *on*, Q_2 *off*), and (Q_1 *off*, Q_2 *on*). When the input voltage is increased to the *upper trigger point*, Q_1 switches *on* and Q_2 switches *off*. When the input is reduced to the *lower trigger point*, Q_1 switches *off* and Q_2 switches *on*.

If, as in Figure 6-1(c), Q_1 is now triggered *on*, the emitter voltage becomes $V_E = V_i - V_{BE}$. Also, the collector current I_{C1} causes a voltage drop across R_{C1}, causing V_{B2} to fall below the level of V_E. Thus, when Q_1 is *on*, Q_2 is biased *off*, and I_{C2} becomes zero. At this point there is no longer any significant voltage drop across R_{C2}, and the output voltage is approximately V_{CC}.

Now, reconsider the conditions illustrated in Figure 6-1(b). It is seen that with Q_2 *on*, $V_E = V_{B2} - V_{BE}$. This is the voltage at the emitter terminal of both transistors, since they are connected together. Transistor Q_1 will not switch *on* until its base voltage becomes greater than V_E. In fact, Q_1 switches *on* approximately at $V_i = V_E + V_{BE}$, which equals V_{B2}. Obviously, if V_i is suddenly made greater than this level, Q_1 would switch *on* rapidly. The lowest level of V_i that causes Q_1 to switch *on* is known as the *upper trigger point* (UTP) for the circuit.

At the moment that Q_1 begins to switch *on*, it starts to pull the common emitter voltage up, the flow of I_C causes V_{C1} to fall, and, consequently, V_{B2} falls. Thus, as Q_1 switches on, it causes a rapid reduction in V_{BE2}. This effect, known as *regeneration*, produces a very rapid switchover from Q_2 *on* to Q_1 *on*.

Now consider the process of switching from Q_1 *on* (Q_2 *off*) to Q_2 *on* (Q_1 *off*). From Figure 6-1(c), with Q_1 *on*, $I_E = (V_i - V_{BE})/R_E$. Thus, a reduction in V_i also reduces I_E, and since $I_C \approx I_E$, I_{C1} also becomes smaller. The voltage drop across R_{C1} is approximately $I_{C1}R_{C1}$, and the collector voltage of Q_1 is $V_{C1} \approx (V_{CC} - I_{C1}R_{C1})$. Therefore, when V_i is reduced, I_{C1} becomes smaller, causing V_{C1} to rise. Because V_{B2} is derived from Q_1 collector, the base voltage of Q_2 rises as V_{C1} rises and V_i falls. If V_i is reduced by a very small amount, the resultant small increase in V_{B2} may leave Q_2 base still below the level of its emitter voltage. In fact, Q_2 switches *on* again only when V_{B2} and V_i become equal. The input voltage at which this occurs is known as the *lower trigger point* (LTP).

In the changeover from Q_1 *on* to Q_2 *on*, regeneration again occurs. When Q_2 starts to switch *on*, it causes Q_1 to begin to switch *off* because of the rise in the common emitter voltage, and Q_1 switching *off* helps Q_2 to turn *on*. Again, the changeover occurs very rapidly.

Speed-up capacitor C_1 [Figure 6-1(a)] is provided solely to improve the circuit switching speed. The effect is essentially as explained in Section 4-4. However, it could also be said that during switching, the voltage change at the collector of Q_1 is potentially divided across R_1 and R_2 before being applied to Q_2 base. The presence of C_1 eliminates the potential division (i.e., $\Delta V_{B2} = \Delta V_{C1}$), and thus speeds up the switching process.

Figure 6-2 shows the effects of various input waveforms on the Schmitt trigger circuit. In each case the output is $(V_{CC} - I_{C2}R_{C2})$ until the upper trigger point is reached. Then, with Q_2 switched *off*, the output becomes approximately V_{CC}. When the signal input falls to the LTP, the output again drops to $(V_{CC} - I_{C2}R_{C2})$ as Q_2 switches *on*. The Schmitt trigger circuit is seen to be essentially a voltage level detector, capable of producing a fast-moving output when either a slow- or fast-moving input arrives at the trigger points.

6-2 DESIGNING FOR THE UTP

Design of any circuit starts from a specification. For the Schmitt trigger circuit, the most important parameters are the upper and lower trigger points, UTP and LTP. A circuit may be designed simply to meet a given UTP, and to completely ignore the

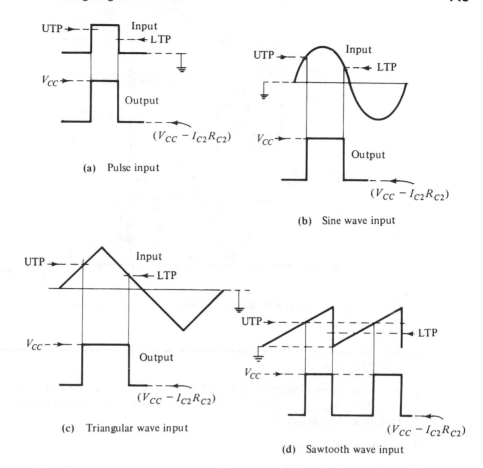

Figure 6-2 Schmitt trigger circuit response to various input waveforms. When the input arrives at the *upper trigger point*, the output switches to V_{CC}. When the input falls to the *lower trigger point*, the output drops to $(V_{CC} - I_{C2}R_{C2})$.

LTP. In this case, the LTP normally is located between the UTP and ground. Such a circuit could function quite satisfactorily with the waveforms illustrated in Figure 6-2.

As already explained, the UTP is equal to the Q_2 base voltage V_{B2} when Q_2 is *on*. Therefore, the circuit is designed to have V_{B2} equal to the specified UTP. Potential divider R_1 and R_2, together with R_{C1}, must provide a stable bias voltage V_{B2} for Q_2 (Figure 6-3). However, R_1 and R_2 must be large enough to avoid overloading R_{C1}. For example, if R_1 and R_2 were made smaller than R_{C1}, then Q_1 collector current would have very little effect on V_{B2} when Q_1 is switched *on*. If R_1 and R_2 are made very large, I_{B2} will cause a large voltage drop across R_1 when Q_2 is switched *on*. Therefore, R_1 and R_2 must be kept as small as possible, but they must be several times larger than R_{C1}. A good rule of thumb here is to take $I_2 \approx I_{E2}/10$ and then calculate R_2 as V_{B2}/I_2. The design procedure is best understood by working through an example.

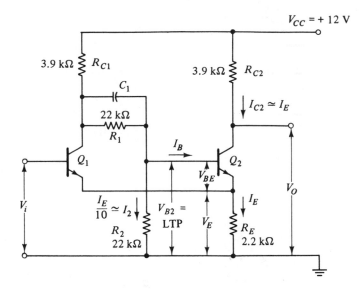

Figure 6-3   Schmitt trigger circuit designed in Example 6-1.

EXAMPLE 6-1

A Schmitt trigger circuit is to have UTP = 5 V. The two silicon transistors to be used have $h_{FE(min)}$ = 100, and I_C is to be 2 mA. The available supply is 12 V. Design a suitable circuit.

Solution   Refer to Figure 6-3.

$$\text{UTP} = V_{B2} = 5 \text{ V}$$

$$V_E = V_{B2} - V_{BE} = 5 \text{ V} - 0.7 \text{ V} = 4.3 \text{ V}$$

$$I_E \approx I_C = 2 \text{ mA}$$

$$R_E = \frac{V_E}{I_E} = \frac{4.3 \text{ V}}{2 \text{ mA}}$$

$$= 2.15 \text{ k}\Omega \quad \text{(use 2.2 k}\Omega \text{ standard value)}$$

Taking Q_2 as saturated, $V_{CE(sat)}$ = 0.2 V typically. The voltage drop across R_{C2} = $I_C R_{C2}$, and

$$I_C R_{C2} = V_{CC} - V_E - V_{CE(sat)} = 12 \text{ V} - 4.3 \text{ V} - 0.2 \text{ V}$$

$$= 7.5 \text{ V}$$

$$R_{C2} = \frac{7.5 \text{ V}}{I_C} = \frac{7.5 \text{ V}}{2 \text{ mA}} = 3.75 \text{ k}\Omega \quad \text{(use 3.9 k}\Omega \text{ standard value)}$$

$$I_2 \approx \frac{1}{10} I_E = \frac{1}{10} \times 2 \text{ mA} = 0.2 \text{ mA}$$

$$R_2 = \frac{V_{B2}}{I_2} = \frac{5 \text{ V}}{0.2 \text{ mA}} = 25 \text{ k}\Omega \quad \text{(use 22 k}\Omega \text{ standard value)}$$

I_2 now becomes 5 V/22 kΩ = 0.227 mA, and

$$I_{B2} = \frac{I_{C2}}{h_{FE(\text{min})}} = \frac{2 \text{ mA}}{100} = 20 \text{ } \mu\text{A}$$

$$I_{B2} + I_2 = 20 \text{ } \mu\text{A} + 0.227 \text{ mA} = 0.247 \text{ mA}$$

$$R_{C1} + R_1 = \frac{V_{CC} - V_{B2}}{I_2 + I_{B2}} = \frac{12 \text{ V} - 5 \text{ V}}{0.247 \text{ mA}}$$

$$= 28.3 \text{ k}\Omega$$

When the LTP is not specified, R_{C1} may be made equal to R_{C2}. *This cannot be done when the circuit is to be designed for a given LTP level.*

$$R_{C1} = R_{C2} = 3.9 \text{ k}\Omega$$

$$R_1 = (R_{C1} + R_1) - R_{C1} = 28.3 \text{ k}\Omega - 3.9 \text{ k}\Omega$$

$$= 24.4 \text{ k}\Omega \quad \text{(use 22 k}\Omega \text{ standard value)}$$

Since standard-value components have been selected in Example 6-1, the UTP will not be exactly as specified. A potentiometer connected between R_1 and R_2, with Q_2 base connected to its moving contact, would provide adjustment to obtain a precise level of UTP.

6-3 DESIGNING FOR THE UTP AND LTP

When V_i decreases to the LTP, the circuit is about to change state, but Q_1 is still *on* and Q_2 is *off*. The situation is illustrated in Figure 6-4. As V_i approaches the LTP, V_{B1} is decreasing and V_{B2} is increasing. The lower trigger point occurs when $V_{B2} = V_{B1} = V_i$. The design procedure, when both UTP and LTP are specified, is exactly the same as in Example 6-1, up to the point at which R_{C1} is chosen.

EXAMPLE 6-2

A Schmitt trigger circuit is to be designed with a UTP of 5 V and an LTP of 3 V. The silicon transistors employed have $h_{FE(\text{min})}$ = 100, and I_C is to be 2 mA. The available supply is 12 V. Design a suitable circuit.

Solution   With the exception of the LTP, the circuit is exactly as specified for Example 6-1. Therefore, from Example 6-1,

$$R_{C2} = 3.9 \text{ k}\Omega \qquad R_E = 2.2 \text{ k}\Omega \qquad R_2 = 22 \text{ k}\Omega$$

$$(R_1 + R_{C1}) = 28.3 \text{ k}\Omega$$

Figure 6-4 shows the circuit conditions when Q_1 is *on* and V_i is exactly at the LTP. $V_i = \text{LTP} = V_{B2} = 3$ V,

$$I_1 = \frac{V_{B2}}{R_2} = \frac{3 \text{ V}}{22 \text{ k}\Omega} = 0.136 \text{ mA}$$

$$I_{C1} \approx I_E = \frac{V_{B1} - V_{BE}}{R_E} = \frac{3 \text{ V} - 0.7 \text{ V}}{2.2 \text{ k}\Omega} = 1.045 \text{ mA}$$

$$V_{CC} = R_{C1}(I_{C1} + I_1) + I_1(R_1 + R_2)$$

$$= R_{C1}(I_{C1} + I_1) + I_1(28.3 \text{ k}\Omega - R_{C1} + R_2)$$

$$= R_{C1}I_{C1} + I_1(28.3 \text{ k}\Omega + R_2)$$

or $\qquad R_{C1} = \dfrac{V_{CC} - I_1(28.3 \text{ k}\Omega + R_2)}{I_{C1}}$

$$= \frac{12 \text{ V} - 0.136 \text{ mA} (28.3 \text{ k}\Omega + 22 \text{ k}\Omega)}{1.045 \text{ mA}}$$

$$= 4.94 \text{ k}\Omega \quad \text{(use 4.7 k}\Omega \text{ standard value)}$$

$$R_1 = 28.3 \text{ k}\Omega - R_{C1}$$

$$= 28.3 \text{ k}\Omega - 4.7 \text{ k}\Omega$$

$$= 23.6 \text{ k}\Omega \quad \text{(use 22 k}\Omega \text{ standard value)}$$

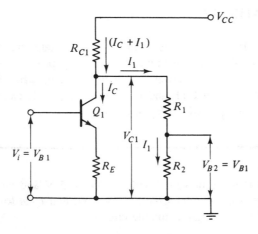

Figure 6-4 To determine the *lower trigger point* of a transistor Schmitt trigger circuit, Q_2 is assumed to be *off*, and the level of V_i is calculated that will give $V_{B_2} = V_i$.

6-4 SPEED-UP CAPACITOR

The effects of a speed-up capacitor (or commutating capacitor) on transistor switching times are discussed in Sec. 4-4. As already explained, the capacitor value can be as large as possible, as long as it is small enough to allow its voltage to return to normal dc levels between switching. For a Schmitt circuit to trigger at the UTP and LTP levels as designed, the capacitor C_1 voltage must settle to the dc level across R_1 in the time interval between triggering.

EXAMPLE 6-3

The Schmitt trigger circuit designed in Example 6-2 is to be triggered at a maximum frequency of 1 MHz. Determine the largest speed-up capacitance (C_1 in Figure 6-3) that may be used.

Solution   The resistance in parallel with the capacitor terminals when Q_1 is *off* is

$$R = R_1 \| (R_{C1} + R_2)$$
$$= 22 \text{ k}\Omega \| (4.7 \text{ k}\Omega + 22 \text{ k}\Omega)$$
$$\approx 12 \text{ k}\Omega$$

(Actually, the input resistance of Q_2 (with emitter resistor R_E) is in parallel with R_2, but is large compared to R_2.)

$$\text{Resolving time} = t_{re} = \frac{1}{\text{Triggering frequency}}$$

$$= \frac{1}{1 \text{ MHz}} = 1 \text{ } \mu\text{s}$$

For C_1 to charge through 90% of its total voltage change,

Eq. 4-4, $$C_{1(max)} = \frac{t_{re}}{2.3R}$$

$$C_1 = \frac{1 \text{ } \mu\text{s}}{2.3 \times 12 \text{ k}\Omega} = 36 \text{ pF}   \text{(use 33 pF standard value)}$$

6-5 OUTPUT/INPUT CHARACTERISTICS

Consider the Schmitt trigger circuit design as finalized in Example 6-2. When Q_2 is *on*, the output voltage is

$$V_o = V_{CC} - I_C R_{C2}$$
$$= 12\ \text{V} - (2\ \text{mA} \times 3.9\ \text{k}\Omega)$$
$$= 4.2\ \text{V}$$

When Q_2 is *off*, the output voltage is

$$V_o = V_{CC} - I_{CO} R_{C2}$$
$$\approx V_{CC} = 12\ \text{V}$$

As designed, the UTP and LTP are approximately 5 V and 3 V, respectively. With the triggering levels and the output voltage levels known, a graph showing output voltage *versus* input voltage may be plotted.

When the input voltage is zero, Q_1 is *off* and Q_2 is *on*. Therefore, $V_o = 4.2$ V. This may be plotted as point A on the output/input characteristics in Figure 6-5(a). As V_i is increased above zero volts, Q_1 remains *off* and Q_2 remains *on* until V_i becomes equal to the UTP, which for this particular circuit is 5 V. Hence, V_o remains at 4.2 V from $V_i = 0$ to $V_i = 5$ V. Point B is plotted at $V_o = 4.2$ V and $V_i = 5$ V.

When the UTP is reached, Q_1 switches *on* and Q_2 switches *off*. Thus, V_o changes from 4.2 V to 12 V. Point C is plotted at $V_o = V_{CC}$ and $V_i =$ UTP. Any further increase in V_i has no effect on V_o. The horizontal line from point C to point D in Figure 6-5(a) shows that V_o remains equal to V_{CC} as V_i increases above the UTP.

Now consider the effect of reducing V_i from a level greater than the UTP. The output voltage V_o remains equal to 12 V until V_i becomes equal to the LTP, at point E on Figure 6-5(b). At the LTP, Q_1 switches *off* and Q_2 switches *on*, returning V_o to 4.2 V. Point F is plotted at $V_o = 4.2$ V and $V_i =$ LTP = 3 V. As V_i is reduced below the LTP, V_o remains at 4.2 V, shown by the line from point F to point A in Figure 6-5(b). The two graphs taken together in Figure 6-5(c) give the complete output/input characteristics for the circuit.

The difference between the UTP and LTP levels is termed the *hysteresis* of the circuit. For many circuit applications, the hysteresis is not very important. In some circumstances, however, circuits with the least possible hysteresis are desirable. Zero hysteresis occurs when the upper and lower trigger points are equal. The hysteresis can be adjusted by altering the ratio of R_1 and R_2, or by adjusting the value of R_{C1}.

6-6 OP-AMP SCHMITT TRIGGER CIRCUITS

Inverting Schmitt Circuit

An IC operational amplifier may be employed as a Schmitt trigger circuit. The design of such a circuit is quite simple. Consider Figure 6-6, which shows a circuit with the input triggering voltage applied to the inverting input terminal. The noninverting terminal is connected to the junction of resistors R_1 and R_2; these resistors operate as a potential

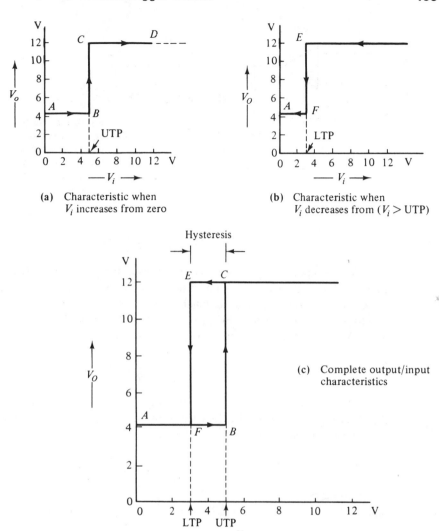

Figure 6-5   The output/input characteristics of a Schmitt trigger circuit show the upper and lower levels of the output, as well as the upper and lower trigger points.

divider from output to ground. The voltage at the noninverting input terminal is the voltage V_2 across R_2, as shown in the figure.

When V_i is less than V_2, the noninverting input terminal voltage V_2 is greater than that at the inverting input. Therefore, the output is saturated in a positive direction. In this case, $V_o \approx (+V_{CC} - 1 \text{ V})$. The voltage at the noninverting input terminal is calculated by using the output voltage V_o and R_1 and R_2:

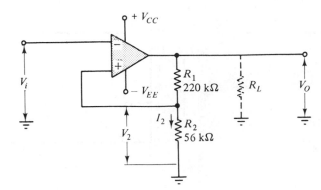

Figure 6-6 An IC operational amplifier can be connected to function as a Schmitt trigger circuit. When V_o is high, the upper trigger point is $+V_o R_2/(R_1 + R_2)$. When V_o is low, the lower trigger point is $-V_o R_2/(R_1 + R_2)$.

$$V_2 \approx \frac{R_2}{R_1 + R_2}(+V_o)$$

When input voltage V_i is raised to the level of V_2, the output begins to go negative, causing V_2 to fall; thus, the noninverting input terminal rapidly becomes negative with respect to the inverting input terminal. When this occurs, the output changes over very rapidly from approximately $+(V_{CC} - 1\text{ V})$ to approximately $-(V_{EE} - 1\text{ V})$.

When the output goes negative, V_2 becomes

$$V_2 \approx \frac{R_2}{R_1 + R_2}(-V_o)$$

Thus, V_2 is a negative voltage, and the output V_o remains negative until the voltage at the inverting input terminal is reduced to the new (negative) level of V_2.

From the above discussion, it can be seen that

$$\text{UTP} \approx \frac{+R_2}{R_1 + R_2}(V_{CC} - 1\text{ V}) \tag{6-1}$$

and

$$\text{LTP} \approx \frac{-R_2}{R_1 + R_2}(V_{EE} - 1\text{ V}) \tag{6-2}$$

EXAMPLE 6-4

A 741 operational amplifier is to be employed as a Schmitt trigger circuit with a UTP of 3 V. Design a suitable circuit, and calculate the actual UTP and LTP when resistors with standard values are selected. Use a supply voltage of ± 15 V.

Solution   From the 741 data sheet in Appendix 1-11,

$$I_{B(\text{max})} = 500 \text{ nA}$$

For a stable level of V_2, $I_2 \gg I_{B(\text{max})}$,

let
$$I_2 = 100 \times 500 \text{ nA}$$
$$= 50 \text{ }\mu\text{A}$$

$$R_2 = \frac{\text{UTP}}{I_2} = \frac{3 \text{ V}}{50 \text{ }\mu\text{A}}$$
$$= 60 \text{ k}\Omega \quad \text{(use 56 k}\Omega \text{ standard value)}$$

I_2 becomes $\dfrac{3 \text{ V}}{56 \text{ k}\Omega} = 53.57 \text{ }\mu\text{A}$

$$V_{R1} = V_o - V_2$$
$$= (V_{CC} - 1 \text{ V}) - V_2$$
$$= 15 \text{ V} - 1 \text{ V} - 3 \text{ V}$$
$$= 11 \text{ V}$$

$$R_1 = \frac{V_{R1}}{I_2} = \frac{11 \text{ V}}{53.57 \text{ }\mu\text{A}}$$
$$= 205 \text{ k}\Omega \quad \text{(use 220 k}\Omega \text{ standard value)}$$

$$\text{Actual UTP} \approx \frac{V_o R_2}{R_1 + R_2}$$
$$= \frac{14 \text{ V} \times 56 \text{ k}\Omega}{220 \text{ k}\Omega + 56 \text{ k}\Omega} = 2.8 \text{ V}$$

$$\text{LTP} \approx -2.8 \text{ V}$$

Slew Rate Effect

The operational amplifier *slew rate* (discussed in Section 5-3) imposes limitations on the performance of the Schmitt circuit. The 741 has a typical slew rate of 0.5 V/μs. (See Appendix 1-11.) For the output voltage to go from -14 to $+14$ V requires a time

$$t = \frac{28 \text{ V}}{0.5 \text{ V/}\mu\text{s}} = 56 \text{ }\mu\text{s}$$

If this time is allowed to occupy 10% of the output pulse width (see the rise time discussion in Section 1-4), then the minimum pulse width is

$$\text{PW} = 10 \times t = 560 \text{ }\mu\text{s}$$

If it is assumed that the output is a square wave, the maximum satisfactory operating frequency is

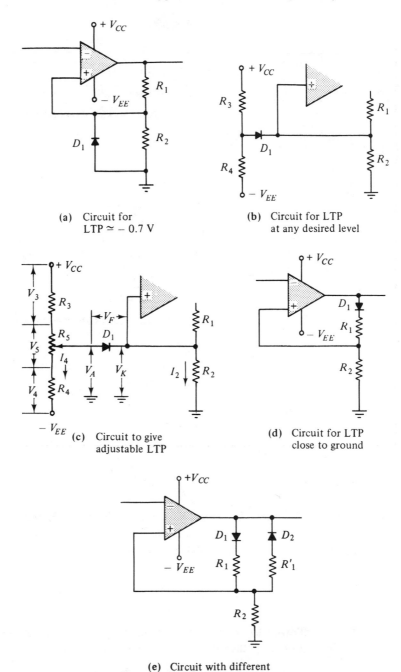

(a) Circuit for
     LTP ≃ − 0.7 V

(b) Circuit for LTP
     at any desired level

(c) Circuit to give
     adjustable LTP

(d) Circuit for LTP
     close to ground

(e) Circuit with different
     UTP and LTP levels

Figure 6-7   Diodes can be employed with op-amp Schmitt trigger circuits to set the trigger points to any desired levels other than $V_o R_2/(R_1 + R_2)$.

$$f_{max} = \frac{1}{2 \times PW} = 893 \text{ Hz}$$

For higher frequency performance, an operational amplifier with a faster slew rate must be employed. The 715, for example, has a slew rate of 100 V/μs. Going through the same calculations as above, the maximum operating frequency for a Schmitt trigger using a 715 is approximately 179 kHz.

Adjusting the Trigger Points

An op-amp Schmitt trigger circuit in which the lower trigger point is clamped to -0.7 V is shown in Figure 6-7(a). When the output voltage becomes negative, diode D_1 is forward biased. Thus, D_1 holds the noninverting input to 0.7 V below ground. When V_i (at the inverting input) drops below -0.7 V, the output again becomes positive. The LTP is now -0.7 V. The UTP is unaffected by the diode, since D_1 is reverse biased when the output is positive.

     The circuit can be designed for any desired UTP. Then, by use of a diode and potential divider, as shown in Figure 6-7(b), the LTP can be fixed at any desired level less than the UTP. A potentiometer placed between R_3 and R_4 [Figure 6-7(c)] provides an adjustable LTP. A diode connected in series with R_1, as illustrated in Figure 6-7(d), gives an LTP which is very close to ground. When the output is negative, D_1 is reverse biased, and only the diode reverse leakage current I_R flows. The LTP now becomes $V_2 = -I_R R_2$. Since I_R is normally very small, and R_2 can be selected as low as a few kilohms, the LTP can be only millivolts from ground. By the reversal of D_1, the UTP can be brought close to ground. Then, the LTP is specified by V_o, R_1, and R_2.

     Figure 6-7(e) shows an arrangement by which the UTP and LTP can be made completely independent of each other at the design stage. When the output is positive, D_2 is reverse biased and D_1 is forward biased. The UTP of the circuit is determined by, R_1 and R_2. When the output of the circuit is negative, D_1 is reverse biased while D_2 is forward biased. The LTP is now determined by R_1' and R_2.

EXAMPLE 6-5

The Schmitt trigger circuit designed in Example 6-4 is to have the LTP adjustable over the range from 2 V to -2 V. Design a suitable circuit.

Solution   From Example 6-4, $R_1 = 220$ kΩ, $R_2 = 56$ kΩ, and $I_2 \approx 50$ μA. Consider the circuit of Figure 6-7(c). For a stable bias voltage V_A, set $I_4 \gg I_2$.

Let $I_4 = 100 \times I_2 = 100 \times 50$ μA

                $= 5$ mA

                $V_{K1} = -2$ V, and $V_{K2} = 2$ V

                $V_{A1} = V_{K1} + V_F = -2$ V $+ 0.7$ V

                $= -1.3$ V

$$V_4 = V_{A1} - V_{EE}$$

$$= -1.3 \text{ V} - (-15 \text{ V}) = 13.7 \text{ V}$$

$$R_4 = \frac{V_4}{I_4} = \frac{13.7 \text{ V}}{5 \text{ mA}}$$

$$= 2.74 \text{ k}\Omega \quad \text{(use 2.7 k}\Omega \text{ standard value)}$$

$$V_{A2} = V_{K2} + V_F = 2 \text{ V} + 0.7 \text{ V}$$

$$= 2.7 \text{ V}$$

$$V_5 = V_{A2} - V_{A1} = 2.7 \text{ V} - (-1.3 \text{ V})$$

$$= 4 \text{ V}$$

$$R_5 = \frac{V_5}{I_4} = \frac{4 \text{ V}}{5 \text{ mA}}$$

$$= 800 \ \Omega \text{ (use 1 k}\Omega \text{ standard potentiometer value)}$$

$$R_3 = \frac{V_{CC} - V_{A2}}{I_4} = \frac{15 \text{ V} - 2.7 \text{ V}}{5 \text{ mA}}$$

$$= 2.46 \text{ k}\Omega \quad \text{(use 2.7 k}\Omega \text{ standard value)}$$

Noninverting Schmitt Circuit

In the circuit shown in Figure 6-8(a) the input voltage V_i is applied to the noninverting input terminal of the operational amplifier via R_1. This means that the circuit output goes positive when V_i is increased to the UTP, and negative when V_i is lowered to the LTP level. Assume that the output voltage of the circuit is negative at a level of approximately $-(V_{EE} - 1 \text{ V})$. If the input voltage is zero, then the voltage across R_1 is

$$V_{R1} = \frac{R_1}{R_1 + R_2}[-(V_{EE} - 1 \text{ V})]$$

[See Figure 6-8(b).] Thus, the voltage at the noninverting input is a negative quantity, and this keeps the output at its negative saturation level.

For the output of the circuit to go positive, V_i must be raised until the voltage at the noninverting terminal goes slightly above ground level (i.e., above the level of the inverting input terminal). When this occurs, the voltage at the right side of R_2 is $-V_o$ and that at the left side is 0 V; i.e., the voltage across R_2 is V_o. [See Figure 6-8(c).] Therefore, the current flowing through R_2 is

$$I_2 = V_o/R_2$$

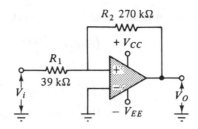

(a) Non-inverting Schmitt trigger circuit

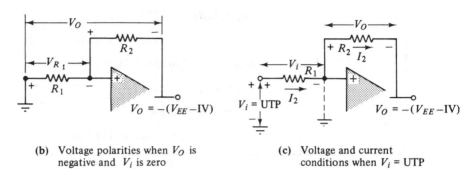

(b) Voltage polarities when V_O is
     negative and V_i is zero

(c) Voltage and current
     conditions when V_i = UTP

Figure 6-8   Noninverting op-amp Schmitt trigger circuit. At the trigger points, the noninverting input terminal is at ground level; V_i appears across R_1, and V_o is developed across R_2.

Since the current flowing into the input terminal of the operational amplifier is negligibly small, the current through R_1 is also I_2. The voltage at the left side of R_1 is V_i, and that at the right side is 0 V. Thus,

$$V_{R1} = V_i = I_2 \times R_1 = \text{UTP}$$

The circuit is designed very simply by first selecting I_2 very much larger than the maximum input current to the amplifier. Then, R_1 is calculated as UTP/I_2, and R_2, as V_o/I_2. The LTP is numerically equal to the UTP, but with reversed polarity.

The modifications to the noninverting Schmitt circuit shown in Figure 6-9 are self-explanatory. Figure 6-9(a) shows a circuit in which the UTP is a positive quantity, but the LTP is very close to ground level. The circuit in Figure 6-9(b) has (numerically) different UTP and LTP values, while that in Figure 6-9(c) has an adjustment to raise or lower the level of both trigger points.

EXAMPLE 6-6

A noninverting Schmitt trigger circuit is to be designed to have trigger points of ±2 V. The available supply is ±15 V. Using a 741 operational amplifier, design a suitable circuit.

Solution   The circuit is shown in Figure 6-8(a).

For the 741, $I_{B(max)} = 500$ nA

$$I_2 \gg I_{B(max)}$$

Let $I_2 = 100 \times I_{B(max)}$

$$= 100 \times 500 \text{ nA} = 50 \text{ μA}$$

$$V_o \approx \pm(V_{CC} - 1 \text{ V})$$

$$\approx \pm(15 \text{ V} - 1 \text{ V}) = \pm 14 \text{ V}$$

$$R_2 = V_o/I_2 = 14 \text{ V}/50 \text{ μA}$$

$$= 280 \text{ kΩ} \quad \text{(use 270 kΩ standard value)}$$

I_2 now becomes $\approx 14 \text{ V}/270 \text{ kΩ} = 51.9 \text{ μA}$

$$R_1 = (\text{UTP or LTP})/I_2$$

$$= 2 \text{ V}/51.9 \text{ μA}$$

$$= 38.5 \text{ kΩ} \quad \text{(use 39 kΩ standard value)}$$

6-7 IC SCHMITT

Schmitt trigger circuits are available as integrated circuits, i.e., without any external components required. The 7413, for example (see Appendix 1-15), contains two complete Schmitt trigger circuits.

The UTP and LTP of the 7413 are identified on the device data sheet as V_{T+} (*positive-going threshold*) and V_{T-} (*negative-going threshold*), respectively. With $V_{CC} = +5$ V, the typical triggering levels are UTP $= 1.7$ V and LTP $= 0.9$ V. The circuit output typically switches between a low level of 0.2 V and a high output voltage of 3.4 V. The switching time between levels is 15 to 18 ns. Input current might be as high as 1 mA, and the output maximum is 16 mA. The input terminals of the 7413 are connected in a way that permits the circuit to function as a *nand gate*. (See Section 10-3.)

Obviously, the 7413 switches very much faster than an IC operational amplifier or discrete-component Schmitt trigger circuit. It is also much less expensive than such other circuits, if only because no external components are required. However, its trigger points are fixed, and where different UTP and LTP levels are required, it is necessary to use a discrete-component circuit, an IC op-amp, or a voltage comparator as discussed in the next section.

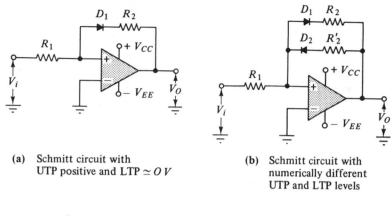

(a) Schmitt circuit with
UTP positive and LTP $\simeq 0\,V$

(b) Schmitt circuit with
numerically different
UTP and LTP levels

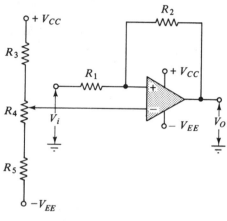

(c) Schmitt circuit with
trigger level adjustment

Figure 6-9 Diodes, or potential divider circuits, can be used to adjust the trigger points on an op-amp noninverting Schmitt trigger circuit.

6-8 IC VOLTAGE COMPARATORS

The output of a *voltage comparator* changes rapidly from one level to another when the input arrives at a predetermined voltage. In this respect the voltage comparator is similar to a Schmitt trigger circuit. Unlike a Schmitt circuit, a voltage comparator has two inputs. The change in output level occurs at the instant that the two input voltages become equal. The input voltages are *compared*; hence the name *comparator*.

Since an IC operational amplifier has two input terminals, it can be employed as a voltage comparator [see the zero-crossing detector in Section 5-4]. But operational amplifiers have much slower response times than circuits designed as comparators (i.e., they are relatively slow in switching from *low* output to *high*, and *vice versa*). This is because the circuitry of a comparator prevents its transistors from going into saturation. In an operational amplifier the transistors do saturate when the output is at one extreme or the other. Voltage comparators can be thought of as fast-switching operational amplifiers, but they cannot be used as amplifiers; like Schmitt trigger circuits, comparator outputs can only be switched from one dc output level to another.

In Appendixes 1-20 and 1-21 partial data sheets are shown for two IC voltage comparators, the 710 and the 311. The 710 is described as a high-speed comparator. Its typical response time is listed as 40 ns. The 311 is much slower, with a response time of 200 ns. (Considering the slew rate of a 741 operational amplifer, the 741 response time could be as large as 50 μs).

The 311 takes a maximum *input bias current* of only 250 nA typically, while the 710 requires 20 μA. Consequently, the 311 offers a much higher input resistance to signals than does the 710. The *input offset voltage* is the voltage difference between input terminals when the comparator detects equality. For the 710 and 311, this quantity is 0.6 mV and 0.7 mV, respectively. The *differential input voltage*, the maximum allowable voltage difference between input terminals, is ±5 V for the 710 and ±14 V for the 311.

The 710 switches between a *low* output of −0.5 V and a *high* output level of +3.2 V. The 311 output can switch between ground and a load voltage (separate from the circuit supply) as high as 40 V. The 710 requires supply voltages of $V_{CC} = +12$ V to +14 V and $V_{EE} = -6$ V to −7 V. The 311 can operate from a single-polarity supply ranging from 5 V to 36 V, or can use the typically available supply of ±15 V.

It is seen that with the single exception of response time, the performance of the 311 comparator is superior to the 710.

Figure 6-10(a) shows a 311 comparator employed as a *zero-crossing detector*. The inverting input is grounded, and the waveform to be monitored is directly connected to the noninverting input. Note that R_1, connected from V_{CC} to the output, must be included because the output of a 311 comparator is an open-circuited transistor collector. At the instant that the input voltage rises above ground, the output voltage goes positive. When the input drops below ground again, the comparator output goes to zero. The output is a pulse wave with its leading and lagging edges occurring exactly at the instants that the input waveform crosses the zero level. The zero-crossing detector could be described as a Schmitt trigger with UTP = LTP = 0 V. Since UTP = LTP, the circuit is said to have zero hysteresis.

A 311 comparator employed as an inverting-type Schmitt trigger is shown in Figure 6-10(b). The upper trigger point for the circuit is

$$\text{UTP} = \frac{V_{CC} \times R_3}{R_1 + R_2 + R_3}$$

and LTP = 0 V, because the *low* output level of the circuit is ground. Design of a comparator Schmitt trigger circuit commences with selection of resistor R_1 to pass a

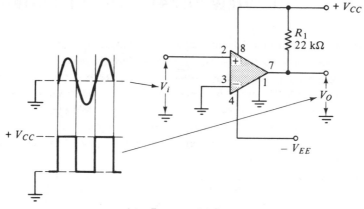

(a)   Zero-crossing detector

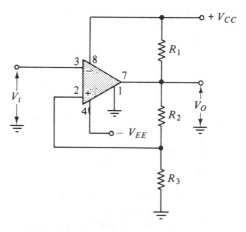

(b) Comparator as a Schmitt trigger

Figure 6-10   When a voltage comparator is used as a zero-crossing detector, the output voltage rapidly changes state each time the input waveform crosses the zero level. The comparator Schmitt trigger circuit shown has an upper trigger point of $V_{CC}R_3/(R_1 + R_2 + R_3)$. The lower trigger point is zero.

suitable collector current through the comparator (internal) output transistor. The current through R_1, R_2, and R_3, with V_o high, is selected to be much larger than the maximum input bias current to the comparator. Then the resistors are calculated for the desired UTP.

REVIEW QUESTIONS AND PROBLEMS

6-1. Sketch a transistor Schmitt trigger circuit, and briefly explain its operation.

6-2. Define the terms *upper trigger point*, *lower trigger point*, *hysteresis*, and *regeneration*.

6-3. Design a transistor Schmitt trigger circuit with UTP = 6 V. Use a 2N3904 transistor with I_C = 1 mA. The available supply is 15 V. Use standard-value resistors.

6-4. The circuit of Problem 6-3 is to have an LTP of 5 V. Make the necessary design modifications and select suitable standard-value resistors.

6-5. Plot the output/input characteristics for the Schmitt trigger circuit designed in Problem 6-4.

6-6. (a) Briefly explain how a *speed-up capacitor* improves the switching time of a Schmitt trigger circuit. (b) The Schmitt trigger circuit designed in Problem 6-4 is to be triggered at a maximum frequency of 800 kHz. Determine the maximum size of the speed-up capacitor that may be employed.

6-7. (a) Sketch the circuit of an operational amplifier employed as a Schmitt trigger circuit. Briefly explain how it functions. (b) Show how this circuit could be modified so that (1) LTP ≈ −0.7 V, (2) LTP ≈ 0 V, and (3) UTP ≈ 0 V.

6-8. Design inverting and noninverting Schmitt trigger circuits using a 741 operational amplifier. The supply voltage is to be ±12 V, and the trigger points ±2 V. Select standard-value resistors and calculate the actual triggering levels.

6-9. Plot the output/input characteristics for the Schmitt trigger circuits designed in Problem 6-8.

6-10. The Schmitt trigger circuits designed in Problem 6-8 are to have an LTP that is adjustable over the range ±1 V. Suitably modify each circuit.

6-11. The circuits of Problem 6-8 have the following input waveforms:
(a) A triangular weveform with an amplitude of ±5 V;
(b) A square wave with an amplitude of 3 V;
(c) A square wave with an amplitude of ±4;
(d) A sine wave with an amplitude of ±10 V;
(e) A sawtooth wave with an amplitude of ±6 V.
Sketch the above waveforms and the resultant output wave from the Schmitt trigger circuit for each case.

6-12. Using a 741 IC operational amplifier, design inverting and noninverting Schmitt trigger circuits to have LTP = −2 V and UTP = 0 V. The available supply is ±9 V.

6-13. Design an inverting Schmitt trigger circuit to have UTP = +4 V and LTP = −3 V. Use a 741 operational amplifier with V_{CC} = ±18 V.

6-14. Explain the operation of a voltage comparator circuit, and discuss how it differs from an operational amplifier.

6-15. Sketch the circuit of a voltage comparator employed as a zero-crossing detector. Also, sketch the input and output waveforms. Briefly explain.

6-16. A voltage comparator is to be employed as a Schmitt trigger circuit with UTP = 2.5 V. The available supply is ±18 V, and the external load (connected between $+V_{CC}$ and the output terminal) is 12 kΩ. Sketch the circuit and determine the values of the other components.

Chapter 7

Monostable and Astable Multivibrators

INTRODUCTION

There are three types of multivibrators: bistable, monostable, and astable. The **bistable multivibrator** (see Chapter 13) has two stable states. The **monostable multivibrator** has one stable state, and may be triggered into another temporary, or quasi-stable, state. When triggered, the circuit generates an output pulse of constant width and amplitude. The **astable multivibrator** has no stable state; the circuit oscillates between two quasi-stable states. Astable circuits are normally designed to operate as square wave generators. IC operational amplifiers and voltage comparators may be employed as monostable and astable multivibrators.

7-1 COLLECTOR-COUPLED MONOSTABLE MULTIVIBRATOR

The *monostable multivibrator* (also known as a *one-shot multivibrator*) has a single stable condition: one transistor is normally *on* and the other transistor is normally *off*. The condition can be reversed by application of a triggering pulse, which temporarily turns *on* the normally-*off* transistor and switches *off* the normally-*on* transistor. This quasi-stable condition lasts only for a brief time period, depending upon the circuit components. During this time, an output pulse is produced.

DC Conditions

Consider the *collector-coupled* monostable circuit shown in Figure 7-1. The circuit is described as collector coupled because the collector terminal of Q_2 is coupled via potential divider R_1 and R_2 to the base terminal of Q_1. In the normal dc condition of the circuit,

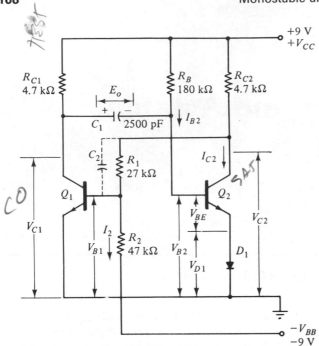

Figure 7-1 Collector-coupled transistor monostable multivibrator circuit. Transistor Q_2 is biased *on* via resistor R_B. Q_1 is biased from Q_2 collector and V_{BB} by means of resistors R_1 and R_2. Consequently, with Q_2 normally *on*, Q_1 is normally *off*.

base current I_{B2} is provided from V_{CC} to Q_2 via resistance R_B. Thus, transistor Q_2 is normally *on*. At this time, diode D_1 is forward biased and has no significant effect on Q_2. The function of D_1 is to protect the base-emitter junction of Q_2 against excessive reverse bias. (See Section 4-5.) With Q_2 *on* in saturation, the collector voltage of Q_2 is $(V_{CE(\text{sat})} + V_{D1})$ above ground level. The base voltage V_{B1} of Q_1 is determined by the negative supply voltage V_{BB} and by R_1 and R_2, as well as by the collector voltage of Q_2. With Q_2 collector near ground level, and V_{BB} a negative quantity, V_{B1} is likely to be negative (i.e., Q_1 base is biased below its grounded emitter). Therefore, with Q_2 normally *on*, Q_1 is normally *off*.

When Q_1 is *off*, its collector current is zero. Therefore, there is no voltage drop across R_{C1}, and the collector of Q_1 is at the supply voltage level V_{CC}. Also, with Q_2 on, the base voltage of Q_2 is $V_{B2} = V_{BE} + V_{D1}$. On the right terminal of capacitor C_1 the voltage is V_{B2}, and on the left terminal it is V_{CC}. Hence the capacitor voltage is $E_o = V_{CC} - V_{B2}$, positive on the left, as shown in Figure 7-1.

Triggered Conditions

Now consider what would occur if Q_1 were triggered *on* to saturation for a brief instant. (This could be made to occur by, for example, capacitor-coupling a positive-going spike to the base of Q_1.) The collector voltage of Q_1 drops almost to ground level. Capacitor C_1 will not lose its charge E_o instantaneously; therefore, when the left terminal of C_1 drops to $V_{CE(\text{sat})}$, the right-hand terminal will drop to $(V_{CE(\text{sat})} - E_o)$. Consequently,

Q_2 base voltage goes to $(V_{CE(sat)} - E_o)$—i.e., Q_2 is biased *off*. With Q_2 *off*, there is no longer a collector current to produce a voltage drop across R_{C2}. Thus, V_{C2} rises rapidly, Q_1 base is biased above ground level, and Q_1 remains *on*. It is seen that when Q_1 is triggered *on* briefly, Q_2 goes *off* and Q_1 remains *on*. As will be seen, Q_1 stays *on* only for a brief time.

The transistor switching process is illustrated by the waveforms in Figure 7-2. Prior to Q_1 being triggered on, the voltages are $V_{B1} = -V$, $V_{C1} = V_{CC}$, $V_{B2} \approx 1.4$ V; and $V_{C2} \approx 0.9$ V. When Q_1 is triggered *on*, $V_{B1} \approx 0.7$ V, $V_{C1} \approx 0.2$ V, $V_{B2} = (V_{C1} - E_o) \approx -E_o$, and $V_{C2} \approx V_{CC}$.

With the exception of V_{B2}, all the above voltages remain constant while Q_2 stays biased *off*. V_{B2} does not remain constant because C_1 discharges via R_B. (See Figure 7-3.) Voltage E_o across C_1 initially is positive on the left side and negative on the right side. Current I flowing into the right side of C_1 will tend to discharge C_1 and then

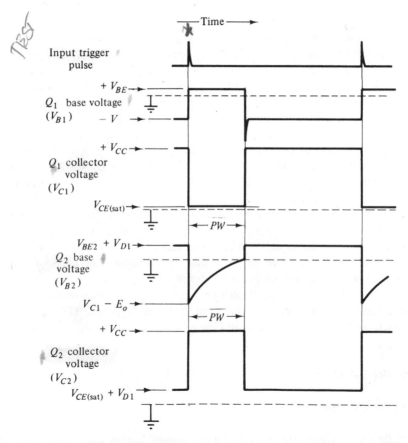

Figure 7-2 Monostable multivibrator circuit waveforms. A positive-going spike at the base of Q_1 switches Q_1 on. The resultant fall in the collector voltage of Q_1 is coupled to the base of Q_2 via capacitor C_1, thus switching Q_2 off and causing its collector voltage to rise. Q_2 remains *off* while its base-emitter junction is reverse biased by C_1.

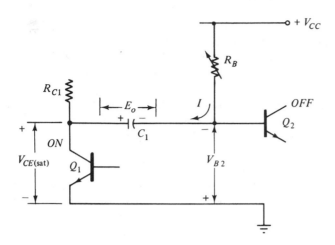

Figure 7-3 Negatively charged capacitor C_1 discharges via resistor R_B when Q_1 is *on* and Q_2 is *off*. Q_2 remains *off* until C_1 discharge allows its base to rise sufficiently above ground level to forward-bias its base-emitter junction.

recharge it with reversed polarity. Thus, V_{B2} begins to rise toward ground level. When C_1 is discharged to $e_{C1} \approx 0$ V, Q_2 base-emitter and D_1 begin to be forward biased again. At this point, I_{C2} again begins to flow and V_{C2} starts to fall. (See Figure 7-2.) When V_{C2} falls, it causes V_{B1} to fall; consequently, V_{C1} rises and causes V_{B2} to rise. The result of this is that Q_1 rapidly switches *off* and Q_2 rapidly comes *on* again, when C_1 is discharged to approximately zero volts. At this time, the negative spike at Q_1 base is due to *speed-up capacitor* C_2 (Figure 7-1) transmitting all the Q_2 collector voltage change to the base of Q_1 and then discharging (see Section 4-4). When Q_1 switches *off* and Q_2 switches *on* again, C_1 is rapidly recharged to E_o via R_{C1} and Q_2 base. The circuit has now returned to its normal stable state and remains in this condition until Q_1 is triggered *on* again.

Refer again to the waveforms in Figure 7-2. It is seen that when the voltage at Q_2 collector is a positive-going pulse, that at Q_1 collector is a negative-going pulse. These two pulses are equal in width, and either or both may be taken as output from the circuit. The pulse width of the output depends upon the values of C_1 and R_B. If R_B is made variable, the output pulse width may be adjusted.

The monostable multivibrator now can be described as a circuit with one stable state, capable of producing an output pulse when triggered. The output pulse width is constant and can be made adjustable by making R_B adjustable. (See Figure 7-3.)

7-2 DESIGNING A MONOSTABLE MULTIVIBRATOR

Resistor Values

The design of a monostable multivibrator usually begins with specification of the output pulse width, the supply voltage, the load, and perhaps the transistors to be employed. As with other circuits, it might be possible to connect the load directly into the circuit as either R_{C1} or R_{C2}. More frequently, the load might be capacitor coupled to Q_2 collector, and R_{C2} would be selected to be much smaller than the load resistance. Alternatively, I_{C2} may be selected to be much larger than the maximum output load current.

I_{C2} and R_{C2} must be chosen so that transistor Q_2 is in saturation, that is, $I_{C2}R_{C2} \approx V_{CC}$. Base current I_{B2} is calculated as $I_{C2}/h_{FE(min)}$; use of the minimum value of h_{FE} ensures that I_{B2} is large enough to drive Q_2 to saturation. The base resistance R_B is then calculated as $(V_{CC} - V_{B2})/I_{B2}$. [See Figure 7-4(a).] R_{C1} is usually made equal to R_{C2}, and so $I_{C1} \approx I_{C2}$.

Resistors R_1 and R_2 provide *on* or *off* bias to Q_1 base. For a stable bias voltage V_{B1}, the current I_2 that flows through R_1 and R_2 should be much larger than the base current to Q_1. When Q_1 is *on*, I_{B1} flows through R_1. If I_{B1} is not much smaller than I_2, variations in I_{B1} may upset the bias voltage at Q_1 base. To achieve the condition $I_2 \gg I_{B1}$, resistors R_1 and R_2 should be selected as small as possible. However, R_1 and R_2 also constitute a load on resistance R_{C2}; thus, to avoid overloading R_{C2}, R_1 and R_2 should be chosen as large as possible. These contradictory requirements are met by applying the rule of thumb that $I_2 \approx I_{C2}/10$. When a design is worked through, it will be seen that this condition results in R_1 and R_2 each being 5 to 10 times R_{C2}. With Q_2 *off*, resistors R_{C2}, R_1, and R_2 constitute a potential divider at the base of Q_1. R_1 and R_2 are calculated, as explained, to set the base of Q_1 at V_{BE} above its emitter voltage [Figure 7-4(b)].

Capacitor

The output pulse width for a monostable circuit is dictated by the time taken for C_1 to discharge from its initial voltage level to approximately zero volts. Therefore, C_1 is calculated from Equation 2-2,

$$e_c = E - (E - E_o)\epsilon^{-t/CR}$$

For the collector-coupled monostable circuit,

$$e_c \approx 0, \qquad E \approx V_{CC},$$

and

$$E_o = -(V_{CC} - V_{B2})$$

Note that C_1 is initially charged positive on the left side and negative on the right side. When Q_1 is *on* and Q_2 is *off*, C_1 tends to charge negative on the left side and positive on the right side. Thus, the initial voltage E_o of C_1 must be taken as negative, and the charging voltage E as positive.

The initial value of Q_2 base voltage, when Q_2 is *on* is

$$V_{B2} = V_{BE} + V_{D1}$$

and

$$E_o = -(V_{CC} - V_{BE} - V_{D1})$$

$$t = \text{Specified PW}$$

$$C = C_1$$

$$R = R_B,$$

which is the resistance through which C_1 is charged when Q_1 is *on* and Q_2 is *off*.

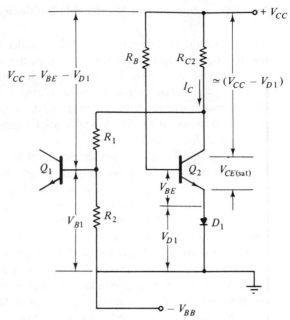

(a) Q_2 *on* in saturation
Q_1 *off*

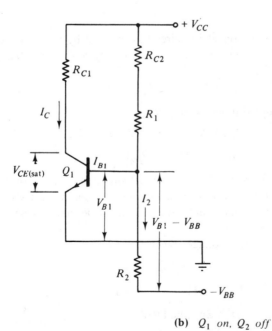

(b) Q_1 *on*, Q_2 *off*

Figure 7-4 (a) In designing a monostable multivibrator, the circuit of Q_2 is first designed with Q_2 *on* and saturated. (b) Q_1 circuit is then designed with Q_1 in a saturated *on* condition. Potential divider current I_2 is selected as approximately $I_C/10$ to avoid over loading Q_2 while still giving stable bias conditions for Q_1.

The speed-up capacitor C_2 is determined by a method similar to that employed for the inverter circuit and the Schmitt trigger circuit.

EXAMPLE 7-1

A collector-coupled monostable multivibrator is to operate from a ± 9 V supply. Transistor collector currents are to be 2 mA, and the transistors used have $h_{FE(min)} = 50$. Neglecting the output pulse width, design a suitable circuit.

Solution   For Q_2 *on* and saturated [Figure 7-4(a)],

$$R_{C2} \approx \frac{V_{CC} - V_{D1} - V_{CE(sat)}}{I_C} = \frac{9\text{ V} - 0.7\text{ V} - 0.2\text{ V}}{2\text{ mA}}$$

$$= 4.05\text{ k}\Omega \qquad \text{(use 4.7 k}\Omega \text{ standard value)}$$

$$I_{B2(min)} = \frac{I_C}{h_{FE(min)}} = \frac{2\text{ mA}}{50} = 40\ \mu\text{A}$$

$$R_B = \frac{V_{CC} - V_{BE} - V_{D1}}{I_{B2}} = \frac{9\text{ V} - 0.7\text{ V} - 0.7\text{ V}}{40\ \mu\text{A}}$$

$$= 190\text{ k}\Omega \qquad \text{(use 180 k}\Omega \text{ standard value)}$$

For Q_1 *on* and saturated [Figure 7-4(b)],

$$R_{C1} = R_{C2} = 4.7\text{ k}\Omega$$

Let

$$I_2 \approx \frac{I_C}{10} = \frac{2\text{ mA}}{10} = 200\ \mu\text{A}$$

$$V_{B1} = V_{BE} = 0.7\text{ V} \qquad \text{(when } Q_1 \text{ is } on)$$

$$V_{R2} = V_{B1} - V_{BB} = 0.7\text{ V} - (-9\text{ V}) = 9.7\text{ V}$$

$$R_2 = \frac{V_{R2}}{I_2} = \frac{9.7\text{ V}}{200\ \mu\text{A}}$$

$$= 48.5\text{ k}\Omega \qquad \text{(use 47 k}\Omega \text{ standard value)}$$

$$I_{B1} + I_2 \approx 200\ \mu\text{A} + 40\ \mu\text{A}$$

$$= 240\ \mu\text{A}$$

$$R_{C2} + R_1 = \frac{V_{CC} - V_{B1}}{I_{B1} + I_2} = \frac{9\text{ V} - 0.7\text{ V}}{240\ \mu\text{A}}$$

$$= 34.6\text{ k}\Omega$$

$$R_1 = (R_{C2} + R_1) - R_{C2} = 34.6\text{ k}\Omega - 4.7\text{ k}\Omega$$

$$= 29.9\text{ k}\Omega \qquad \text{(use 27 k}\Omega \text{ standard value)}$$

The circuit design is now complete (ignoring the PW). V_{B1} should be calculated with Q_2 *on* to determine that Q_1 is *off* at this time, and to check that the reverse bias is not excessive on the Q_1 base-emitter junction.

When Q_2 is *on*,

$$V_{C2} = V_{D1} + V_{CE(\text{sat})} \approx 0.7 \text{ V} + 0.2 \text{ V}$$

$$= 0.9 \text{ V}$$

$$V_{R1} = \frac{R_1}{R_1 + R_2}(V_{C2} - V_{BB})$$

$$= \frac{27 \text{ k}\Omega}{27 \text{ k}\Omega + 47 \text{ k}\Omega}[0.9 \text{ V} - (-9 \text{ V})]$$

$$\approx 3.6 \text{ V}$$

$$V_{B1} = V_{C2} - V_{R1}$$

$$V_{B1} = 0.9 \text{ V} - 3.6 \text{ V}$$

$$= -2.7 \text{ V}$$

This level of V_{B1} is sufficient to ensure that Q_1 is biased *off* when Q_2 is *on*. Also, -2.7 V is less than the typical limit of -5 V for a reverse-biased base-emitter junction.

EXAMPLE 7-2

For the circuit designed in Example 7-1, select a suitable capacitor to give an output pulse of 250 μs.

Solution

From Eq. 2-9, $\quad C = \dfrac{t}{R \ln\left(\dfrac{E - E_o}{E - e_c}\right)}$

$$e_c = 0 \text{ V}$$

$$E = V_{CC} = 9 \text{ V}$$

$$E_o = -(V_{CC} - V_{BE} - V_{D_1})$$

$$= -(9 \text{ V} - 0.7 \text{ V} - 0.7 \text{ V}) = -7.6 \text{ V}$$

$$t = 250 \text{ μs}$$

$$R = R_B = 180 \text{ k}\Omega$$

$$C_1 = \frac{250 \ \mu s}{180 \ k\Omega \ \ln \left[\dfrac{9 \ V - (-7.6 \ V)}{9 \ V - 0 \ V} \right]}$$

$$= 2300 \ pF \qquad \text{(use 2500 pF standard capacitor value)}$$

7-3 TRIGGERING MONOSTABLE MULTIVIBRATORS

Monostable multivibrator triggering can be effected either by switching *off* the normally *on* transistor, or by turning *on* the normally *off* transistor. Figure 7-5(a) shows a positive-going spike capacitor-coupled to the base of normally-*off* transistor Q_1. This raises Q_1 base above its grounded emitter, thus switching it *on*. Q_1 switch-*on* then causes Q_2 to switch *off*. The input spike "sees" resistances R_1 and R_2 in parallel with the transistor input resistance. Therefore, the spike has to supply current through R_1 and R_2 as well as base current to Q_1. To ensure that Q_1 switches *on* and Q_2 switches *off*, the input current must be supplied for a time t equal to the turn-on time for Q_1 added to the turn-off time for Q_2.

The arrangement in Figure 7-5(b) provides for Q_2 (the normally *on* transistor) to be switched *off*. In this case, the negative-going spike pulls Q_2 base below ground for the transistor turn-*off* time. During this brief time, C_1 behaves as a short circuit, so that the load "seen" by the input spike is $R_B \| R_{C_1}$. This is greater than the load "seen" by the positive-going spike in Figure 7-5(a). Therefore, triggering by a negative-going spike at Q_{2B} requires a larger input current than triggering by a positive-going spike at Q_{1B}.

Perhaps the most effective monostable triggering circuit is that shown in Figure 7-5(c), in which an additional transistor Q_3 is employed. Q_3 is normally biased *off* by means of resistor R_3 shorting its base and emitter terminals together. Coupling capacitor C_c and resistor R_3 operate as a differentiating circuit (see Section 2-6), so that the pulse input is differentiated, as illustrated in the figure. Only the positive spike will turn *on* Q_3. In the event that the negative spike is too large for Q_3 base-emitter, diode D_2 may be used to clip it off. When Q_3 switches *on*, its collector current causes a voltage drop across R_{C1}, and the charge on capacitor C_1 causes Q_2 to be biased *off*. Thus, Q_3 switch-*on* has the same effect as Q_1 switch-*on*. To correctly trigger the circuit of Figure 7-5(c), the input spike must hold Q_3 *on* for the turn-*off* of Q_2.

The design procedure for the triggering circuit of Figure 7-5(c) is similar to the capacitor-coupled inverter design in Example 4-9, except that t is made equal to the turn-off time for Q_2 instead of the input pulse width. This is also the procedure followed for selecting C_c in the circuit of Figure 7-5(a). Design of the circuit of Figure 7-5(b) is similar to the design given in Example 4-8.

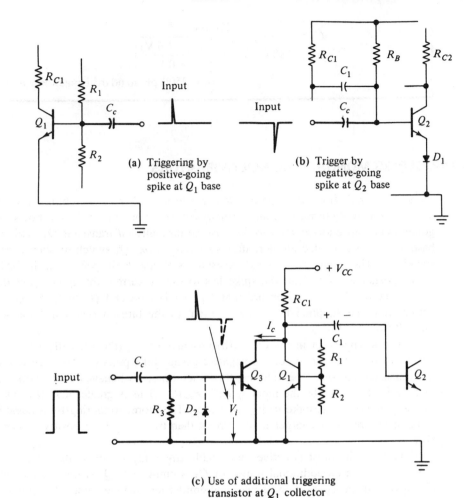

(a) Triggering by
positive-going
spike at Q_1 base

(b) Trigger by
negative-going
spike at Q_2 base

(c) Use of additional triggering
transistor at Q_1 collector

Figure 7-5   A monostable multivibrator circuit may be triggered into its quasi-stable condition by (a) a positive-going spike at the base of Q_1, (b) a negative-going spike at the base of Q_2, or (c) an additional transistor which produces a negative-going spike at the collector of Q_1.

7-4 EMITTER-COUPLED MONOSTABLE MULTIVIBRATORS

In the emitter-coupled monostable multivibrator circuit in Figure 7-6, a resistance R_E connects both transistor emitter terminals to ground. Also, instead of R_2 being connected to a negative supply voltage, it now is connected to ground. The negative supply voltage is no longer required, and it is seen that one advantage of the emitter-coupled circuit is that it operates from a single supply voltage. Another advantage of this circuit is that the presence of R_E makes it easy to maintain the transistors unsaturated. Consequently,

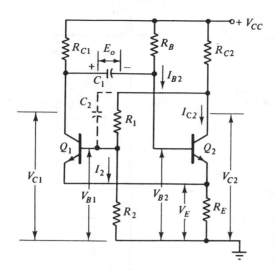

Figure 7-6   Emitter-coupled monostable multivibrator circuit. Emitter resistor R_E, common to Q_1 and Q_2, eliminates the need for a negative bias supply. When Q_2 is *on*, the base of Q_1 is biased below the level of the common emitter voltage.

the transistors can be made to switch faster than in the case of the collector-coupled multivibrator.

Reference to Figure 7-6 shows that transistor Q_2 is normally *on*, because it is supplied with base current via R_B. At this time, there is a voltage drop V_E across resistor R_E, as illustrated. Also, the voltage drop across R_{C2} makes Q_2 collector voltage lower than the supply voltage level. Q_1 base is biased from Q_2 collector via potential divider R_1 and R_2; their ratio is such that with Q_2 on, V_{B1} is lower than V_E. Therefore, Q_1 base voltage is less than its emitter voltage, and Q_1 is biased *off*. With Q_1 off, its collector voltage equals the supply voltage. The initial voltage across C_1 at this time is $V_{CC} - V_{B2}$.

When Q_1 is triggered *on*, its collector voltage drops, and the charge on C_1 causes the base voltage of Q_2 to drop. When Q_2 begins to turn *off*, its collector voltage starts to rise, thus raising the base voltage of Q_1. With Q_1 on, the new level of V_E is $V_{B1} - V_{BE1}$, and because the base of Q_2 is pushed below this level (by the charge on C_1) Q_2 is biased *off*. Then, Q_2 remains *off* until C_1 has discharged enough to allow V_{B2} to rise above V_E.

The design procedure for a saturating emitter-coupled monostable multivibrator is not very different from that for the collector-coupled circuit. When the circuit is designed for nonsaturating operation, a minimum V_{CE} level must be selected, and $h_{FE(max)}$ must be used in the calculations. Triggering methods for the emitter-coupled circuit are exactly the same as those for the collector-coupled multivibrator.

7-5 OP-AMP MONOSTABLE MULTIVIBRATORS

An IC operational amplifier connected to function as a monostable multivibrator is shown in Figure 7-7(a). The inverting input terminal is grounded via resistance R_3, and the noninverting terminal is biased above ground by resistances R_1 and R_2. Since

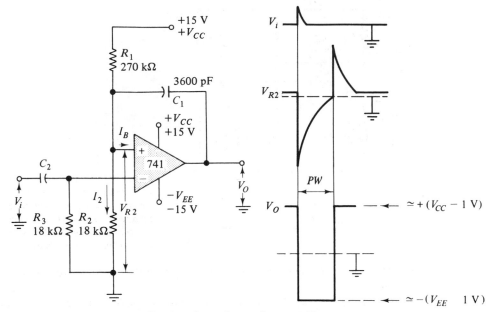

(a) Circuit and waveforms of monostable
multivibrator using IC operational amplifier

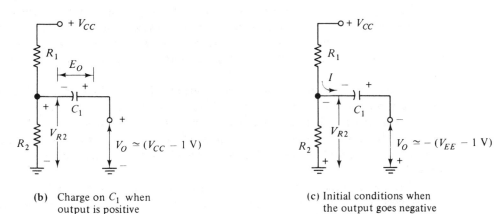

(b) Charge on C_1 when
output is positive

(c) Initial conditions when
the output goes negative

Figure 7-7   IC operational amplifier as a monostable multivibrator. The noninverting input terminal
is biased to a slightly higher voltage than the inverting input terminal; thus, the output is normally
high. When the inverting input is briefly triggered to a higher level, the output switches *low*, and
the charge on capacitor C_1 holds the noninverting terminal below ground level until C_1 discharges.

the noninverting terminal has a positive input, the output is saturated near the V_{CC}
level. In Figure 7-7(b), it is seen that capacitor C_1 is normally charged positive on the
right side and negative on the left side.

When a large enough positive-going step voltage input is coupled to the inverting
terminal via C_2, the inverting terminal voltage is raised above the level of the noninverting

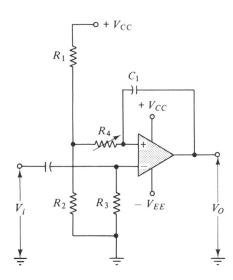

Figure 7-8 IC op-amp monostable circuit with pulse-width control. Variation of resistor R_4 alters the rate of discharge of C_1, and thus adjusts the output pulse width.

terminal. The output then switches rapidly to approximately $-(V_{EE} - 1 \text{ V})$. [See Figure 7-7(c).] This pushes the noninverting terminal down to $-(V_{EE} - 1 \text{ V}) - E_o$, thus ensuring that the output remains negative until C_1 discharges. C_1 begins to discharge via R_1 and R_2 as soon as the output goes negative. Eventually, C_1 will charge positive on the left side and negative on the right side. When the voltage on the left side of C_1 rises above the voltage level at the inverting terminal, the noninverting terminal again has a positive input. Now, V_o rapidly returns to approximately $(V_{CC} - 1 \text{ V})$, and the circuit has returned to its original condition.

The output voltage of the circuit moved from its normal positive level of $(V_{CC} - 1 \text{ V})$ to a negative level of $-(V_{EE} - 1 \text{ V})$, and then returned to $(V_{CC} - 1 \text{ V})$. Thus, a negative output pulse is generated when the circuit is triggered. The output pulse width depends upon C_1, the values of R_1 and R_2, and the bias level at the inverting terminal.

Figure 7-8 shows a modification to the operational amplifier monostable circuit to facilitate pulse width control. If R_1 and R_2 are made much smaller than R_4, then R_4 is effectively the charge and discharge resistance for C_1. When R_4 is adjustable, as illustrated, the output pulse width can be controlled.

EXAMPLE 7-3

Design a monostable multivibrator using a 741 operational amplifier with $V_{CC} = \pm 15$ V. The circuit is to be triggered by a 1.5 V input spike, and the output pulse width is to be 200 μs.

Solution   To use a triggering input of 1.5 V, let $V_{R2} = 1$ V [See Figure 7-7(a).]

and let $I_2 = 100 \times I_{B(\text{max})}$

$$= 100 \times 500 \text{ nA} = 50 \text{ μA}$$

Then $\qquad$ $R_2 = \dfrac{V_{R2}}{I_2} = \dfrac{1 \text{ V}}{50 \text{ }\mu\text{A}}$

$\qquad\qquad\quad = 20 \text{ k}\Omega \qquad \text{(use 18 k}\Omega \text{ standard value)}$

I_2 becomes $\qquad$ $I_2 = \dfrac{1 \text{ V}}{18 \text{ k}\Omega}$

$\qquad\qquad\quad \approx 56 \text{ }\mu\text{A}$

$\qquad V_{R1} = V_{CC} - V_{R2} = 15 \text{ V} - 1 \text{ V}$

$\qquad\qquad\quad = 14 \text{ V}$

$\qquad R_1 = \dfrac{V_{B1}}{I_2} = \dfrac{14 \text{ V}}{56 \text{ }\mu\text{A}}$

$\qquad\qquad\quad = 250 \text{ k}\Omega \qquad \text{(use 270 k}\Omega \text{ standard value)}$

$\qquad R_3 = R_1 \| R_2 = 18 \text{ k}\Omega \| 270 \text{ k}\Omega$

$\qquad\qquad\quad \approx 16.9 \text{ k}\Omega \qquad \text{(use 18 k}\Omega \text{ standard value)}$

When V_o is positive, the initial charge on C_1 is

$$E_o = V_{R2} - V_o$$
$$\approx V_{R2} - (V_{CC} - 1 \text{ V})$$
$$\approx 1 \text{ V} - 15 \text{ V} + 1 \text{ V}$$
$$\approx -13 \text{ V}$$

When V_o is negative, the final charge on C_1 at switchover is

$$e_c \approx +(V_{EE} - 1 \text{ V})$$
$$= +14 \text{ V}$$

$$\text{Charging voltage} = E = V_{R2} - (-V_o)$$
$$= 1 \text{ V} + 14 \text{ V}$$
$$= 15 \text{ V}$$

$$\text{Charging resistance} = R_1 \| R_2$$
$$= 18 \text{ k}\Omega \| 270 \text{ k}\Omega$$
$$\approx 16.9 \text{ k}\Omega$$

From Eq. 2-9, $\qquad$ $C_1 = \dfrac{t}{R \ln \left(\dfrac{E - E_o}{E - e_c}\right)}$

$$= \frac{200 \ \mu s}{16.9 \ k\Omega \ \ln \left[\dfrac{15V - (-13 \ V)}{15 \ V - 14 \ V} \right]}$$

$$= 3550 \ pF \qquad \text{(use 3600 pF standard value)}$$

7-6 IC MONOSTABLE MULTIVIBRATORS

Monostable multivibrators are available as integrated circuit components. The 74121 shown in Figure 7-9 is typical of such components. Operating from a 5 V supply, the unit provides two complementary outputs (Q and $\overline{Q}$) and has three input terminals. Two of the inputs (A_1 and A_2) typically require -1.4 V to effect triggering. The other input triggers the circuit when a typical level of $+1.55$ V is applied. The output pulse width is around 30 to 35 ns when no external components are employed and R_{int} (terminal 9) is connected directly to V_{CC}. With R_{int} left unconnected, and with an external timing capacitor and resistor (R_T and C_T in Figure 7-9), the output pulse width is

$$PW \approx 0.7 \ C_T R_T \tag{7-1}$$

Pulse widths of up to 28 s are possible.

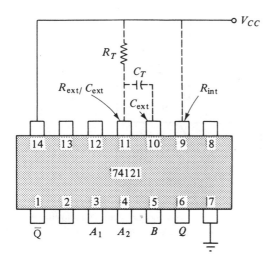

Figure 7-9 The 74121 IC monostable multivibrator. The output pulse width is determined by selection of resistor R_T and capacitor C_T.

7-7 ASTABLE MULTIVIBRATORS

The *astable multivibrator* [Fig. 7-10(a)] has no stable state, but has two quasi-stable states. The circuit oscillates between the states (Q_1 *on*, Q_2 *off*) and (Q_2 *on*, Q_1 *off*). The output at the collector of each transistor is a square wave; therefore, the circuit is applied as a *square wave generator*.

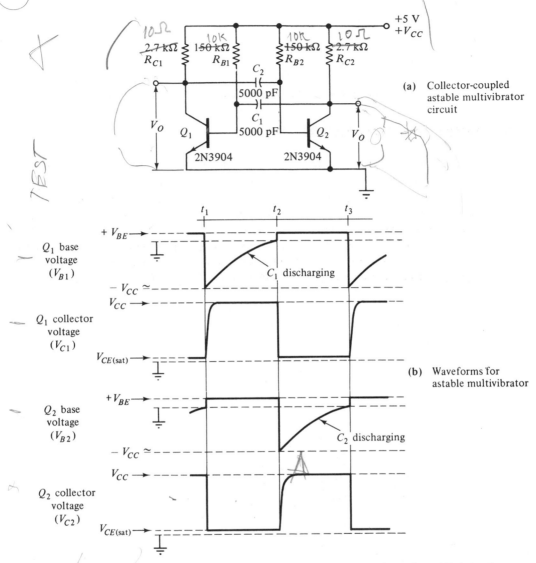

Figure 7-10 Collector-coupled astable multivibrator circuit and waveforms. Both transistors are biased to be normally *on*, but each is biased *off* by the charge on the capacitor at its base when the other transistor is *on*.

Consider the *collector-coupled astable multivibrator* shown in Figure 7-10(a). Each transistor has a bias resistance R_B, and each base is capacitor-coupled to the collector of the other transistor. This is similar to the arrangement of the normally-*on* transistor in a monostable multivibrator. Consequently, each transistor in an astable circuit functions in the same way as the normally-*on* transistor in a monostable circuit. When Q_1 is *on* and Q_2 is *off*, capacitor C_1 is charged to $(V_{CC} - V_{BE1})$, positive on the right side. For Q_2 *on* and Q_1 *off*, C_2 is charged to $(V_{CC} - V_{BE2})$, positive on the left side.

Refer to the circuit waveforms in Figure 7-10(b); it is seen that just prior to time t_1, transistor Q_1 is *on*, and its collector voltage is $V_{CE(sat)}$. Also, Q_2 is *off*, and its collector voltage is V_{CC}. Thus, capacitor C_1 is charged to $(V_{CC} - V_{BE1})$. At t_1, the base voltage of transistor Q_2 rises above ground, causing Q_2 to switch *on*. The collector current I_{C2} now causes Q_2 collector voltage to fall to $V_{CE(sat)}$. Since C_1 will not discharge instantaneously, the base voltage of Q_1 becomes

$$V_{B1} = V_{C2} - (\text{Charge on } C_1)$$

$$= V_{CE(sat)} - (V_{CC} - V_{BE1})$$

$$\approx -V_{CC}$$

With its emitter grounded, and its base at $-V_{CC}$, transistor Q_1 is biased *off*. Therefore, at time t_1, the collector voltage of Q_1 rises to V_{CC}. The rise of V_{C1} is not instantaneous, because capacitor C_2 charges via R_{C1} as Q_1 switches *off*.

Between times t_1 and t_2, the base voltage of Q_2 remains at V_{BE}, and Q_2 remains biased *on*. During this time, C_1 discharges via resistance R_{B1}. Therefore, the voltage at Q_1 base rises from $-V_{CC}$ toward V_{CC}. When Q_1 base rises above ground, the transistor begins to switch *on*. The falling collector voltage of Q_1 is coupled to Q_2 base via capacitor C_2, thus causing Q_2 to switch *off*. As Q_2 turns *off*, its collector voltage rises, and C_1 is recharged via R_{C2} and Q_1 base. This pumps a large current into the base of Q_1, making it switch *on* very fast. Consequently, the collector voltage of Q_1 falls very rapidly at switch-*on*. The switchover process is reversed when C_2 discharges sufficiently to allow Q_2 base to rise above ground.

The output pulse width from either transistor is equal to the time during which the transistor is *off*. This is the time taken by the capacitor to discharge from approximately V_{CC} to zero volts, as given by Eq. 2-9:

$$t = CR \ln \left(\frac{E - E_o}{E - e_c} \right)$$

In this equation,

$$t = \text{PW}, \qquad C = C_1 = C_2, \qquad R = R_{B1} = R_{B2}$$

and E is equal to the supply voltage V_{CC}; E_o, the initial capacitor charge, is equal to $-V_{CC}$. This is taken as negative, because the capacitor would eventually charge with reversed polarity to approximately $+V_{CC}$ if transistor switchover did not occur. The final capacitor charge at switchover is $e_c = 0$ V.

Using Eq. 2-9, $$\text{PW} = CR \ln \left[\frac{V_{CC} - (-V_{CC})}{V_{CC} - 0} \right]$$

$$= CR \ln \left[\frac{2V_{CC}}{V_{CC}} \right]$$

$$= CR \ln 2$$

or $PW \approx 0.69\, CR$ (7-2)

For $C = 0.1\ \mu F$ and $R_B = 100\ k\Omega$,

$$PW = 0.69 \times 0.1\ \mu F \times 100\ k\Omega$$

$$= 6.9\ ms$$

and the output frequency is

$$f = \frac{1}{2\ PW} = \frac{1}{2 \times 6.9\ ms}$$

$$\approx 72.5\ Hz$$

EXAMPLE 7-4

Design an astable multivibrator as in Fig. 7-10(a) to generate a 1 kHz square wave. The supply voltage is 5 V, and the output load current is to be 20 μA.

Solution

Let $I_C \approx 100 \times$ (load current)

$$= 100 \times 20\ \mu A = 2\ mA$$

Use 2N3904 transistors, which, from the data sheet in Appendix 1-4, have $h_{FE(min)} = 70$.

$$R_C \approx \frac{V_{CC}}{I_C} = \frac{5\ V}{2\ mA}$$

$$= 2.5\ k\Omega \qquad \text{(use 2.7 k\Omega\ standard value)}$$

$$I_{B(min)} = \frac{I_C}{h_{FE(min)}} = \frac{2\ mA}{70}$$

$$\approx 28.6\ \mu A$$

$$R_B = \frac{V_{CC} - V_{BE}}{I_B} = \frac{5\ V - 0.7\ V}{28.6\ \mu A}$$

$$= 150\ k\Omega \qquad \text{(standard value)}$$

$$PW = \frac{1}{2f}$$

$$= \frac{1}{2 \times 1\ kHz} = 0.5\ ms$$

From Eq. 7-2 $C_1 = \dfrac{PW}{0.69\ R_B} = \dfrac{0.5\ ms}{0.69 \times 150\ k\Omega}$

$$= 4800\ pF \qquad \text{(use 5000 pF standard capacitor value)}$$

The circuit of Figure 7-11 shows several possible modifications to the simple astable multivibrator circuit of Figure 7-10(a). Each transistor has its base biased to approximately $-V_{CC}$ when *off*. Consequently, if V_{CC} is greater than the maximum base-emitter reverse voltage, the transistors may be destroyed. Inclusion of diodes D_1 and D_2 (Figure 7-11) affords protection for the transistor base-emitter junctions, as explained in Section 7-1.

If C_1 and C_2 are unequal, or if R_{B1} and R_{B2} are unequal, one transistor will remain *off* for a longer time than the other one. In this case, the output is a pulse waveform instead of a square wave. The output frequency of the circuit may be made adjustable by including a variable resistor R_1 in series with one of the base bias resistors. In Figure 7-11, R_1 controls the rate of discharge of capacitor C_1; thus, R_1 can be used to adjust the *off*-time of Q_1.

Occasionally, the frequency of an astable multivibrator has to be synchronized to some external frequency. Figure 7-11 shows a negative-going synchronizing spike capacitor-coupled to the base of Q_2. When the spike input is applied, Q_2 is switched *off* and C_2 is recharged to its maximum voltage. Q_2 then remains *off* for its normal pulse width.

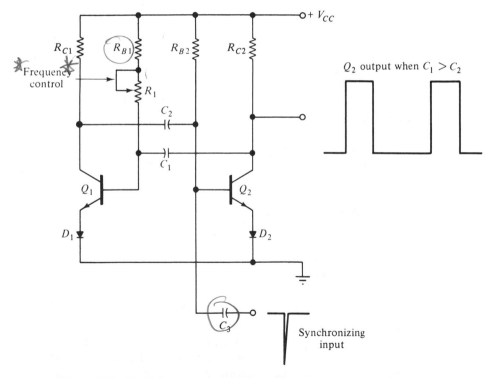

Figure 7-11   Modified astable multivibrator circuit. Variable resistor R_1 provides frequency control by adjusting the *off*-time of Q_1. Diodes D_1 and D_2 protect the transistor base-emitter junctions against excessive reverse bias. Capacitor C_3 couples a negative-going spike to the base of Q_2, to synchronize the astable frequency to the spike frequency.

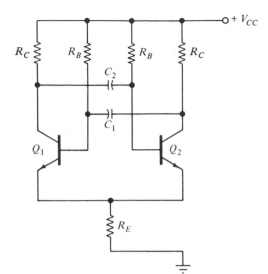

Figure 7-12   Emitter-coupled astable multivibrator circuit. The presence of emitter resistor R_E ensures that the circuit oscillates, by eliminating the possibility of both transistors settling in an *on* condition.

One problem with the collector-coupled astable circuit is that it may not always start oscillating when the supply voltage is switched *on*. Because of the circuit symmetry, it can happen that both transistors switch *on* and remain *on*. Oscillation can be started by shorting one of the transistor bases to its emitter terminal for a brief instant. However, this usually is not practical. The emitter-coupled astable multivibrator circuit shown in Figure 7-12 solves the problem. In this circuit, when one transistor begins to conduct, the other transistor has its emitter voltage raised and its base voltage reduced. Thus, it is almost impossible for the two to remain *on* at one time. Approximately half of the supply voltage might be dropped across R_E, giving $R_E = V_{CC}/(2I_C)$. Then, using $V_{CC}/2$, the other components are determined as already discussed.

7-8 OP-AMP ASTABLE MULTIVIBRATORS

An inverting Schmitt trigger circuit using an IC operational amplifier can be easily converted into an astable multivibrator with the addition of a resistor and capacitor. In the circuit shown in Figure 7-13, R_2, R_3, and the operational amplifier constitute a Schmitt trigger circuit. The waveforms in the illustration help to explain the circuit operation. When the output is high, current flows through R_1, charging C_1 positively until it reaches the UTP of the Schmitt. The Schmitt output then switches negative, and current commences to flow out of C_1 via R_1 until its voltage reaches the Schmitt LTP. The Schmitt output then switches positive once again, and the cycle recommences.

Design of this circuit merely involves the Schmitt circuit design, followed by selection of suitable R and C values for the desired charge and discharge time and trigger voltage levels.

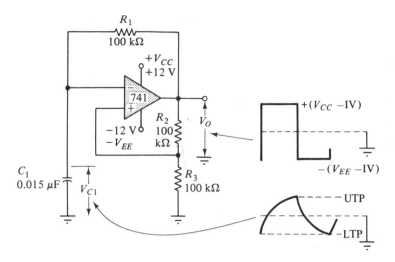

Figure 7-13   IC op-amp astable multivibrator circuit. The op-amp with resistors R_2 and R_3 constitutes a Schmitt trigger circuit. (See Section 6-6.) C_1 and R_1 are selected to have C_1 charge from LTP to UTP, and *vice versa*, in the desired time.

EXAMPLE 7-5

Using a 741 operational amplifier, design an astable multivibrator as in Fig. 7-13, to have an output frequency of 300 Hz and an output amplitude of ±11 V.

Solution

$$V_{CC} \approx \pm(V_o + 1 \text{ V})$$

$$\approx \pm(11 \text{ V} + 1 \text{ V}) = \pm12 \text{ V}$$

Let the Schmitt trigger voltage $= V_o/2 = \pm5.5$ V.

$$V_{R3} = \pm5.5 \text{ V}$$

Let $I_2 = 100 \times I_{B(\text{max})} = 100 \times 500 \text{ nA} = 50 \text{ }\mu\text{A}$.

$$R_3 = \frac{V_{R3}}{I_2} = \frac{5.5 \text{ V}}{50 \text{ }\mu\text{A}}$$

$$= 110 \text{ k}\Omega \qquad \text{(use 100 k}\Omega \text{ standard value)}$$

$$V_{R2} = V_{R3}$$

$$R_2 = R_3 = 100 \text{ k}\Omega$$

Let $I_{R1(\text{min})} = 100 \times I_{B(\text{max})} = 50 \text{ }\mu\text{A}$.

$$V_{R1(min)} = V_o - \text{trigger voltage}$$

$$= 11 \text{ V} - 5.5 \text{ V} = 5.5 \text{ V}$$

$$R_1 = \frac{5.5 \text{ V}}{50 \text{ }\mu\text{A}} \approx 100 \text{ k}\Omega$$

$$T = \frac{1}{300 \text{ Hz}} = 3.3 \text{ ms}$$

C_1 charging time $t = \frac{1}{2} T = 1.67 \text{ ms}.$

From Eq. 2-9,
$$C = \frac{t}{R_1 \ln \left[\dfrac{E - E_o}{E - e_c}\right]}$$

$$= \frac{1.67 \text{ ms}}{100 \text{ k}\Omega \ln \left[\dfrac{11 \text{ V} - (-5.5 \text{ V})}{11 \text{ V} - 5.5 \text{ V}}\right]}$$

$$= 0.015 \text{ }\mu\text{F}$$

7-9 VOLTAGE COMPARATOR AS MONOSTABLE AND ASTABLE

Comparator Monostable

A voltage comparator connected to function as a monostable multivibrator is shown in Figure 7-14. This configuration is similar to op-amp monostable circuits, except that, for the 311 comparator, a resistor R_5 must be included between $+V_{CC}$ and the output terminal. The resistance of R_5 is selected to pass a suitable level of current through the collector of the comparator (internal) output transistor. A minimum of 1 mA is usually satisfactory. Apart from R_5, design procedure for this circuit is similar to that for the op-amp monostable. Because the comparator has a much faster response time than an operational amplifier, comparators are more suitable for monostable circuits that are to produce very short duration output pulses.

Comparator Astable

Like an op-amp astable multivibrator, the voltage comparator astable circuit illustrated in Figure 7-15(a) consists of a Schmitt trigger and a CR circuit. Capacitor C_1 is charged and discharged from the comparator output via resistor R_1. Resistor R_2 completes the collector circuit of the comparator output transistor. The comparator, together with resi-

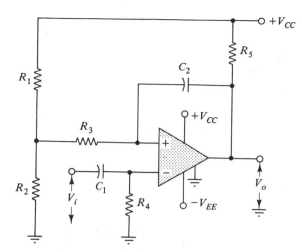

Figure 7-14   Voltage comparator as a monostable multivibrator. This is similar to the op-amp monostable circuit, except that a 311 comparator must have a resistor R_5 from $+V_{CC}$ to the output terminal.

stors R_2 through R_5, constitutes a Schmitt trigger circuit. This particular Schmitt circuit is designed to have UTP $= 2\ V_{CC}/3$ and LTP $= V_{CC}/3$. Thus, the capacitor charges from $V_{CC}/3$ to $2\ V_{CC}/3$ and back again, to produce a square wave output, as illustrated.

Consider Figures 7-15(b) and (c). When the comparator output is higher (open circuited), the resistor network is as in Figure 7-15(b). Series-connected resistors R_2 and R_3 are in parallel with R_5; R_4 is in series with the parallel combination. For this condition, the voltage across R_4 is the UTP for the Schmitt. Figure 7-15(c) shows the network arrangement when the comparator output is low (at ground level). Here, parallel-connected R_3 and R_4 are in series with R_5, and V_{R4} is now the LTP of the Schmitt. To give $V_{R4} = 2\ V_{CC}/3$ in one case, and $V_{R4} = V_{CC}/3$ in the other case, resistors R_3, R_4, and R_5 are made equal, and R_2 is selected very much smaller tha R_3.

When V_o is high,

$$V_{R4} = V_{CC} \times \frac{R_4}{R_4 + R_5 \| (R_2 + R_3)}$$

With $R_3 = R_4 = R_5$, and $R_2 \ll R_4$,

$$V_{R4} \approx \frac{2\ V_{CC}}{3}$$

When V_o is low,

$$V_{R4} = V_{CC} \times \frac{R_4 \| R_3}{R_5 + R_4 \| R_3}$$

With $R_3 = R_4 = R_5$,

$$V_{R4} = \frac{V_{CC}}{3}$$

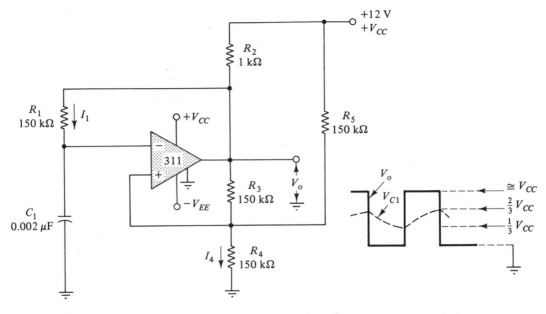

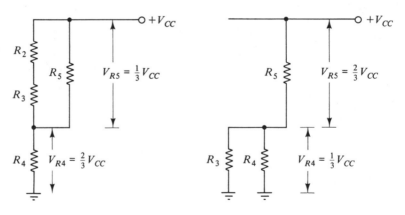

(a) Circuit and waveforms

(b) When V_o is *high*, $V_{R4} = \frac{2}{3} V_{CC}$ (c) When V_o is *low*, $V_{R4} = \frac{1}{3} V_{CC}$

Figure 7-15 Voltage comparator as an astable multivibrator. Resistors R_2 through R_5 are selected to make the comparator perform as a Schmitt trigger circuit with LTP = $V_{CC}/3$ and UTP = $2 V_{CC}/3$. R_1 and C_1 are determined so as to allow C_1 to charge between LTP and UTP over the desired time.

Also, from the development of Eq. 7-2,

$$C_1 = \frac{t}{R_1 \ln 2} \qquad (7\text{-}3)$$

EXAMPLE 7-6

Using a 311 comparator and a supply of ± 12 V, design an astable multivibrator to have an output frequency of 3 kHz.

Solution   From Appendix 1-20, $I_{B(\text{max})} = 250$ nA for the 311.

Let
$$I_4 \approx 100 \times I_{B(\text{max})}$$
$$= 25 \ \mu\text{A}$$

With V_o low,
$$V_{R4} = \text{LTP} = \frac{V_{CC}}{3} = \frac{12 \text{ V}}{3}$$
$$= 4 \text{ V}$$
$$R_4 = \frac{V_{R4}}{I_4} = \frac{4 \text{ V}}{25 \ \mu\text{A}}$$
$$= 160 \text{ k}\Omega \text{ (use 150 k}\Omega \text{ standard value)}$$
$$R_3 = R_5 = R_4 = 150 \text{ k}\Omega$$

For $R_2 \ll R_4$, let $R_2 = 1$ kΩ.

When V_o is low,
$$I_{R2} = \frac{V_{CC}}{R_2} = \frac{12 \text{ V}}{1 \text{ k}\Omega}$$
$$= 12 \text{ mA (which is satisfactory)}$$

Let
$$I_{R1(\text{min})} \approx 100 \times I_{B(\text{max})}$$
$$= 25 \ \mu\text{A}$$
$$R_1 = \frac{V_{R1(\text{min})}}{I_{R1(\text{min})}} = \frac{4 \text{ V}}{25 \ \mu\text{A}}$$
$$= 160 \text{ k}\Omega \text{ (use 150 k}\Omega)$$

$$\text{Pulse width} = t = \frac{T}{2} = \frac{1}{2f}$$
$$= \frac{1}{2 \times 3 \text{ kHz}} = 167 \ \mu\text{s}$$

Eq. 7-3,
$$C_1 = \frac{t}{R_1 \ln 2} = \frac{167 \ \mu\text{s}}{150 \text{ k}\Omega \times \ln 2}$$
$$= 1600 \text{ pF (standard value)}$$

REVIEW QUESTIONS AND PROBLEMS

7-1. Sketch the circuit of a collector-coupled monostable multivibrator. Sketch the waveforms, and explain the operation of the circuit. Also, explain the function of each component.

7-2. A collector-coupled monostable multivibrator is to operate from a ± 12 V supply. The transistor collector currents are to be 3 mA, and the transistors have $h_{FE(min)} = 70$. Neglecting the output pulse width, design a suitable circuit.

7-3. Select suitable capacitors for the circuit in Problem 7-2 to give an output pulse of 330 μs.

7-4. The monostable multivibrator designed for Problems 7-2 and 7-3 is to be triggered at 10 μs intervals between output pulses. Calculate the maximum speed-up capacitance that may be employed.

7-5. Sketch and explain the various methods of triggering a monostable multivibrator. For the multivibrator in problem 7-2, design a triggering system using an additional transistor. The triggering input is a 3 V pulse with a source resistance of 3.3 kΩ. The turn-off time for Q_2 is 1 μs.

7-6. Sketch the circuit of an emitter-coupled monostable multivibrator. Carefully explain how the circuit operates. Discuss the relative advantages and disadvantages of emitter-coupled and collector-coupled monostable multivibrators. Show how the emitter-coupled circuit may be modified to provide pulse-width control.

7-7. Sketch the circuit of a monostable multivibrator employing an IC operational amplifier. Explain how the circuit operates, and show how the output pulse width may be controlled.

7-8. Design a monostable multivibrator using a 741 operational amplifier with $V_{CC} = \pm 9$ V. The circuit is to be triggered by a 0.5 V input spike, and the output pulse width is to be 300 μs.

7-9. A 74121 IC monostable multivibrator is to have an output pulse width of 2 μs. Select suitable external components, and show how they should be connected to the circuit.

7-10. Sketch the circuit of a collector-coupled astable multivibrator. Also, sketch the waveforms of the collector and base voltages. Carefully explain how the circuit operates.

7-11. Derive an expression for the output pulse width of an astable multivibrator. Design an astable multivibrator to have a 5 kHz output square wave. The available supply is ± 9 V, and the load current is 50 μA.

7-12. Sketch the circuit of an astable multivibrator in which the output frequency may be adjusted. Also, show how the multivibrator frequency may be synchronized with an external frequency.

7-13. Sketch the circuit of an emitter-coupled astable multivibrator. Explain its operation, and discuss its advantages compared to a collector-coupled circuit.

7-14. Design an emitter-coupled monostable multivibrator to have $V_{CC} = 9$ V, $I_L = 50$ μA, and PW = 100 μs.

7-15. Sketch the circuit of an IC operational amplifier employed as an astable multivibrator. Briefly explain how the circuit functions.

7-16. Using a 741 IC operational amplifier, design an astable multivibrator to produce a square wave output with an amplitude of approximately ± 9 V and frequency of approximately 500 Hz.

7-17. Sketch the circuit of an IC voltage comparator employed as a monostable multivibrator. Briefly explain how the circuit operates.

7-18. Design a monostable multivibrator using a 311 voltage comparator. The supply voltage is ±15 V, the circuit is to be triggered by a 1 V input pulse with PW = 1 μs, and the output pulse width is to be 500 μs.

7-19. Sketch the circuit of an IC voltage comparator employed as an astable multivibrator. Briefly explain how the circuit operates.

7-20. Design an astable multivibrator using a 311 voltage comparator. The supply voltage is ±15 V, and the output frequency is to be 7 kHz.

Chapter 8

IC Timer Circuits

INTRODUCTION

Integrated circuit timers are made up of voltage comparators, flip-flops, and low-impedance output stages, all contained in a single package. Using additional externally connected resistors and capacitors, these devices can be made to function as monostable multivibrators, astable multivibrators, square wave generators, sequential timers, etc. The design procedures for calculating the values of the external components are extremely simple.

8-1 THE 555 IC TIMER

The 555 integrated circuit *timer* can be applied to a myriad of timing applications: monostable multivibrators, astable multivibrators, ramp generators, sequential timers, etc. Although at first glance the functional block diagram (Figure 8-1) may look a little complex, it is really quite easily understood by anyone who has a basic understanding of flip-flops and comparators. Calculation of the values of external components for various applications is also very simple.

Referring to the function block diagram, the 555 timer is seen to consist of a *potential dividing network* R_1, R_2, and R_3; two *voltage comparators*; a *set-reset flip-flop*; an inverting buffer *output stage*, and two transistors. The 555 data sheet in Appendix 1-16 indicates that the circuit functions satisfactorily with supply voltages V_{CC} ranging from 4.5 V to 18 V. The set-reset flip-flop is explained in Chapter 13. For now, note that its output switches to *low* when a positive input is applied to the *set* terminal, and switches to *high* when a positive input appears at the *reset* terminal.

The potential divider provides a bias voltage to the inverting input terminal of comparator 1, and a different bias voltage to the noninverting terminal of comparator

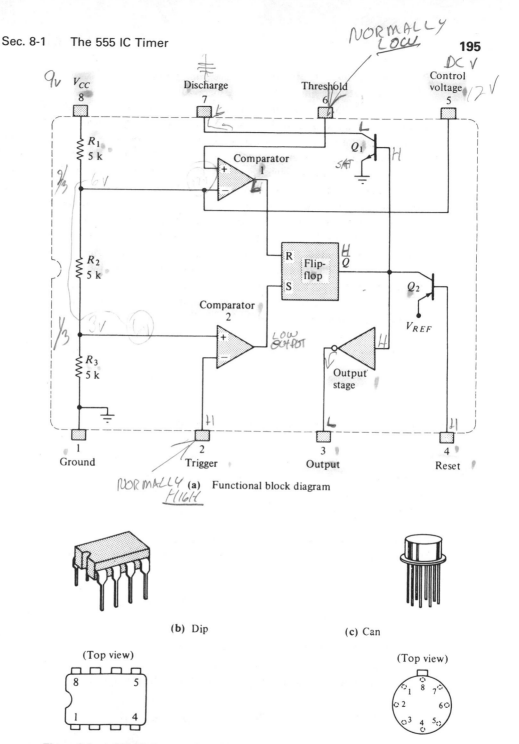

(a) Functional block diagram

(b) Dip

(c) Can

(Top view)

8			5
1			4

(Top view)

Figure 8-1 A 555 IC timer consists of a potential divider, two voltage comparators, a flip-flop, an inverting buffer output stage, a discharge transistor, and a reset transistor.

2. Access to the other inputs of the comparators is available via terminals 2 and 6, identified as *trigger* and *threshold*, respectively.

The comparator output levels control the flip-flop, and the flip-flop output is fed to the output stage and to the base of *npn* transistor Q_1. When the flip-flop output is high, Q_1 is biased *on*. In this condition, the transistor is typically used to *discharge* a capacitor connected to terminal 7. Q_1 is *off* when the flip-flop output is *low*. The *output stage* provides a low output resistance and also inverts the output level of the flip-flop. The voltage at terminal 3 is *low* when the flip-flop out is *high*, and *high* when the flip-flop output is *low*. The output stage can *sink* or *source* (at output terminal 3) a maximum current of 200 mA. (See the 555 data sheet in Appendix 1-16).

Transistor Q_2 is a *pnp* device with its emitter connected to an internal reference voltage V_{REF} which is always less than V_{CC}. If *reset* terminal 4 is connected to V_{CC}, the base-emitter junction of Q_2 is reverse biased, causing the transistor to remain *off*. When terminal 4 is pulled below V_{REF} (i.e., towards ground level), Q_2 switches *on*. This turns Q_1 *on*, causes the output at terminal 3 to go to ground level, and resets the flip-flop to its *high* output state.

The complete functioning of the timer circuit is best understood by considering a typical application, such as the monostable multivibrator in the next section.

8-2 555 AS A MONOSTABLE MULTIVIBRATOR

A basic 555 monostable circuit is shown in Figure 8-2. The supply voltage is connected across terminal 8 (+ V_{CC}) and terminal 1 (ground). Terminal 2 (*trigger*) is directly connected to a trigger pulse source. C_A is a capacitor which charges from V_{CC} via resistor R_A when *npn* transistor Q_1 (see Figure 8-1) is *off*. Terminal 4 is connected directly to V_{CC} to ensure that *pnp* transistor Q_2 (see Figure 8-1) remains *off* at all times. Terminal 5 is left open circuited, and the output is taken from terminal 3.

Operation of the 555 monostable circuit is explained below in point form. Refer to Figures 8-1 and 8-2.

Initial state

- Terminal 2 is *high* because trigger source level is normally *high*.
- Comparator 2 output is *low*, because terminal 2 is *high* (inverting input) and the voltage at the noninverting input of the comparator is V_{R3},

  where $$V_{R3} = V_{CC} \times \frac{R_3}{R_1 + R_2 + R_3} = \tfrac{1}{3} V_{CC}$$

- Comparator 1 output is *low*, because terminal 6 is *low* (noninverting input) and the inverting input of the comparator is at $V_{(R_2+R_3)}$,

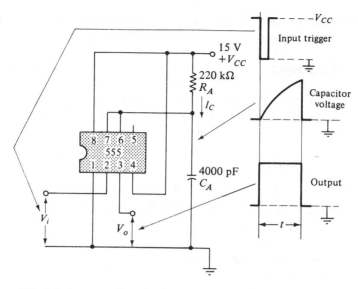

Figure 8-2   A basic monostable multivibrator can be constructed by connecting a resistor and a capacitor to a 555 timer. The resistor is calculated as $R_A = V_{CC}/(3\ I_{C(min)})$, and the capacitor is determined from $t = 1.1\ C_A R_A$.

where
$$V_{(R2+R3)} = V_{CC} \times \frac{R_2 + R_3}{R_1 + R_2 + R_3} = \tfrac{2}{3} V_{CC}$$

- Flip-flop output $\overline{Q}$ is *high*, in its *reset* condition.
- Q_1 is *on* because the flip-flop output is *high*.
- Capacitor C_A is in its discharged state because Q_1 is *on*.
- Terminals 6 and 7 are at a *low* voltage level because Q_1 is *on*.
- Output voltage at terminal 3 is *low* because the flip-flop output is *high*.

Triggered state

- The trigger input causes terminal 2 to go below V_{R3}; i.e., the inverting input to comparator 2 is driven below the voltage level at the noninverting input.
- Comparator 2 output goes *high* because of the trigger input.
- The flip-flop is driven into its *set* condition (output level *low*) by the *high* output of comparator 2.
- Q_1 is switched *off* by the flip-flop output switching to *low*.
- Terminal 3 *output* switches to *high* because of the flip-flop output switching to *low*.
- With Q_1 *off*, C_A (connected to terminals 6 and 7) commences to charge exponentially via R_A.

- When the trigger input at terminal 2 switches to *high* once again, comparator 2 output switches to *low*. The flip-flop remains in its *set* condition.

Final state

- Comparator 1 output remains *low* until the capacitor voltage (connected to terminal 6) becomes equal to $V_{(R2+R3)} = \frac{2}{3} V_{CC}$. Then comparator 1 output switches to *high*.
- The flip-flop is driven into its *high*-output *reset* condition once again by the *high* from comparator 1.
- Q_1 is switched *on* by the *high* output from the flip-flop.
- C_A is rapidly discharged by Q_1, and the voltage level at terminals 6 and 7 falls.
- The output voltage at terminal 3 switches to a *low* level because the flip-flop output is *high*.
- Comparator 1 output switches to *low* once again as terminal 6 voltage falls below $V_{(R2+R3)}$. The flip-flop remains in its *reset* condition.
- The final state of the monostable multivibrator is the same as its initial state. The circuit is now ready for triggering once again.

The 555 monostable circuit gives an output pulse each time it is triggered. The output pulse width depends upon the values of R_A and C_A, and upon the internal voltage levels of the 555 circuit.

8-3 DESIGNING A 555 MONOSTABLE CIRCUIT

Design of the monostable circuit in Fig. 8-2 involves nothing more than selection of R_A and C_A. The supply voltage V_{CC} can be anything from 4.5 V to 18 V. (See Appendix 1-16.) Regardless of the value of V_{CC}, $V_{R3} = \frac{1}{3} V_{CC}$ and $V_{(R2+R3)} = \frac{2}{3} V_{CC}$, as already shown above. Also, as already seen, when the circuit is triggered, C_A charges up to $\frac{2}{3} V_{CC}$, and then the circuit returns to its initial state. The time t required for C_A to charge from zero to $\frac{2}{3} V_{CC}$ determines the output pulse width. This time can be readily calculated from Equation 2-9.

$$t = CR \ln \left[\frac{E - E_o}{E - e_c} \right]$$

For the circuit in Figure 8-2:

$$C = C_A, \qquad R = R_A, \qquad E = V_{CC}, \qquad E_o = 0,$$

and $e_c = \frac{2}{3} V_{CC}$ is the final capacitor voltage. Substituting the quantities into Eq. 2-9 gives

$$t = 1.1 \, C_A \, R_A \tag{8-1}$$

C_A should normally be chosen as small as possible to ensure that Q_1 (see Figure 8-1) has no difficulty in discharging it rapidly. However, C_A should not be so small that it is affected by stray capacitance. If C_A is to be as small as possible, then the charging current must also be as small as possible. The minimum level of charging current occurs when the capacitor voltage is at its maximum level, i.e., when $e_c = \frac{2}{3} V_{CC}$. At this instant the voltage across R_A is

$$V_{RA} = V_{CC} - \tfrac{2}{3} V_{CC} = \tfrac{1}{3} V_{CC}$$

and the capacitor charging current is

$$I_{C(min)} = \frac{\frac{1}{3} V_{CC}}{R_A}$$

or

$$R_A = \frac{V_{CC}}{3 \, I_{C(min)}} \qquad (8\text{-}2)$$

$I_{C(min)}$ should be chosen much greater than the threshold current I_{th} which flows into terminal 6. This is to ensure that I_{th} does not divert a significant amount of I_C away from the capacitor.

The design procedure now becomes:

1. Note the value of I_{th} from the 555 specification sheet.
2. Select $I_{C(min)} \gg I_{th}$.
3. Calculate R_A, using Eq. (8-2).
4. Calculate C_A, using Eq. (8-1).

EXAMPLE 8-1

Design a 555 monostable circuit to have a 1 ms output pulse width. The supply voltage is to be $V_{CC} = 15$ V.

Solution   From the 555 data sheet, Appendix 1-16, maximum $I_{th} = 0.25$ μA.

$$I_{C(min)} \gg I_{th}$$

$$\text{let } I_{C(min)} = 100 \times I_{th}$$

$$= 100 \times 0.25 \text{ μA} = 25 \text{ μA}$$

From Eq. 8-2,
$$R_A = \frac{V_{CC}}{3 \, I_{C(min)}} = \frac{15 \text{ V}}{3 \times 25 \text{ μA}}$$

$$= 200 \text{ kΩ} \qquad \text{(use 220 kΩ standard value)}$$

From Eq. 8-1,
$$C_A = \frac{t}{1.1 \, R_A} = \frac{1 \text{ ms}}{1.1 \times 220 \text{ kΩ}}$$

$$\approx 4000 \text{ pF} \qquad \text{(use 4000 pF standard value)}$$

8-4 MODIFICATIONS TO THE BASIC 555 MONOSTABLE CIRCUIT

Figure 8-3 shows a monostable circuit with the trigger input *capacitor coupled* to terminal 2. Terminal 2 is connected to V_{CC} via resistor R_B. This is to ensure that the inverting input (trigger) terminal of comparator 2 remains above the noninverting input voltage until the trigger input is applied. The input waveform is differentiated by C_1 and R_B, and diode D_1 clips off the unwanted positive spike. (See the waveforms in Figure 8-3.)

In Figure 8-3, R_B performs exactly the same function as R_B does in the *normally-off capacitor-coupled inverter* discussed in Section 4-6; i.e., it holds a transistor (inside the comparator) in the *off* condition. As explained in Section 4-6, a 22 kΩ resistor is a reasonable maximum value to use for R_B.

The minimum value of C_1 is determined by considering the trigger input current (specified in Appendix 1-16 as $I_T = 0.5$ μA typical). The current that flows through R_B when the trigger voltage is present must also be considered. C_1 is then calculated in a similar way to that used for the capacitor-coupled inverter circuit (see Example 4-8), except that t is made equal to the 555 *output rise time* instead of the input pulse width. According to Appendix 1-16, the output rise time is 100 ns.

Decoupling capacitor C_2 (usually 0.01 μF) minimizes pickup of unwanted noise signals at the inverting input of comparator 1 when there is no other external connection

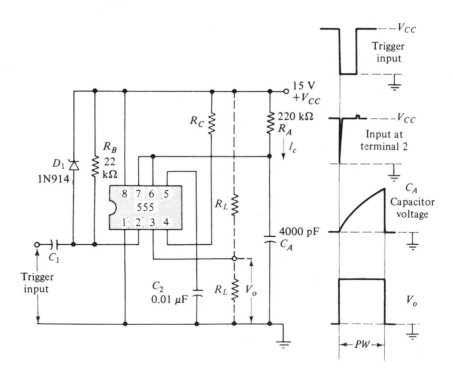

Figure 8-3 555 monostable multivibrator circuit with capacitor-coupled triggering (C_1, D_1, and R_B), load resistance (R_L), and noise decoupling capacitor (C_2).

to the *control voltage* terminal (terminal 5). C_2 also functions as a bypass capacitor to maintain the dc voltage constant across internal resistors R_2 and R_3 at high triggering frequencies.

As illustrated in Figure 8-3, the load R_L can be connected from *output* terminal 3 to either $+V_{CC}$ or ground level. This has no effect on the voltage level at the 555 output terminal, but it obviously affects the current direction through the load.

A resistor R_C may be connected between the *reset* terminal (terminal 4) and V_{CC}. This allows a reset voltage to be capacitor coupled to terminal 4, in order to reset the circuit back to its initial state, i.e., prior to the end of the normal pulse width. A suitable maximum value for R_C is 22 kΩ. (See Section 4-6.)

8-5 555 ASTABLE MULTIVIBRATOR

The monostable circuit is converted into an astable circuit simply by connecting the *trigger* terminal (terminal 2) directly to the *threshold* terminal (terminal 6). [See Figure 8-4(a).] The charging resistor is replaced by the two resistors R_A and R_B, and the *discharge* terminal (terminal 7) is connected to the junction of the two.

When the capacitor voltage (connected to terminal 6 and terminal 2) goes below $\frac{1}{3} V_{CC}$, the inverting input of comparator 2 is below the level of the noninverting input (which is $V_{R3} = \frac{1}{3} V_{CC}$). Comparator 2 output goes *high*, and triggers the flip-flop into its *set* condition, in which its output is a *low* level. Q_1 is now *off*, and C_A charges via R_A and R_B.

C_A continues to charge until it reaches $\frac{2}{3} V_{CC}$, at which point the noninverting input of comparator 1 (connected to C_A via terminal 6) is raised above the level of the inverting input (at $V_{(R2+R3)} = \frac{2}{3} V_{CC}$). The output of comparator 1 now goes *high*, triggering the flip-flop into its *reset* state (output *high*), and causing Q_1 to switch *on*. Capacitor C_A is now discharged by Q_1 via resistor R_B. Discharge of C_A continues until its voltage falls below $\frac{1}{3} V_{CC}$. At this point the output of comparator 2 goes *high*, triggering the flip-flop to its *low* output state and switching Q_1 *off* once again. The cycle has now recommenced, and it continues repetitively.

Design of a 555 astable multivibrator involves only the calculation of R_A, R_B, and C_A. Note that I_C should be selected much larger than both the trigger current and the threshold current. Capacitor C_2 is usually 0.01 μF, as already explained. Since C_A charges via $(R_A + R_B)$ from $\frac{1}{3} V_{CC}$ to $\frac{2}{3} V_{CC}$, the initial capacitor voltage is $E_o = \frac{1}{3} V_{CC}$, and the final level is $e_c = \frac{2}{3} V_{CC}$. Also, the charging voltage is $E = V_{CC}$. Substituting the values into Equation 2-9 gives

$$t_1 = 0.693 \ C_A(R_A + R_B) \tag{8-3}$$

Similarly, for the discharge period: $E_o = \frac{2}{3} V_{CC}$, $e_c = \frac{1}{3} V_{CC}$, and $E = 0$. Using these quantities, Equation 2-9 yields

$$t_2 = 0.693 \ C_A \ R_B \tag{8-4}$$

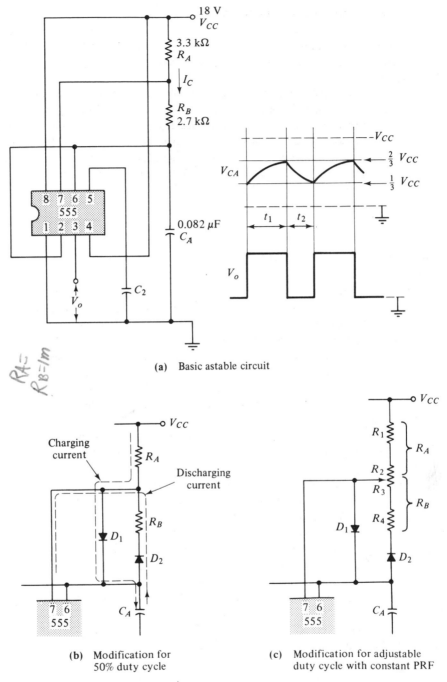

(a) Basic astable circuit

$R_A = R_B = 1m$

(b) Modification for
50% duty cycle

(c) Modification for adjustable
duty cycle with constant PRF

Figure 8-4 A 555 astable multivibrator is basically a self-triggering monostable circuit (a). The ouput is a pulse waveform with a duty cycle greater than 50%. Each time period is governed by the equation $t = 0.69\, C_A R$. The duty cycle can be altered by the use of diodes, as in (b) and (c).

EXAMPLE 8-2

Design a 555 astable multivibrator to give a pulse output with PRF = 2 kHz and duty cycle = 66%. Use V_{CC} = 18 V.

Solution   Referring to the waveform in Figure 8-4(a),

$$t_1 + t_2 = \frac{1}{PRF} = \frac{1}{2 \text{ kHz}}$$

$$= 500 \text{ μs}$$

$$t_1 = (\text{duty cycle}) \times (t_1 + t_2)$$

$$= \frac{66}{100} \times 500 \text{ μs}$$

$$= 330 \text{ μs}$$

$$t_2 = (t_1 + t_2) - t_1 = 500 \text{ μs} - 330 \text{ μs} = 170 \text{ μs}$$

From data sheet,

$$I_{C(\text{min})} \gg I_{th} = 0.25 \text{ μA}$$

and

$$I_{C(\text{min})} \gg I_{trig} = 0.5 \text{ μA}$$

$$\text{Let } I_{C(\text{min})} = 1 \text{ mA}.$$

From Eq. 8-2,

$$R_A + R_B = \frac{V_{CC}}{3 I_{C(\text{min})}} = \frac{18 \text{ V}}{3 \times 1 \text{ mA}}$$

$$= 6 \text{ k}\Omega$$

From Eq. 8-3,

$$C_A = \frac{t_1}{0.693(R_A + R_B)} = \frac{330 \text{ μs}}{0.693 \times 6 \text{ k}\Omega}$$

$$\approx 0.08 \text{ μF} \qquad (\text{use } 0.082 \text{ μF standard value})$$

From Eq. 8-4,

$$R_B = \frac{t_2}{0.693 \, C_A} = \frac{170 \text{ μs}}{0.693 \times 0.08 \text{ μF}}$$

$$= 3.07 \text{ k}\Omega \qquad (\text{use } 2.7 \text{ k}\Omega \text{ standard value})$$

$$R_A = (R_A + R_B) - R_B$$

$$= 6 \text{ k}\Omega - 2.7 \text{ k}\Omega$$

$$= 3.3 \text{ k}\Omega \qquad (\text{standard value})$$

EXAMPLE 8-3

Analyze the circuit designed in Example 8-2 to determine the actual PRF and the duty cycle.

Solution   From Eq. 8-3,

$$t_1 = 0.693 \, C_A(R_A + R_B)$$

$$= 0.693 \times 0.082 \, \mu F \times (3.3 \text{ k}\Omega + 2.7 \text{ k}\Omega)$$

$$= 341 \, \mu s$$

From Eq. 8-4,

$$t_2 = 0.693 \, C_A R_B$$

$$= 0.693 \times 0.082 \, \mu F \times 2.7 \text{ k}\Omega$$

$$= 153 \, \mu s$$

$$t_1 + t_2 = 341 \, \mu s + 153 \, \mu s$$

$$= 494 \, \mu s$$

$$PRF = \frac{1}{(t_1 + t_2)} = \frac{1}{494 \, \mu s}$$

$$= 2.02 \text{ kHz}$$

$$\text{Duty cycle} = \frac{t_1}{t_1 + t_2} \times 100\% = \frac{341 \, \mu s}{153 \, \mu s + 341 \, \mu s} \times 100\%$$

$$= 69\%$$

8-6 MODIFICATIONS TO THE BASIC 555 ASTABLE CIRCUIT

50% Duty Cycle Astable

When Eqs. 8-3 and 8-4 are applied to the circuit of Figure 8-4(a), it is seen that t_1 must always be greater than t_2 (i.e., the duty cycle of the pulse output is greater than 50%). This is because Equation 8-3 involves $(R_A + R_B)$, whereas Equation 8-4 for t_2 involves only R_B. Clearly, for $t_1 = t_2$, $(R_A + R_B)$ must be equal to R_B, and (since R_A cannot be zero) this is impossible.

Figure 8-4(b) shows a modification which allows the duty cycle to be made 50% or less. During the charging cycle, diode D_2 is reverse biased and D_1 is forward biased. C_A is charged from V_{CC} via R_A and D_1. When Q_1 (in Figure 8-1) is *on*, terminal 7 is close to ground level, and C_A is discharged via D_2 and R_B. Equation 8-3 now becomes

$$t_1 = 0.693 \, C_A R_A$$

while Equation 8-4 remains

$$t_2 = 0.693 \ C_A R_B$$

Now, with $R_A = R_B$, t_1 and t_2 are equal, and the duty cycle is 50%. When R_A is less than R_B, t_1 is less than t_2, and the duty cycle becomes less than 50%.

Variable Duty Cycle Circuit

The further modification in Figure 8-4(c) allows the duty cycle to be adjusted without altering the PRF of the circuit. The duty cycle depends upon the relative values of R_A and R_B. Since the variable resistor $(R_2 + R_3)$ can be adjusted to increase R_A while decreasing R_B, and *vice versa*, the duty cycle is variable.

The pulse repetition frequency is the reciprocal of $t_1 + t_2$;

thus,
$$PRF = \frac{1}{t_1 + t_2}$$

and
$$t_1 + t_2 = 0.693 \ C_A R_A + 0.693 \ C_A R_B$$
$$= 0.693 \ C_A \ (R_A + R_B)$$
$$= 0.693 \ C_A \ [(R_1 + R_2) + (R_3 + R_4)]$$

or
$$PRF = \frac{1}{[0.693 \ C_A(R_1 + R_2 + R_3 + R_4)]}$$

Clearly, the PRF remains constant regardless of the distribution of the variable resistance $(R_2 + R_3)$ between R_A and R_B.

Variable-Frequency Square Wave Generator

A common requirement is a square wave generator with a frequency control that does not affect the duty cycle of the square wave output. This could be very simply achieved by making C_A a variable capacitor. A wide range of output frequencies could be produced by including a multiposition switch to select different fixed capacitor values. A variable capacitor could also be permanently connected for continuous frequency adjustment between switched ranges. [See Figure 8-5(a).] This arrangement could be employed with the basic astable circuit in Figure 8-4(a), or with either of the modifications shown in Figures 8-4(b) and (c).

Another method of constructing a variable-frequency square wave generator with an almost constant duty cycle uses the basic astable circuit in Figure 8-4(a). R_B is replaced with a variable resistor in series with a fixed-value resistor. [See Figure 8-5(b).] R_A is made very much smaller than the minimum value of R_B. The output frequency is adjusted by altering R_B. Because $(R_A + R_B)$ is always a little larger than R_B, the duty cycle is always a little greater than 50%, but is not significantly changed when the frequency is adjusted. When calculating component values, R_A should be not

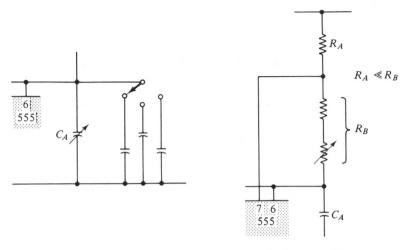

(a) Capacitor value adjustment   (b) Resistor adjustment

Figure 8-5   A variable-frequency square wave generator can be constructed by providing a 555 astable circuit with a range of selectable capacitor values and an adjustable charging resistance.

less than about 1 kΩ. This is to avoid overloading discharge transistor Q_1. (See Figure 8-1.)

Another Square Wave Generator Circuit

In Figure 8-6(a) yet another method of constructing a 50% duty cycle astable multivibrator is shown. In this case capacitor C_A is charged via R_A and R_B from the low-impedance output terminal 3. As in Figure 8-4, terminals 2 and 6 detect the *low* and *high* limits of capacitor voltage. Discharge terminal 7 is left unconnected.

When the output (at terminal 3) is high, C_A charges positively until its voltage arrives at $\frac{2}{3} V_{CC}$ (detected at terminal 6). The output then switches to a low level, and C_A is discharged through the same two resistors (R_A and R_B) until it falls to $\frac{1}{3} V_{CC}$ (detected at terminal 2). Once again, the output switches to a high level and the cycle repeats.

The one problem with this circuit is that V_o at terminal 3 does not go all the way up to V_{CC}; instead, it is usually about 1 V below the level of the supply. Equation 8-3, which should be applicable here, is not quite correct, because C_A is being charged to $\frac{2}{3} V_{CC}$ and discharged to $\frac{1}{3} V_{CC}$, but is being charged from a voltage $V_o \approx V_{CC} - 1$ V. The use of resistor R_C connected at terminal 5 can correct the error. Referring to Figure 8-1, it is seen that a resistor from terminal 5 to ground shunts R_2 and R_3. Selection of a suitable value for R_C can make $V_{R3} = \frac{1}{3} V_o$, and $(V_{R2} + V_{R3}) = \frac{2}{3} V_o$. The output from the circuit of Figure 8-6 now has a 50% duty cycle, and adjustment of R_B affords

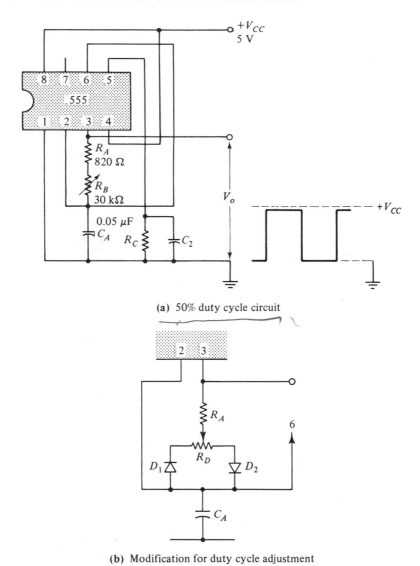

(a) 50% duty cycle circuit

(b) Modification for duty cycle adjustment

Figure 8-6   Another circuit for a 555 square wave generator. In (a), the capacitor is charged and discharged using output terminal 3, instead of V_{CC}. In (b), diodes and variable resistors are included to provide duty cycle adjustment.

output frequency control without affecting the duty cycle. Figure 8-6(b) shows a method of adding duty cycle control to the circuit of Figure 8-6(a). Diodes D_1 and D_2 and potentiometer R_D provide different (variable-resistance) charge and discharge paths for capacitor C_A.

8-7 MORE 555 TIMER APPLICATIONS

Delay Timers

The two circuits in Figure 8-7 provide time-delayed outputs from the instant of supply switch-*on*. The *high* output from circuit (a) can be employed to turn *on* a circuit or device for a time t after V_{CC} is first applied. Circuit (b) produces a *low* output for a time t after the supply is turned *on*. Thus, it would be used to delay switch-*on* of a circuit for the desired time period.

Both circuits are essentially self-triggering monostable multivibrators. Self-triggering is achieved by connecting *trigger* terminal 2 directly to the capacitor. Note that in both cases, *discharge* terminal 7 is left unconnected.

When V_{CC} is switched *on* to the circuit in Figure 8-7(a), the capacitor is initially in a discharged state. Thus, as illustrated, $V_{CA} = 0$, and this *low* level at terminal 2 triggers the output to *high*. The output remains *high* until V_{CA} exceeds 2 $V_{CC}/3$, at which point the input voltage to *threshold* terminal 6 resets the output voltage to *low*. In Figure 8-7(b), C_A and R_A are interchanged. In this case, the discharged state of C_A provides a *high* level (V_{CC}) at terminals 2 and 6 at the instant of switch-*on*. The *high* voltage at the *threshold* terminal causes the output to be held at a *low* level for the timing period. As the capacitor charges, the voltage across the resistor decreases. When V_{RA} falls below $V_{CC}/3$, this voltage level at terminal 2 triggers the output to high.

The design procedure for these circuits is exactly the same as for a 555 monostable multivibrator.

Sequential Timer

Figure 8-8 shows three 555 monostable multivibrators cascaded to make a *sequential timer*. An input pulse V_i triggers *monostable 1* to produce an output with pulse width *PW1*. When the output from *monostable 1* switches to low, its negative-going trailing edge triggers *monostable 2*, which then generates an output with pulse width *PW2*. Similarly, at the end of *PW2*, *monostable 3* is triggered to produce *PW3*. Each monostable circuit is designed individually for the desired output pulse width.

In circuits requiring several 555 timers, it may be desirable for reasons of economy of space and cost to use 556 dual timers. Figure 8-9 shows the terminal connections for the 556, which consists of two 555 timers in a single package. The two circuits are completely independent of each other, except for supply and ground terminals.

Pulsed-Tone Oscillator

The pulsed-tone oscillator circuit in Figure 8-10 generates a repetitive succession of groups of high-frequency square waves. Both sections of the circuit are astable multivibrators. The output of low-frequency *astable 1* controls *astable 2*. When output terminal 3 on *astable 1* is *low*, it holds C_A in *astable 2* in an uncharged state via the interconnecting diode D_1. When the output of *astable 1* switches to *high*, D_1 is reverse biased. This allows C_A to charge, and thus, as illustrated by the waveforms, *astable 2* oscillates for the duration of the output pulse from *astable 1*.

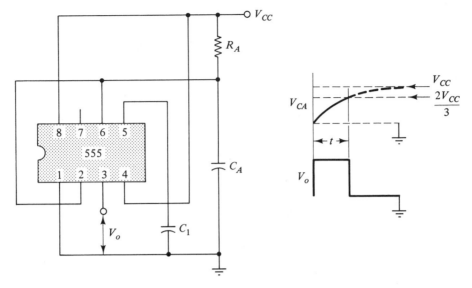

(a) High output for time t

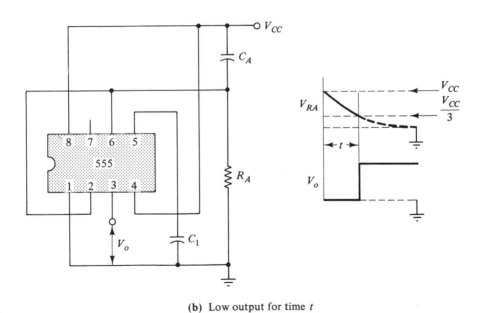

(b) Low output for time t

Figure 8-7   Delay timers are similar to monostable circuits, except that *trigger* terminal 2 is connected directly to the capacitor. In (a), the circuit is triggered at switch-*on*, to provide a *high* output for time t. In circuit (b), triggering occurs when V_{RA} falls to $V_{CC}/3$. The output is *low* for time t.

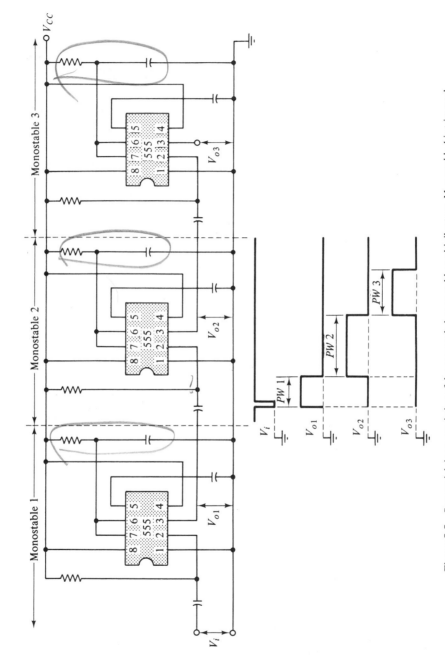

Figure 8-8 Sequential timer consisting of three cascaded monostable multivibrators. *Monostable 1* is triggered by V_i. *Monostable 2* is triggered by the trailing edge of V_{o1}, and *monostable 3* is triggered by the trailing edge of V_{o2}.

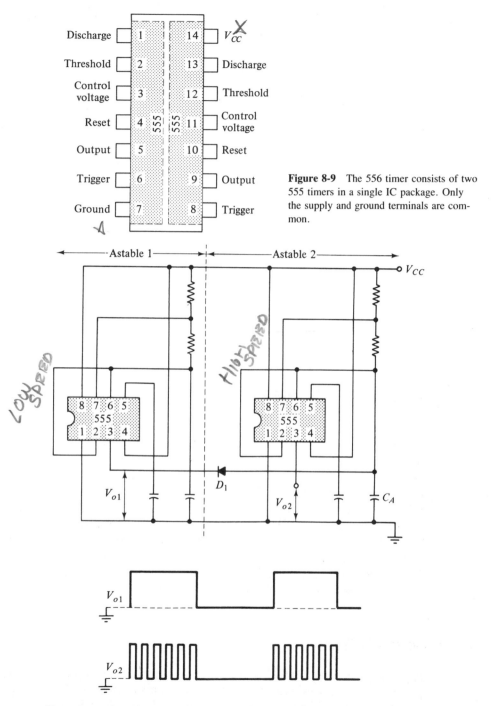

Discharge 1 14 V_CC

Threshold 2 13 Discharge

Control voltage 3 12 Threshold

Reset 4 555 555 11 Control voltage

Output 5 10 Reset

Trigger 6 9 Output

Ground 7 8 Trigger

Figure 8-9 The 556 timer consists of two 555 timers in a single IC package. Only the supply and ground terminals are common.

Astable 1 | Astable 2

V_{CC}

8 7 6 5
555
1 2 3 4

8 7 6 5
555
1 2 3 4

V_{o1}

D_1

V_{o2}

C_A

V_{o1}

V_{o2}

Figure 8-10 Pulsed-tone oscillator consisting of two astable multivibrators. High-frequency circuit *astable 2* oscillates for the duration of the output pulse from low-frequency *astable 1*.

211

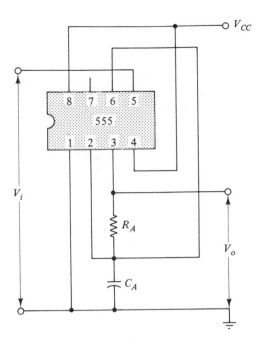

Figure 8-11 555 voltage-controlled oscillator. The circuit is an astable multivibrator with a control voltage V_i connected to terminal 5. Adjustment of V_i alters the capacitor upper and lower voltage levels, thus controlling the time period and frequency of the output waveform.

Voltage-Controlled Oscillator

The circuit in Figure 8-11 is a square wave generator, as in Figure 8-6(a), except that terminal 5 (the *control voltage* terminal) has an external voltage V_i applied to it. Recall that terminal 5 is connected internally to the junction of potential divider resistors R_1 and R_2. When no external voltage is applied, the voltage at terminal 5 is $V_5 = 2\,V_{CC}/3$, and that at the junction of R_2 and R_3 is $V_{CC}/3$, or $V_5/2$. As already explained, when the 555 is connected to function as an astable circuit, the capacitor charges up to $2\,V_{CC}/3$ and discharges to $V_{CC}/3$. With V_i present, the capacitor charges to V_i and discharges to $V_i/2$. Adjusting V_i alters the capacitor charge and discharge times, thus varying the output frequency and duty cycle. Hence, the circuit is a *voltage-controlled oscillator*. The internal potential divider network constitutes a load of $R_1 \| (R_2 + R_3)$ on the signal source. The signal source resistance should be much smaller than $R_1 \| (R_2 + R_3)$.

8-8 CMOS TIMER CIRCUIT

The ICM7555 manufactured by Intersil is typical of timers using CMOS technology. The 7555 can be substituted in all situations for a bipolar 555 IC. However, the 7555 typically draws only 80 μA from the supply, whereas the 555 requires 10 mA. The 7555 can also operate from a supply voltage as low as 2 V, compared to 4.5 V minimum for the 555. One other advantage of the CMOS 7555 is that its output voltage (at terminal 3) exhibits very low offset, i.e., the output effectively swings from ground to $+V_{CC}$. This means, for example, that the square wave generator circuit of Figure 8-6, using a 7555 timer, could operate satisfactorily without resistor R_C.

The maximum levels of trigger and threshold current for the 7555 are 50 pA, compared to 0.5 µA and 0.25 µA, respectively, for the 555 timer. Because of these very low current levels, the usual approach to calculation of resistor values (using a current which is one hundred times the input current) gives extremely high resistances. The high resistances result in very low capacitance values. Consider Example 8-1. If a 7555 device is used, and the same design procedure is followed, R_A is calculated as 100 MΩ, and C_A is approximately 10 pF. The low capacitance can be affected by stray capacitances, and the high resistance may allow pickup of unwanted noise spikes.

To design satisfactory circuits using a 7555 timer, start by selecting a capacitor value very much larger than stray capacitance, say, a minimum of 1000 pF. Then, use the equation relating C and R to determine a suitable resistance value.

The 7556 is a dual 7555. Pin connections for the 7555 and 7556 are the same as for 555 and 556 devices.

REVIEW QUESTIONS AND PROBLEMS

8-1. Sketch the functional block diagram of a 555 IC timer. Briefly explain the function of each component.

8-2. Sketch the circuit and waveforms generated for a 555 monostable multivibrator. Referring to the monostable circuit and the 555 functional block diagram, explain how the 555 operates as a monostable circuit.

8-3. Using a supply of 18 V, design a 555 monostable circuit to produce a 0.5 ms output pulse.

8-4. Show how a basic 555 monostable circuit should be modified to (a) use capacitor-coupled triggering, (b) prevent unwanted signals from being picked up at the control voltage terminal, and (c) permit a reset signal to be coupled to the reset terminal. Sketch the various waveforms and briefly explain.

8-5. Sketch the circuit of a 555 timer employed as an astable multivibrator. Show the capacitor and output waveforms, and explain how the circuit functions.

8-6. Show how a basic 555 astable circuit may be modified (a) to produce a 50% duty cycle, (b) to provide an adjustable duty cycle with a constant PRF, and (c) to create a variable-frequency square wave generator. Briefly explain in each case.

8-7. Design a 555 astable multivibrator to generate an output with PRF = 5 kHz and a duty cycle of 75%. Use V_{CC} = 15 V.

8-8. Analyze the circuit designed in Problem 8-7 to determine the actual PW and duty cycle.

8-9. Sketch circuit diagrams to show how 555 timers may be used to construct (a) a sequential timer, and (b) a pulsed-tone oscillator. Explain how each circuit operates.

8-10. Design the 555 square wave generator illustrated in Fig. 8-6(a). Output amplitude is to be ≈ 10 V, and frequency is to be adjustable over the range 1 kHz to 10 kHz. Assume that V_o is approximately 1 V less than V_{CC}.

8-11. Modify the circuit designed for Problem 8-10, as in Figure 8-6(b), to provide duty cycle adjustment of 20% to 80% at $f = 10$ kHz.

8-12. Sketch a circuit which uses a 555 timer to produce a *high* output level for a given time period from the instant of supply switch-*on*. Explain the operation of the circuit.

8-13. Using a 10 V supply, design the circuit in Question 8-12 to have a *high* output level for a time of 15 ms.

8-14. Sketch a circuit which uses a 7555 timer to produce a *low* output level for a given time period from the instant of supply switch on. Explain the operation of the circuit.

8-15. Using a 12 V supply, design the circuit in Question 8-14 to have a *low* output level for a time of 25 ms.

8-16. Using 7555 timers, design a four-stage sequential timer to produce outputs of 0.5 ms, 1 ms, 2 ms, and 4 ms. Use $V_{CC} = 15$ V.

8-17. Using 555 timers, design a pulsed-tone oscillator to produce 5 ms output pulses of a 5 kHz tone with 3 ms intervals. Use $V_{CC} = 12$ V.

8-18. Sketch the circuit of a voltage-controlled oscillator using a 555 timer. Explain the operation of the circuit.

8-19. Design a 555 voltage controlled oscillator to have an output amplitude of approximately 18 V, and an output frequency ranging from approximately 2 kHz to 3 kHz.

8-20. Compare 7555/7556 CMOS timers to 555/556 bipolar devices. Design a square wave generator using a 7555 timer to have an output amplitude of 5 V and output frequency ranging from 500 Hz to 15 kHz.

Chapter 9

Ramp, Pulse, and Function Generators

INTRODUCTION

A simple *ramp generator* can be constructed using a capacitor charged via a resistor in conjunction with a discharge transistor. To improve the ramp linearity, the charging resistor may be replaced with a transistor constant-current circuit. When a UJT is employed as the discharge transistor, the circuit becomes a relaxation oscillator. Bootstrap ramp generators and Miller integrators produce nearly linear output ramps. *Pulse generator* circuits can be constructed to have variable frequency and pulse width, as well as amplitude and dc level adjustment. Integrated circuit *function generators* operate with several additional external components.

9-1 CR RAMP GENERATOR

The simplest ramp generator circuit is a capacitor charged via a series resistance. A transistor must be connected in parallel with the capacitor to provide a discharge path, as shown in the circuit of Figure 9-1(a). Capacitor C_1 is charged from V_{CC} via R_1. Q_1 is biased *on* via R_B, so that the capacitor is normally in a discharged state. When a negative-going input pulse is coupled by C_2 to Q_1 base, the transistor switches *off*. With Q_1 *off*, C_1 begins to charge, thus producing an approximate ramp output until the input pulse ends. [See Figure 9-1(b).] At this point, Q_1 switches *on* again and rapidly discharges the capacitor.

The output from a simple CR circuit is exponential rather than linear. For voltages very much less than the supply voltage, however, the output is approximately linear. When the transistor is *on*, the capacitor is discharged to $V_{CE(sat)}$. Hence, $V_{CE(sat)}$ is the starting level of the output ramp. Output amplitude control can be provided by making charging resistance R_1 adjustable.

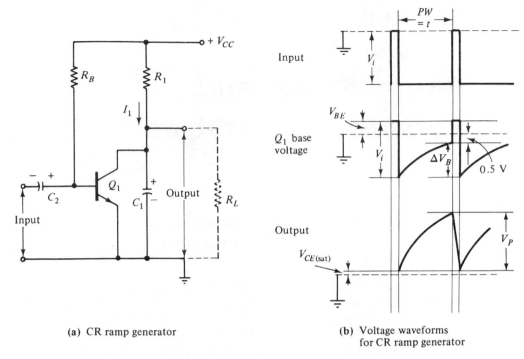

(a) CR ramp generator

(b) Voltage waveforms
for CR ramp generator

Figure 9-1 Simple ramp generator circuit consisting of a capacitor charged via a resistor, and a discharge transistor connected as a capacitor-coupled normally-*on* inverter. The output waveform is exponential.

The capacitance of C_2, which couples the input pulse to the transistor base, should be selected as low as possible, for both minimum cost and the smallest possible physical size. As discussed in Section 4-6, the minimum capacitor size can be determined by allowing the base voltage of Q_1 to rise during the input pulse time. [See Figure 9-1(b).] The base voltage starts approximately at 0.7 V when Q_1 is *on*. Then, V_{B2} is pulled down by the input pulse, but starts to rise again as C_2 is charged through R_B. To ensure that Q_1 is still *off* at the end of the pulse time, V_{B2} should not rise above approximately -0.5 V.

EXAMPLE 9-1

Design a simple CR ramp generator as in Figure 9-1(a) to give an output that peaks at 5 V. The supply voltage is 15 V, and the load to be connected at the output is 100 kΩ. The ramp is to be triggered by a negative-going pulse with an amplitude of 3 V, PW of 1 ms, and a time interval between pulses of 0.1 ms. Take the transistor $h_{FE(min)}$ as 50.

Solution   Maximum output current is

$$I_{L(\max)} = \frac{V_P}{R_L}$$

$$= \frac{5\ \text{V}}{100\ \text{k}\Omega} = 50\ \mu\text{A}$$

At peak output voltage, let

$$I_1 = 100 \times I_{L(\max)} = 100 \times 50\ \mu\text{A}$$

$$= 5\ \text{mA}$$

$$R_1 = \frac{V_{CC} - V_P}{I_1} = \frac{15\ \text{V} - 5\ \text{V}}{5\ \text{mA}}$$

$$= 2\ \text{k}\Omega \qquad \text{(use 2.2 k}\Omega \text{ standard value)}$$

The voltages for capacitor C_1 are

$$\text{Initial voltage} = E_o = V_{CE(\text{sat})} \approx 0.2\ \text{V}$$

$$\text{Final voltage} = e_c \approx V_P = 5\ \text{V}$$

$$\text{Charging voltage} = E = V_{CC} = 15\ \text{V}$$

From Eq. 2-9, $$C_1 = \frac{t}{R \ln \dfrac{E - E_o}{E - e_c}} = \frac{1\ \text{ms}}{2.2\ \text{k}\Omega \ln \dfrac{15\ \text{V} - 0.2\ \text{V}}{15\ \text{V} - 5\ \text{V}}}$$

$$\approx 1.2\ \mu\text{F}$$

The discharge time for C_1 is 0.1 ms, which is one-tenth of the charging time. For Q_1 to discharge C_1 in one-tenth of the charging time,

$$I_C \approx 10 \times (C_1 \text{ charging current})$$

$$= 10\, I_1 = 50\ \text{mA}$$

$$I_B = \frac{I_C}{h_{FE(\min)}} = \frac{50\ \text{mA}}{50}$$

$$= 1\ \text{mA}$$

$$R_B = \frac{V_{CC} - V_{BE}}{I_B} = \frac{15\ \text{V} - 0.7\ \text{V}}{1\ \text{mA}}$$

$$= 14.3\ \text{k}\Omega \qquad \text{(use 12 k}\Omega \text{ standard value)}$$

For Q_1 to remain biased *off* at the end of the input pulse, let $V_B = -0.5$ V.

$$\Delta V_B = V_i - V_{BE} - V_B \qquad \text{[see Figure 9-1(b)]}$$

$$= 3\ \text{V} - 0.7\ \text{V} - 0.5\ \text{V} = 1.8\ \text{V}$$

The charging current for C_2 is equal to the current through R_B when Q_1 is *off*:

$$I \approx \frac{V_{CC} - V_i}{R_B} = \frac{15 \text{ V} - (-3 \text{ V})}{12 \text{ k}\Omega}$$

$$= 1.5 \text{ mA}$$

From Eq. 2-8,

$$C_2 = \frac{It}{\Delta V} = \frac{1.5 \text{ mA} \times 1 \text{ ms}}{1.8 \text{ V}}$$

$$= 0.83 \text{ }\mu\text{F} \qquad \text{(use 1 }\mu\text{F standard value)}$$

9-2 CONSTANT-CURRENT RAMP GENERATORS

Bipolar Transistor Constant-Current Circuits

The major disadvantage of the simple CR ramp generator is its nonlinearity. To produce a linear ramp, the capacitor charging current must be held constant. This can be achieved by replacing the charging resistance with a *constant-current circuit*.

A basic transistor constant-current circuit is shown in Figure 9-2(a). The potential divider (R_1 and R_2) provides a fixed voltage V_B at the base of *pnp* transistor Q_2. The voltage across emitter resistor R_3 remains constant at ($V_B - V_{BE}$). Thus, the emitter current is also constant, $I_E = (V_B - V_{BE})/R_3$. Since $I_C \approx I_E$, the collector current remains constant. Figure 9-2(b) shows an arrangement that allows the level of constant current

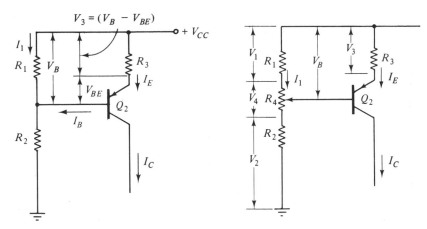

(a)   Constant current circuit   (b)   Adjustable constant current circuit

Figure 9-2   Constant-current circuits are constructed by providing a constant bias voltage V_B at the base of a transistor. The transistor emitter voltage is constant at $V_E = V_B - V_{BE}$. This gives a constant current $I_E = V_E/R_3$.

to be adjusted. R_4 provides adjustment of V_B. Since $V_3 = (V_B - V_{BE})$, V_3 also is adjustable by R_4, and I_E can be set to any desired level over a range dependent upon R_4. To maintain a constant level of I_E (and I_C), the voltage across R_3 should be several times larger than the base emitter voltage V_{BE}. This ensures that changes in V_{BE} do not significantly affect I_E.

Constant-Current Ramp Generator Circuit

Figure 9-3 shows a ramp generator that employs the constant-current circuit. Note that because I_C of Q_2 is a constant charging current for C_1, the (output) capacitor voltage grows linearly. The simpler capacitor-charging equation, Eq. 2-8, may now be used for C_1 calculations. The circuit of Figure 9-3 functions like the simple CR ramp generator of Figure 9-1, with R_1 replaced by the constant-current circuit.

The output voltage from the constant-current ramp generator remains linear only if a sufficient collector-emitter voltage is maintained across Q_2 for it to operate in the active region of its characteristics. If Q_2 reaches saturation, the output voltage is clipped. Therefore, V_{CE2} should not fall below about 3 V. Because of this, and because of the constant voltage V_3 across resistor R_3, the maximum ramp output voltage obtainable from the circuit of Figure 9-3 is approximately $V_P = V_{CC} - V_3 - 3$ V.

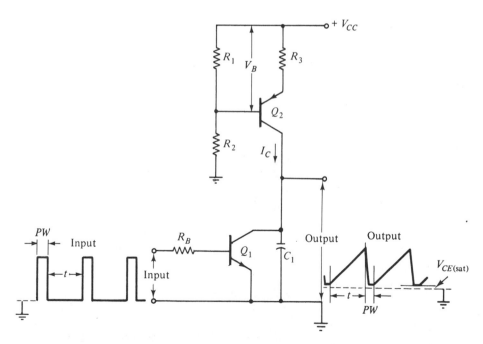

Figure 9-3   A linear ramp generator is created by substituting a constant-current circuit in place of the charging resistor (of Figure 9-1). In this circuit, a positive pulse directly coupled to the base of Q_1 causes the capacitor to be discharged.

In Figure 9-3, a direct-coupled transistor Q_1 is employed to discharge C_1, instead of the capacitor-coupled circuit of Figure 9-1. (A capacitor-coupled transistor could, of course, be used.) The input pulse is directly connected to the base of transistor Q_1. When the input is at ground level, Q_1 is *off* and capacitor C_1 charges via Q_2. When a positive input is applied, Q_1 is switched *on* and C_1 is rapidly discharged. Q_1 remains *on* during the positive input pulse; thus, C_1 is held in a discharged condition, and the ramp generator output voltage remains at the $V_{CE(\text{sat})}$ of Q_1.

EXAMPLE 9-2

Using a constant-current circuit, modify the ramp generator designed in Example 9-1 to produce a linear ramp output.

Solution   From the circuit of Figure 9-3,

$$\text{Output} = V_P = 5 \text{ V}$$

$$V_3 + V_{CE2} = V_{CC} - 5 \text{ V} = 10 \text{ V}$$

Let

$$V_{CE2} = 3 \text{ V, minimum}$$

Then

$$V_3 = 10 \text{ V} - 3 \text{ V} = 7 \text{ V}$$

For $C_1 = 1 \ \mu\text{F}$, $V_P = 5$ V, and $t = 1$ ms;

$$I_C = \frac{C_1 \Delta V}{t} = \frac{1 \ \mu\text{F} \times 5 \text{ V}}{1 \text{ ms}} = 5 \text{ mA}$$

$$R_3 \approx \frac{7 \text{ V}}{5 \text{ mA}} = 1.4 \text{ k}\Omega \qquad \text{(use 1.2 k}\Omega \text{ standard value)}$$

For $R_3 = 1.2$ kΩ,

$$V_3 = 5 \text{ mA} \times 1.2 \text{ k}\Omega$$

$$= 6 \text{ V}$$

$$V_B = V_3 + V_{BE2} = 6 \text{ V} + 0.7 \text{ V}$$

$$= 6.7 \text{ V}$$

V_B must be a stable bias voltage that is unaffected by I_{B2}. Make $I_1 \approx I_E = 5$ mA. Then

$$R_1 = \frac{V_B}{I_1} = \frac{6.7 \text{ V}}{5 \text{ mA}}$$

$$= 1.34 \text{ k}\Omega \qquad \text{(use 1.2 k}\Omega \text{ standard value)}$$

I_1 then becomes

$$I_1 = \frac{6.7 \text{ V}}{1.2 \text{ k}\Omega} = 5.58 \text{ mA}$$

$$V_2 = V_{CC} - V_1 = 15 \text{ V} - 6.7 \text{ V} = 8.3 \text{ V}$$

$$R_2 \approx \frac{V_2}{I_1} = \frac{8.3 \text{ V}}{5.58 \text{ mA}}$$

$$= 1.49 \text{ k}\Omega \qquad \text{(use 1.5 k}\Omega\text{ standard value)}$$

EXAMPLE 9-3

Redesign the circuit of Example 9-2 to make the ramp amplitude adjustable from 3 V to 5V.

Solution   The circuit modification is shown in Figure 9-2(b). The charging current, with $V_P = 3$ V, is

$$I_C = \frac{C_1 \Delta V}{t} = \frac{1 \text{ }\mu\text{F} \times 3 \text{ V}}{1 \text{ ms}}$$

$$= 3 \text{ mA}$$

For $V_P = 5$ V,

$$I_C = \frac{1 \text{ }\mu\text{F} \times 5 \text{ V}}{1 \text{ ms}} = 5 \text{ mA}$$

For $I_C = 3$ mA, $I_E \approx 3$ mA and

$$V_3 = I_E \times R_3 = 3 \text{ mA} \times 1.2 \text{ k}\Omega$$

$$= 3.6 \text{ V}$$

$$V_B = V_1 = V_3 + V_{BE} = 3.6 \text{ V} + 0.7 \text{ V}$$

$$= 4.3 \text{ V}$$

(At this point, the moving contact on the potentiometer is at the upper end.) For $R_1 = 1.2$ kΩ,

$$I_1 = \frac{4.3 \text{ V}}{1.2 \text{ k}\Omega} \approx 3.6 \text{ mA}$$

For $I_C = 5$ mA,

$$V_3 = 5 \text{ mA} \times 1.2 \text{ k}\Omega$$

$$= 6 \text{ V}$$

and

$$V_B = V_3 + V_{BE} = 6.7 \text{ V}$$

(At this point, the moving contact on the potentiometer is at the lower end.)
Now,

$$V_B = V_1 + V_4$$

$$V_4 = 6.7\ V - 3.6\ V = 3.1\ V$$

and $$R_4 = \frac{V_4}{I_1} = \frac{3.1\ V}{3.6\ mA}$$

$$= 0.86\ k\Omega \qquad \text{(use a 1 k}\Omega \text{ standard potentiometer value)}$$

Then V_4 becomes

$$V_4 = I_1 R_4 = 3.6\ mA \times 1\ k\Omega = 3.6\ V$$

and $$V_2 = V_{CC} - V_1 - V_4$$

$$= 15\ V - 4.3\ V - 3.6\ V$$

$$= 7.1\ V$$

$$R_2 = \frac{V_2}{I_1} = \frac{7.1\ V}{3.6\ mA}$$

$$= 1.97\ k\Omega \qquad \text{(use 2.2 k}\Omega \text{ standard value)}$$

FET Constant-Current Circuits

A field effect transistor with a single source resistance can function as a constant-current circuit. A *p*-channel FET is shown in Figure 9-4(a) with a resistor connected between the source terminal and V_{CC}. With the gate terminal also connected to V_{CC}, the gate-source voltage is the voltage drop across R_S, which is $I_S R_S$, or $I_D R_S$. Referring to the FET transconductance characteristics in Figure 9-4(b), the desired drain/source current I_S can be selected and the corresponding gate-source voltage V_{RS} determined as illustrated. Then $R_S = V_{RS}/I_S$.

This approach is satisfactory only when the transconductance characteristic for the particular FET has been plotted. For any given FET type, there are two possible extreme characteristics, as shown by the broken lines in Figure 9-4(b). These characteristics occur because of the spread in levels of *drain-source saturation current* [$I_{DSS(max)}$ and $I_{DSS(min)}$] and *pinch-off voltage* [$V_{P(max)}$ and $V_{P(min)}$]. Normally, only the FET maximum and minimum characteristics are known. Thus, to set I_S in Figure 9-4(a) to a desired level, R_S should be made adjustable. The range of adjustment of R_S is determined by drawing a horizontal line on the transconductance characteristics at the desired current. Then, the extremes of R_S are calculated from I_S and the corresponding V_{GS} levels.

One important caution that must be observed when using a FET constant-current

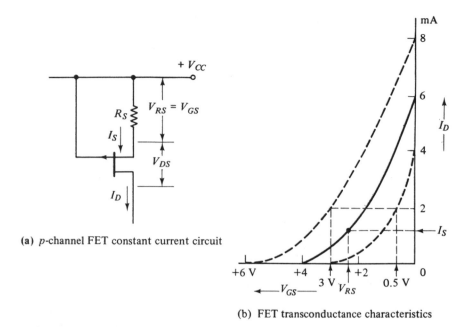

(a) *p*-channel FET constant current circuit

(b) FET transconductance characteristics

Figure 9-4   A field effect transistor can be connected to function as a constant-current circuit. For a given level of current I_S, V_{RS} is determined from the FET transconductance characteristics. Then, $R_S = V_S/I_S$.

circuit is that to keep the FET from saturating, the drain-source voltage V_{DS} must not be allowed to fall below the level $V_{DS(min)} = V_{P(max)} - V_{GS} + 1$ V. $V_{P(max)}$ is the maximum pinch-off voltage, or gate-source cutoff voltage, for the device at $V_{GS} = 0$. Just as a bipolar transistor cannot be expected to function linearly if its collector-emitter voltage approaches the saturation level, so, too, a FET will not function correctly if its drain-source voltage falls below this minimum.

A *constant-current diode* (or *field effect diode*) is essentially a FET and a resistor connected as illustrated in Figure 9-4(a) and contained in a single package. These devices can be purchased with various constant-current levels.

EXAMPLE 9-4

Determine the range of adjustment of R_S for the FET circuit and characteristics shown in Figure 9-4 to give a constant current level of 2 mA. Also, calculate the minimum drain-source voltage.

Solution   Draw a horizontal line at $I_D = 2$ mA on the transconductance characteristics in Figure 9-4(b), as illustrated.

From the intersection of the I_D line and the transconductance characteristics,

$$V_{GS(max)} \approx 3 \text{ V and } V_{GS(min)} \approx 0.5 \text{ V}$$

$$R_{S(max)} = \frac{V_{GS(max)}}{I_D} \approx \frac{3 \text{ V}}{2 \text{ mA}} = 1.5 \text{ k}\Omega$$

$$R_{S(min)} = \frac{V_{GS(min)}}{I_D} \approx \frac{0.5 \text{ V}}{2 \text{ mA}} = 250 \text{ }\Omega$$

From the maximum characteristics, $V_{P(max)} = 6$ V, and $V_{GS} = 3$ V,

therefore, $V_{DS(min)} = V_{P(max)} - V_{GS} + 1 \text{ V} = 6 \text{ V} - 3 \text{ V} + 1 \text{ V}$

$$= 4 \text{ V}$$

9-3 UJT RELAXATION OSCILLATORS

UJT Circuit

A unijunction transistor can be used in conjunction with a capacitor and a charging resistor to construct an oscillator with an approximate ramp-type output. Figure 9-5(a) shows the simplest form of such a circuit, which is called a *UJT relaxation oscillator*. The UJT remains *off* until its emitter voltage V_{EB_1} approaches the firing voltage V_P for the particular device. At this point, the UJT switches *on*, and a large emitter current I_E flows. This causes capacitor C_1 to discharge rapidly. When the capacitor voltage falls to the emitter saturation level, the UJT switches *off*, allowing C_1 to begin to charge again.

The frequency of a relaxation oscillator can be made variable by switched selection of capacitors and/or by adjustment of the charging resistance. [See Figure 9-5(b).] The resistance R_2, in series with UJT terminal B_1, allows synchronizing input pulses to be applied. When an input pulse pulls B_1 in a negative direction, V_{EB1} is increased to the level at which the UJT fires. Once the UJT fires, it will not switch *off* again until the capacitor is discharged.

In the design of a UJT relaxation oscillator, the charging resistance R_1 must be selected between certain upper and lower limits. Resistance R_1 must not be so large that the emitter current is less than the *peak point current* I_P when V_{EB1} is at the firing voltage; otherwise, the device may not switch *on*. If R_1 is very small, then when V_{EB1} is at the emitter saturation level, a current greater than the *valley point current* I_V might flow into the emitter terminal. In this case, the UJT may not switch *off*. Thus, for correct UJT operation, R_1 must be selected between two limits that allow the emitter current to be a minimum of I_P and a maximum of I_V.

The UJT oscillator circuits shown in Figures 9-5(a) and (b) produce exponential output waveforms, because the capacitors are charged via resistances. Constant-current circuits could be used in place of R_1 to generate linear ramp output waveforms.

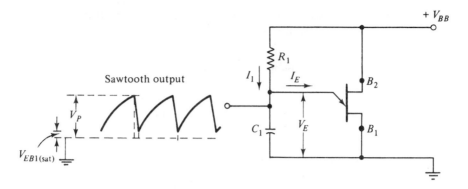

(a)   UJT relaxation oscillator

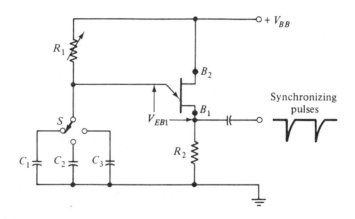

(b)   Variable frequency UJT relaxation
oscillator

Figure 9-5   A UJT relaxation oscillator is created when a unijunction transistor is connected to discharge the capacitor in a simple CR circuit. The capacitor charges from the UJT valley voltage to the device firing voltage. The output frequency may be altered by selection of C and/or adjustment of R.

EXAMPLE 9-5

The circuit of Figure 9-5(a) is to use a 2N3980 UJT. The supply voltage V_{BB} is 20 V, and output frequency is to be 5 kHz. Design a suitable circuit, and calculate the typical peak-to-peak output amplitude.

Solution   Capacitor C_1 charges from $V_{EB1(sat)}$ to the firing voltage $V_P = V_F + \eta V_{BB}$. The data sheet for the 2N3980 (Appendix 1-13) gives the following specifications:

$$V_{EB1(sat)} = 3 \text{ V maximum}, \qquad I_P = 2 \text{ μA}, \qquad I_V = 1 \text{ mA}, \qquad \eta = 0.68 \text{ to } 0.82$$

$$\eta \approx 0.75 \text{ average}$$

$$V_P = 0.7 \text{ V} + (\eta \times V_{BB}) = 0.7 + (0.75 \times 20 \text{ V})$$

$$= 15.7 \text{ V}$$

Therefore, for the capacitor,

$$E = \text{Supply voltage} = V_{BB} = 20 \text{ V}$$

$$E_o = \text{Initial charge} = V_{EB1(sat)} = 3 \text{ V}$$

$$e_c = \text{Final charge} = V_P = 15.7 \text{ V}$$

$$R_{1(max)} = \frac{V_{BB} - V_P}{I_P} = \frac{20 \text{ V} - 15.7 \text{ V}}{2 \text{ μA}}$$

$$\approx 2.15 \text{ MΩ}$$

$$R_{1(min)} = \frac{V_{BB} - V_{EB1(sat)}}{I_V} = \frac{20 \text{ V} - 3 \text{ V}}{1 \text{ mA}}$$

$$\approx 17 \text{ kΩ}$$

R_1 must be in the range 17 kΩ to 2.15 MΩ. If R_1 has a very high resistance, C_1 must have a very small capacitance. Let $R_1 = 22$ kΩ.

$$t = \frac{1}{\text{Output frequency}} = \frac{1}{5 \text{ kHz}} = 200 \text{ μs}$$

From Eq. 2-9,
$$C_1 = \frac{t}{R_1 \ln \left(\dfrac{E - E_o}{E - e_c} \right)} = \frac{200 \text{ μs}}{22 \text{ kΩ} \ln \left(\dfrac{20 \text{ V} - 3 \text{ V}}{20 \text{ V} - 15.7 \text{ V}} \right)}$$

$$= 6600 \text{ pF} \qquad \begin{array}{l} \text{[use 6800 pF standard capacitor} \\ \text{(see Appendix 2-2)]} \end{array}$$

$$\text{Typical output amplitude} = V_P - V_{EB1(sat)} = 15.7 \text{ V} - 3 \text{ V}$$

$$= 12.7 \text{ V}$$

Programmable UJT Circuit

The *programmable unijunction transistor* (PUT) is a four-layer device used in a particular way to simulate a UJT. The interbase resistances r_{B1} and r_{B2} and the intrinsic standoff ratio η may be programmed to any desired values by selecting two resistors. This means that the device firing voltage V_P can also be programmed.

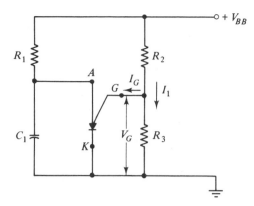

Figure 9-6   Programmable UJT relaxation oscillator. The firing voltage is programmed by selection of resistors R_2 and R_3.

Consider Figure 9-6. The *gate G* of the *pnpn* device is connected to the junction of resistors R_2 and R_3. The gate voltage is $V_G = V_{BB}R_3/(R_2 + R_3)$. The device will trigger *on* when the capacitor voltage V_{AK} makes the anode positive with respect to the gate. When this occurs, the anode-to-cathode voltage rapidly drops to a low level, and the device conducts heavily from anode to cathode. This situation continues until the current becomes too low to sustain conduction. With the anode used as an emitter terminal, the circuit action simulates a UJT.

A data sheet for 2N6027 and 2N6028 PUT devices is included in Appendix 1-14. For the 2N6027, the value of I_P is given as 1.25 μA typical, and I_V as 18 μA typical. The offset voltage, which is equivalent to $V_{EB1(sat)}$, is typically 0.7 V. To provide a stable gate bias voltage, the current through the potential divider (R_2 and R_3) must be much larger than the gate current at switch-*on*:

9-4 BOOTSTRAP RAMP GENERATORS

Transistor Bootstrap Ramp Generator

The circuit of a transistor *bootstrap ramp generator* is shown in Figure 9-7(a). The ramp is generated across capacitor C_1, which is charged via resistance R_1. The discharge transistor Q_1 holds the capacitor voltage V_1 down to $V_{CE(sat)}$ until a negative input pulse is applied. Transistor Q_2 is an emitter follower that provides a low-output impedance. Emitter resistor R_E is connected to a negative supply level, rather than to ground. This is to ensure that Q_2 remains conducting when its base voltage V_1 is close to ground. Capacitor C_3, known as the *bootstrapping capacitor*, has a much higher capacitance than C_1. The function of C_3, as will be shown, is to maintain a constant voltage across R_1, and thus maintain the charging current constant.

To understand the operation of the bootstrap ramp generator, first consider the dc voltage levels before an input signal is applied. Transistor Q_1 is *on*; its voltage is $V_{CE(sat)}$, which is typically 0.2 V. This level is indicated as point A on the graph of voltage V_1 in Figure 9-7(b). The emitter of Q_2 is now at $(V_1 - V_{BE2})$, which is also the output

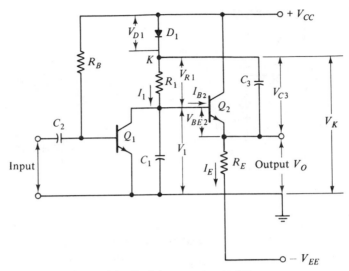

(a)   Bootstrap ramp generator

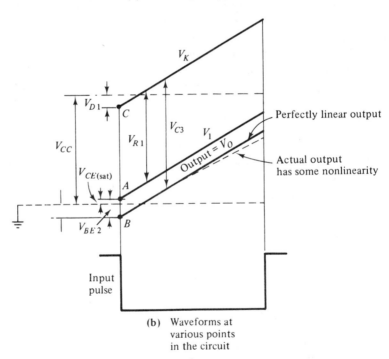

(b)   Waveforms at
various points
in the circuit

Figure 9-7   A transistor bootstrap ramp generator is basically a simple CR ramp generator (as in Figure 9-1) with the addition of emitter follower Q_2, diode D_1, and bootstrapping
capacitor C_3. These components maintain a constant voltage across R_1 and thus produce
a linear output ramp.

voltage V_o (point B on the V_o graph). At this time, the voltage at the cathode of diode D_1 is $V_K = V_{CC} - V_{D1}$, where V_{D1} is the diode forward voltage drop. The voltage $V_{CC} - V_{D1}$ is shown at point C on the graph of V_K [Figure 9-7(b)]. The voltage across capacitor C_3 is the difference between V_K and V_o.

When Q_1 is switched *off* by a negative-going input pulse, C_1 starts to charge via R_1. Voltage V_1 now increases, and the emitter voltage V_o of Q_2 (the emitter follower) also increases. Thus, as V_1 grows, V_o also grows, remaining only V_{BE} below V_1. [See Figure 9-7(b).] As V_o increases, the lower terminal of C_3 is *pulled up*. Because C_3 has a high capacitance, it retains its charge, and, as V_o increases, the voltage at the upper terminal of C_3 also increases. Thus, V_K increases as V_1 increases, and V_K remains V_{C3} volts above V_o. (In fact, V_K goes above the level of V_{CC}, causing D_1 to be reverse biased.) The constant voltage across C_3 maintains the voltage V_{R1} constant across R_1. Therefore, the charging current through R_1 is held constant, and consequently, C_1 charges linearly, giving a linear output ramp.

During the ramp time, D_1 is reverse biased as already explained, and the charging current through R_1 is provided by capacitor C_3. If C_3 has a very high capacitance, and I_1 is small, then C_3 will discharge by only a very small amount. When the input pulse is removed and C_1 is discharged rapidly by Q_1, V_o drops to its initial level. Also, V_K drops, allowing D_1 to become forward biased. At this time, a current pulse through D_1 replaces the small charge lost from C_3. The circuit is then ready to generate another output ramp.

In addition to producing a very linear output ramp, another advantage of the bootstrap generator is that the amplitude of the ramp can approach the level of the supply voltage. Note that the output ramp amplitude may be made adjustable over a fixed time period by making R_1 adjustable.

The broken line on the graph of output voltage [Figure 9-7(b)] shows that, instead of being perfectly linear, the output may be slightly nonlinear. If the difference between the actual output and the ideal output is 1% of the output peak voltage, then the ramp may be said to have 1% nonlinearity. Some nonlinearity results from the slight discharge of C_3 that occurs during the ramp time. Another source of nonlinearity is the Q_2 base current I_{B2}. As the capacitor voltage grows, I_{B2} increases. Since I_{B2} is part of I_1, the capacitor charging current decreases slightly as I_{B2} increases. Thus, the charging current does not remain perfectly constant, and consequently, the ramp is not perfectly linear. The design of a bootstrap ramp generator begins with a specification of ramp linearity. This dictates the charging current and the capacitance of C_3. The percentage of nonlinearity usually is allocated in equal parts to ΔI_{B2} and ΔV_{C3}.

Note that the preceding reasoning about ramp linearity assumes that there is no significant leakage current through the capacitors. This requires that the capacitors not be electrolytic.

EXAMPLE 9-6

Design a transistor bootstrap ramp generator to provide an output amplitude of 8 V over a time period of 1 ms. The ramp is to be triggered by a negative-going pulse with an amplitude of 3 V, a pulse width of 1 ms, and a time interval between pulses of 1 ms. The load resistor to be supplied has a value of 1 kΩ, and the ramp is to be linear within 2%. The supply voltage is to be ±15 V. Take $h_{FE(min)} = 100$.

Solution   The circuit is as shown in Figure 9-7(a).

$$R_E = R_L = 1 \text{ k}\Omega$$

When $V_o = 0$, $I_E \approx \dfrac{V_{EE}}{R_E} = \dfrac{15 \text{ V}}{1 \text{ k}\Omega} = 15 \text{ mA}$

When $V_o = V_P$, $I_E \approx \dfrac{V_P - V_{EE}}{R_E} = \dfrac{8 \text{ V} - (-15 \text{ V})}{1 \text{ k}\Omega}$

$$= 23 \text{ mA}$$

At $V_o = 0$, $I_{B2} = \dfrac{I_{E2}}{h_{FE}} = \dfrac{15 \text{ mA}}{100} = 0.15 \text{ mA}$

At $V_o = V_P$, $I_{B2} = \dfrac{23 \text{ mA}}{100} = 0.23 \text{ mA}$

$$\Delta I_{B2} = 0.23 \text{ mA} - 0.15 \text{ mA} = 80 \text{ }\mu\text{A}$$

Allow 1% nonlinearity due to ΔI_{B2} (that is, ΔI_{B2} represents a loss of charging current to C_1):

$$I_1 = 100 \times \Delta I_{B2}$$

$$= 100 \times 80 \text{ }\mu\text{A}$$

$$= 8 \text{ mA}$$

$$C_1 = \dfrac{I_1 t}{\Delta V} = \dfrac{I_1 \times (\text{Ramp time})}{V_P}$$

$$= \dfrac{8 \text{ mA} \times 1 \text{ ms}}{8 \text{ V}}$$

$$= 1 \text{ }\mu\text{F} \qquad \text{(standard capacitor value)}$$

$$V_{R1} = V_{CC} - V_{D1} - V_{CE(sat)}$$

$$= 15 \text{ V} - 0.7 \text{ V} - 0.2 \text{ V}$$

$$= 14.1 \text{ V}$$

$$R_1 = \dfrac{V_{R1}}{I_1} = \dfrac{14.1 \text{ V}}{8 \text{ mA}}$$

$$= 1.76 \text{ k}\Omega \qquad \text{(use 1.8 k}\Omega \text{ standard value)}$$

For 1% nonlinearity due to C_3 discharge,

$$\Delta V_{C3} = 1\% \text{ of initial } V_{C3} \text{ level}$$

$$V_{C3} \approx V_{CC} = 15 \text{ V}$$

$$\Delta V_{C3} = \frac{15 \text{ V}}{100} = 0.15 \text{ V}$$

and C_3 discharge current is equal to $I_1 = 8$ mA.

Thus, $$C_3 = \frac{I_1 t}{\Delta V_{C3}} = \frac{8 \text{ mA} \times 1 \text{ ms}}{0.15 \text{ V}}$$

$$= 53 \text{ } \mu\text{F} \qquad \text{(use 56 } \mu\text{F standard capacitance value)}$$

R_B and C_2 are calculated in the same way as for Example 9-1.

Note that the recharge path for C_3 is via D_1 and R_E in the circuit of Figure 9-7(a). Using Eq. 2-2 and the component values from Example 9-6, it is found that the time required to recharge C_3 by a ΔV_{C3} of 0.15 V is approximately 0.6 ms. This means that the time interval between ramp outputs (and between input pulses) should be not less than 0.6 ms. If Q_2 is replaced by a *voltage follower* (see the next circuit), the recharge time for C_3 is usually small enough to be ignored.

IC Op-amp Bootstrap Ramp Generator

An IC operational amplifier connected as a *voltage follower* (see Section 5-2) forms part of the bootstrap ramp generator in Figure 9-8. The circuit of the IC operational amplifier bootstrap generator is almost exactly like that of the transistor bootstrap circuit. The voltage follower takes the place of the emitter follower. Note that, although a plus-and-minus supply is still required, load resistance R_L can now be connected to ground, instead of to $-V_{EE}$. Note also that the output ramp starts at $V_{CE(sat)}$ instead of at $V_{CE(sat)} - V_{BE}$. The low input current to the operational amplifier has an almost negligible effect on the charging current to C_1. In fact, the reverse leakage current of D_1 (when it is reverse biased) can be more significant than the input bias current of the amplifier. Using a 1N914 diode (Appendix 1), I_R might be as large as 3 μA. For the 741, the maximum input bias current is 500 nA. (Note that for the transistor bootstrap circuit, I_R of D_1 is very much smaller than I_B of transistor Q_2.) The leakage current of D_1 can be the starting point for the IC bootstrap circuit design. This results in a lower charging current to C_1 and in lower capacitance values for C_1, C_2, and C_3.

If D_1 leakage current is extremely small, the preceding approach may result in a very small charging current and consequently in a very low capacitance value for C_1. The typical input capacitance for an oscilloscope is 30 pF. So C_1 should not be made so small as to affect the circuit performance when an oscilloscope is connected to any part of it. As a minimum, C_1 should be selected approximately 1000 times greater

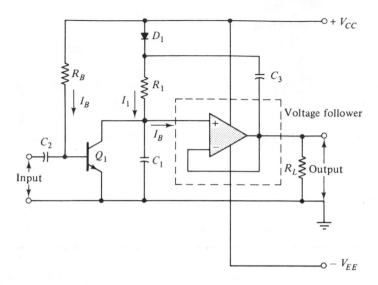

Figure 9-8   The use of an op-amp voltage follower instead of an emitter follower improves the performance of a transistor bootstrap ramp generator.

than the typical 30 pF C_{in} of an oscilloscope. This will also ensure that C_1 is not affected by the *stray capacitance* of wiring, etc.

EXAMPLE 9-7

Design a bootstrap ramp generator using a 741 operational amplifier. The specifications for the circuit are the same as those for the circuit of Example 9-6, with the exception that the time interval between input pulses is 0.1 ms.

Solution   The circuit is as illustrated in Figure 9-8.

$$R_L = 1 \text{ k}\Omega$$

$$\text{Maximum } I_R = 3 \text{ }\mu A \qquad \text{(when } D_1 \text{ is reverse biased)}$$

Allow 1% nonlinearity due to I_R:

$$I_1 = 100 \times I_R$$

$$= 100 \times 3 \text{ }\mu A = 300 \text{ }\mu A$$

$$C_1 = \frac{I_1 t}{\Delta V} = \frac{I_1 \times (\text{Ramp time})}{V_P}$$

$$= \frac{300 \text{ }\mu A \times 1 \text{ ms}}{8 \text{ V}}$$

$$= 0.0375 \text{ }\mu F \qquad \text{(use } 0.039 \text{ }\mu F \text{ standard value)}$$

$$V_{R1} = V_{CC} - V_{D1} - V_{CE(sat)}$$

$$= 15 \text{ V} - 0.7 \text{ V} - 0.2 \text{ V}$$

$$= 14.1 \text{ V}$$

$$R_1 = \frac{V_{R1}}{I_1} = \frac{14.1 \text{ V}}{300 \text{ }\mu\text{A}}$$

$$= 47 \text{ k}\Omega \qquad \text{(standard value)}$$

For 1% nonlinearity due to C_3 discharge:

$$\Delta V_{C3} = 1\% \text{ of initial } V_{C3}$$

$$V_{C3} \approx V_{CC} = 15 \text{ V}$$

$$\Delta V_{C3} = \frac{15 \text{ V}}{100} = 0.15 \text{ V}$$

C_3 discharge current $= I_1 = 300 \text{ }\mu\text{A}$. So

$$C_3 = \frac{I_1 t}{\Delta V_{C3}} = \frac{300 \text{ }\mu\text{A} \times 1 \text{ ms}}{0.15 \text{ V}} = 2 \text{ }\mu\text{F} \qquad \text{(standard value)}$$

(Compare this to $C_3 = 56 \text{ }\mu\text{F}$ for the transistor circuit of Example 9-6.) The discharge time of C_1 is equal to one-tenth of the charge time. Therefore, the discharge current of C_1 is ten times greater than the charge current.

$$\text{Minimum } I_C \text{ of } Q_1 = 10 \times I_1$$

$$= 10 \times 300 \text{ }\mu\text{A} = 3 \text{ mA}$$

$$I_B = \frac{I_C}{h_{FE}} = \frac{3 \text{ mA}}{100}$$

$$= 30 \text{ }\mu\text{A}$$

$$R_B = \frac{V_{CC} - V_{BE}}{I_B}$$

$$= \frac{15 \text{ V} - 0.7 \text{ V}}{30 \text{ }\mu\text{A}}$$

$$= 477 \text{ k}\Omega \qquad \text{(use 470 k}\Omega \text{ standard value)}$$

During the input pulse, $\Delta V_{C2} = 1.8 \text{ V}$ (see Example 9-1) and the charging current of C_2 can be expressed by

$$I = \frac{V_{CC} - V_i}{R_B} = \frac{15 \text{ V} - (-3 \text{ V})}{470 \text{ k}\Omega}$$

$$= 38 \text{ }\mu\text{A}$$

Thus,
$$C_2 = \frac{It}{\Delta V} = \frac{38\ \mu A \times 1\ ms}{1.8\ V}$$

$$= 0.02\ \mu F \qquad \text{(standard value)}$$

9-5 FREE-RUNNING RAMP GENERATORS

A bootstrap ramp generator may be made free-running by employing a Schmitt trigger circuit to detect the output peak level and generate a capacitor discharge pulse. In the circuit shown in Figure 9-9(a), *pnp* transistor Q_1 discharges C_1 when the Schmitt circuit output is negative. Diode D_2 protects the base-emitter junction of Q_1 against excessive reverse bias when the Schmitt output is positive.

Consider the circuit waveforms shown in Figure 9-9(b). During the time that the Schmitt circuit output is positive, Q_1 remains *off* and C_1 charges. This provides a positive-going ramp output. When the ramp amplitude arrives at the UTP of the Schmitt circuit, the Schmitt output switches to negative. This causes I_{B1} to flow, biasing Q_1 *on* and rapidly discharging C_1. As the voltage of capacitor C_1 falls, the ramp output also falls rapidly, and this continues until the Schmitt LTP is reached. The presence of D_3 makes the Schmitt circuit have an LTP close to ground. (See Sec. 6-6.) Therefore, when the ramp output falls to ground level, the Schmitt output goes positive again, switching Q_1 *off* and allowing ramp generation to recommence.

The free-running ramp generator can be synchronized with another waveform by means of negative pulses coupled via capacitor C_3. The presence of the negative pulse lowers the UTP of the Schmitt circuit, so that the Schmitt output becomes negative, causing the ramp to go to zero when the synchronizing pulse is applied.

Potentiometer R_7 [Figure 9-9(c)] allows the charging current to C_1 to be adjusted, thus controlling the ramp length and the output frequency. In Figure 9-9(d), R_6 affords adjustment of the Schmitt UTP. This provides control of the ramp amplitude.

EXAMPLE 9-8

Design a free-running ramp generator with an output frequency of 1 kHz and an output amplitude in the range 0 V to 8 V. Use 741 operational amplifiers and a supply voltage of ± 15 V.

Solution   *Schmitt circuit.* For an output of 0 V to 8 V, the Schmitt circuit must have an LTP of 0 V and a UTP of 8 V. Design the Schmitt circuit as explained in Section 6-6.

Bootstrap circuit. The bootstrap output should go from 0 V to 8 V over a time period of 1/1 kHz (i.e., 1 ms). Design the circuit as in Example 9-7, substituting a *pnp* transistor for Q_1.

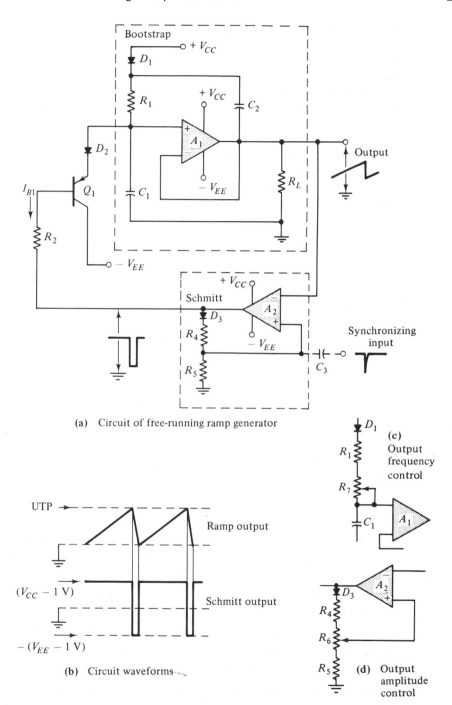

(a)   Circuit of free-running ramp generator

(c) Output frequency control

(b)   Circuit waveforms

(d) Output amplitude control

Figure 9-9   Free-running ramp generator circuit, using a bootstrap ramp generator and an inverting Schmitt trigger. When the ramp voltage level equals the Schmitt UTP, the Schmitt output switches to negative and discharges C_1 via Q_1.

9-6 MILLER INTEGRATOR RAMP GENERATORS

The circuit in Figure 9-10 shows a Miller integrator (see Section 5-7) operating as a ramp generator. The negative pulse generates the positive ramp by producing current I_1 in the direction shown. At this time, (n-channel FET) Q_1 is biased *off* by the negative input pulse. When the input switches to ground level, I_1 goes to zero and Q_1 is switched *on*. Q_1 rapidly discharges C_1 and keeps it discharged until the input becomes negative again. If C_1 is to be discharged in one-tenth of the charge time, then Q_1 must be able to pass a current ten times greater than the charge current I_1. To ensure that Q_1 is biased *off* when the input pulse is present, the input pulse must have a negative amplitude greater than the FET pinchoff voltage. Because the capacitor is completely discharged by the action of the FET, there is no need to include resistor R_3 [Figure 5-11(a)] in this circuit.

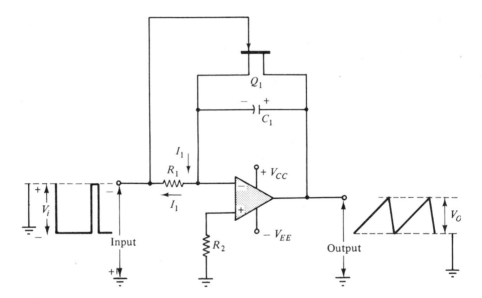

Figure 9-10   Miller integrator used as a ramp generator. The FET remains *off* during the negative input pulse, and the capacitor charges to produce a linear output ramp. When the input goes to zero, the FET switches *on* and discharges the capacitor.

EXAMPLE 9-9

Design a Miller integrator ramp generator circuit as in Figure 9-10. The output is to have 5 V peak, with a time period of 1 ms and a discharge time of 50 μs. Use a 2N4857 FET, specify the input pulse, and calculate the FET drain current.

Solution   Select C_1 very much larger than stray and instrument capacitance:

Let $C_1 = 1000 \times 30 \text{ pF} = 0.03 \text{ μF}$

From Eq. 2-8,
$$I = \frac{C \, \Delta V}{t} = \frac{0.03 \, \mu F \times 5 \, V}{1 \, ms}$$

$$= 150 \, \mu A$$

From Appendix 1-8, for the 2N4857,

$$V_{P(max)} = V_{GS(off)} = 6 \, V$$

To ensure that Q_1 is switched off,

let
$$V_i = V_{P(max)} + 1 \, V$$

$$= 7 \, V$$

Then
$$R_1 = \frac{V_i}{I_1} = \frac{7 \, V}{150 \, \mu A}$$

$$= 47 \, k\Omega \qquad \text{(standard value)}$$

The input pulse has $V_i = -7$ V, PW = 1 ms, and the space width = 50 μs.

$$I_{D1} = C_1 \text{ discharge current} = \frac{I_1 \times \text{PW}}{\text{space width}} = \frac{150 \, \mu A \times 1 \, ms}{50 \, \mu s}$$

$$= 3 \, mA$$

Compensate the op-amp for $A_V = 1$ (see Section 5-2)

9-7 TRIANGULAR WAVEFORM GENERATORS

A free-running triangular waveform generator can be constructed using the output of a Miller integrator to generate its own square wave input. Consider the circuit in Figure 9-11(a). The Miller integrator circuit used is exactly as discussed in the last section. The output of the Miller circuit is fed directly to a noninverting Schmitt trigger circuit. The Schmitt is designed to have a positive UTP and a negative LTP (see Section 6-6), and its output is applied as an input to the Miller circuit.

   Operation of the circuit is easily understood by considering the waveforms in Figure 9-11(b). At time t_1 the integrator output has reached the UTP (a positive voltage), and the noninverting Schmitt circuit output is positive at approximately $+(V_{CC} - 1 \, V)$. The positive voltage from the Schmitt causes current I_1 to flow into the Miller circuit, charging C_1 positive on the left side. As C_1 charges in this direction, the integrator output is a negative-going ramp. The integrator continues to produce a negative-going ramp while its input is a positive voltage. At time t_2, the integrator output arrives at the LTP (negative voltage). The Schmitt trigger circuit output now becomes negative and reverses the direction of I_1. Thus, the integrator output becomes a positive-going ramp. This positive-going ramp generation continues until the integrator output arrives at the UTP of the Schmitt circuit once again. Synchronizing pulses may be applied via

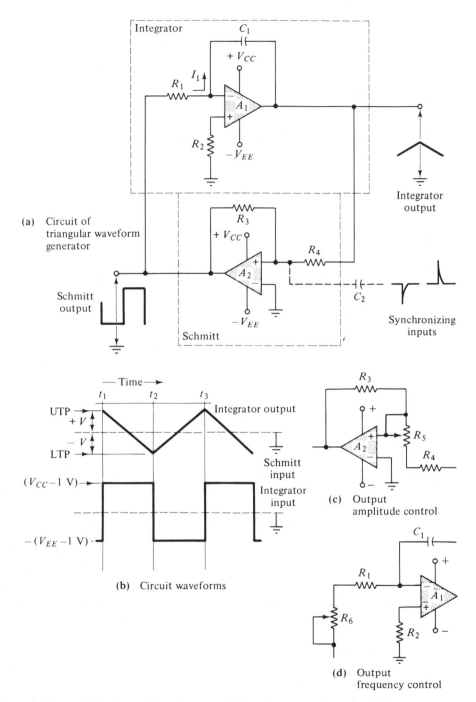

(a) Circuit of triangular waveform generator

(b) Circuit waveforms

(c) Output amplitude control

(d) Output frequency control

Figure 9-11 Free-running triangular waveform generator consisting of a Miller integrator and a noninverting Schmitt. When the ramp equals the UTP or LTP of the Schmitt, the Schmitt output switches polarity and reverses the direction of the capacitor charging current.

C_2 to lower the trigger point of the Schmitt circuit, causing it to trigger before the ramp arrives at its normal peak level.

The circuit described above generates a triangular waveform with a constant peak-to-peak output amplitude and a constant frequency. The modifications shown in Figures 9-11(c) and (d) allow both frequency and amplitude adjustment. R_5 adjusts the UTP and LTP of the Schmitt circuit, thus controlling the peak-to-peak output amplitudes. R_6 affords adjustment of the input current to the integrator, to control the rate of charge of C_1. This means that the ramp time period is controlled by adjusting R_6.

For short rise and fall times at the output of the Schmitt, an operational amplifier with a high slew rate is required. Alternatively, in the case of the Schmitt circuit, a voltage comparator might be employed. Although the output of the integrator changes relatively slowly, rounding of the triangular waveform peaks may occur if the op-amp does not have a high enough frequency response. The integrator operational amplifier should be compensated as for a voltage follower.

EXAMPLE 9-10

Design a free-running triangular waveform generator to have a peak-to-peak output of 4 V at a frequency of 250 Hz. Use 741 operational amplifiers and a supply voltage of ± 15 V.

Solution *Schmitt circuit.* For 4 V p-to-p, the Schmitt circuit UTP = 2 V and LTP = -2 V. A Schmitt circuit can be designed as in Example 6-6 to give these desired trigger points.
Miller integrator circuit. The input to the Miller circuit is the Schmitt output; that is, $V_i \approx \pm 14$ V.

$$\text{Ramp amplitude} = 4 \text{ V}$$

$$\text{Ramp time period} = \frac{1}{2f} = \frac{1}{2 \times 250 \text{ Hz}}$$

$$= 2 \text{ ms}$$

Design the Miller circuit as in Example 5-5.

9-8 PULSE GENERATOR CIRCUIT

Circuit Operation

A three-section pulse generator circuit consisting of a square wave generator, a monostable multivibrator, and an attenuator is illustrated in Figure 9-12(a). The circuit waveforms are shown in Figure 9-12(b).

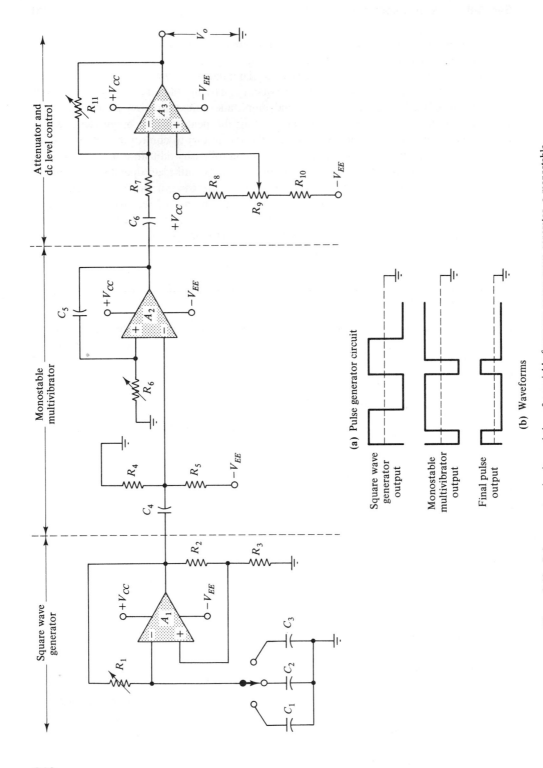

Figure 9-12 Pulse generator circuit consisting of a variable-frequency square wave generator, a monostable multivibrator with adjustable pulse width, and an output stage with dc level and amplitude controls.

The square wave generator, which is an op-amp astable multivibrator, produces a waveform at the desired frequency for triggering the monostable multivibrator. The frequency range can be changed by selection of a capacitor from among C_1, C_2, and C_3, and continuous frequency adjustment is possible by adjustment of R_1. The op-amp monostable generates a pulse of the desired width each time it is triggered. Pulse-width adjustment is afforded by variable resistor R_6, and the range of the pulse width can be switched by using switch-selectable capacitors in place of C_5.

An op-amp inverting amplifier is employed as an attenuator to provide output pulse amplitude adjustment, and to offer a low output impedance. The amplifier voltage gain is R_{11}/R_7. Thus, if R_{11} is less than R_7, the output pulse amplitude is attenuated. With R_{11} adjustable, the output pulse amplitude can be adjusted as desired. In other inverting amplifier applications, the noninverting input terminal of the op-amp is usually grounded. In this case, the noninverting input terminal can be adjusted above or below ground by means of the potential divider constituted by R_8, R_9, and R_{10}. Since the inverting input terminal assumes the dc level of the noninverting terminal, and since R_{11} directly connects the output to the inverting input terminal, the dc output level of the inverting amplifier can be adjusted by means of R_9.

The waveforms in Figure 9-12(b) show that each positive-going edge of the square wave triggers the monostable to produce a negative-going pulse. The pulse is attenuated and inverted by the output stage, and its dc level can be adjusted with respect to ground.

Pulse Generator Design

Design of the pulse generator illustrated in Figure 9-12 commences with the specification of maximum and minimum pulse repetition frequency, pulse width, and pulse amplitude. High-frequency operational amplifiers should be used for all three sections, to give fast rise times. Alternatively, voltage comparators (or 555 timers) may be employed for the square wave generator and the monostable multivibrator to achieve the fastest possible rise times. The output attenuator is operating as a linear circuit, consequently an operational amplifier must be used for this part of the pulse generator.

The square wave generator is designed as in Example 7-5 using an op-amp, or as in Example 7-6 if a voltage comparator is employed. The output frequency ranges might be 100 Hz to 1 kHz, 1 kHz to 10 kHz, and 10 kHz to 100 kHz. The circuit should be designed first for the *lowest operating frequency within the highest range*— 10 kHz in the above specification. This gives the lowest capacitance value for C_1 and the highest resistance $R_{1(max)}$. The minimum resistance for R_1 is then calculated as $R_{1(min)} = R_{1(max)}/10$. This gives a ten times reduction in time period, and hence a ten times increase in output frequency, from 10 kHz to 100 kHz. The capacitance of C_2 is selected ten times larger than that of C_1 to give a frequency range ten times lower. By the same reasoning, $C_3 = 10 C_2$.

The design procedure for the monostable multivibrator is similar to that in Example

7-3, although the monostable circuit in Figure 9-12 is slightly different from the circuit designed in the example. The values of R_6 and C_5 should first be determined for the longest output pulse width. Then, using C_5, the minimum value for R_6 should be calculated to give the shortest required pulse width. In some circumstances, it might be best to include switched capacitors in the monostable (as in the square wave generator) to give several ranges of adjustable pulse width.

The capacitance of coupling capacitor C_4 is determined similarly to that for a capacitor-coupled inverter circuit. C_4 and the input resistance differentiate the output of the square wave generator. The capacitance of C_4 must be large enough to hold the inverting input terminal of A_2 above the level of the (grounded) noninverting terminal until the output of A_2 changes state.

The attenuator is designed exactly as any other inverting amplifier. R_7 is determined in the usual way, using the output from the monostable as an input signal and making the current through R_7 very much larger than the op-amp input bias current. The maximum resistance of R_{11} is calculated for maximum gain as $A_v R_7$. A maximum voltage gain of 1 might be appropriate, giving $R_{11(max)} = R_7$. With R_{11} adjustable from its maximum resistance to zero, the output pulse can be reduced to any desired amplitude.

The dc level control (R_8, R_9, and R_{10}) is designed simply by determining the appropriate resistor voltages, selecting a current very much larger than the op-amp input bias current, and then applying Ohm's law. If, for example, the dc output level is to be adjustable between $+2$ V and -2 V, then those voltage levels should appear at the top and bottom, respectively, of R_9.

Coupling capacitor C_6 is determined using the longest pulse width to be passed from the monostable to the output stage, and the acceptable maximum tilt on the output pulse. (See Section 3-5.) From Eq. 3-2, the coupling capacitor equation is

$$C_6 = \frac{2 \, V_{CC} \, \text{PW}_{(max)}}{\Delta V \, R_7}$$

where $\text{PW}_{(max)}$ is the maximum output pulse width, and ΔV is the acceptable tilt in volts.

Amplifier A_3 should be frequency compensated as a voltage follower. When R_{11} is adjusted to zero, the inverting amplifier circuit has the same feedback characteristics as a voltage follower. Also, A_3 should have a slew rate which allows the output voltage to go from its minimum to maximum level in a time less than or equal to one-tenth the minimum output pulse width. This gives an output rise time which is not greater than one-tenth the minimum pulse width. In some circumstances a longer rise time may be acceptable.

Whether it is an operational amplifier or a comparator, A_2 should have a slew rate smaller than that of A_3. (Review Eq. 1-6 to recall how rise times accumulate.) The rise time of the output from the monostable multivibrator can be reduced by the use of a clipper: if the output from A_2 is clipped in half, the rise and fall times are halved.

9-9 OP-AMP FUNCTION GENERATOR

Function Generator Circuit

The triangular waveform generator illustrated in Figure 9-11 has a square wave output (from the Schmitt trigger circuit), as well as a triangular output. If it could also produce a sinusoidal waveform, it would be a *function generator*. In general, a function generator has sine, square, triangular, pulse, and sawtooth outputs.

Figure 9-13 shows a modified version of the circuit in Figure 9-11. One obvious modification is the *sine wave converter*, shown in block form. This is explained shortly. The other major modification is that resistor R_1 of Figure 9-11 is replaced with two resistors and two diodes. When the Schmitt output is high, current flows via R_1 and D_1 to charge capacitor C_1, positively on the left, negatively on the right. As explained in Section 9-7, this action produces a negative-going ramp output from the integrator. When the Schmitt output is low, the capacitor current direction is reversed, flowing

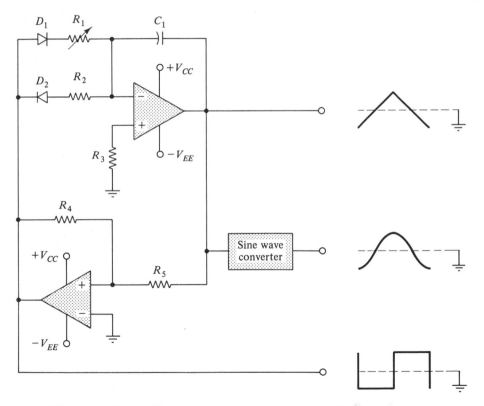

Figure 9-13 Function generator, consisting of a triangular waveform generator, as in Figure 9-11, and a sine wave converter. Adjustable resistors R_1 and R_2, together with diodes D_1 and D_2, facilitate duty cycle adjustment.

through D_2 and R_2 to produce a positive-going output ramp. The different charge and discharge paths, and the fact that R_1 and R_2 are adjustable, allow the capacitor charge and discharge times to be independently adjusted. Thus, instead of a triangular output waveform, a sawtooth wave can be produced. With a sawtooth output from the integrator, the Schmitt output no longer has a 50% duty cycle. Instead of a square wave, it is now a pulse waveform.

Apart from the sine wave converter circuit, the function generator is designed exactly as discussed in Section 9-7. The maximum and minimum resistances of R_1 determine the maximum and minimum (positive) pulse widths at the output of the Schmitt. Thus, $R_{1(max)}$ and $R_{1(min)}$ are calculated in the same way as R_1 in Figure 9-11, allowing for the diode voltage drop. $R_{2(max)}$ and $R_{2(min)}$ are also determined in the same way, for maximum and minimum space widths.

Sine Wave Converter

A widely used method for converting a triangular wave into an approximate sinusoidal waveform is illustrated in Figure 9-14. If diodes D_1 and D_2, and resistors R_3 and R_4, were not present in the circuit of Figure 9-14(a), R_1 and R_2 would simply behave as a voltage divider. In this case, the output from the circuit would be an attenuated version of the triangular wave input:

$$V_o = V_i \frac{R_2}{R_1 + R_2}$$

With D_1 and R_3 in the circuit, R_1 and R_2 still behave as a voltage divider until V_{R2} exceeds $V_1 + V_F$, where V_F is the diode forward voltage drop. At this point, D_1 becomes forward biased and R_3 is effectively in parallel with R_2, so that

$$V_o \approx V_1 + V_F + V_i \frac{R_2 \| R_3}{R_1 + R_2 \| R_3}$$

Output voltage levels above $V_1 + V_F$ are attenuated to a greater extent than levels below $V_1 + V_F$. Consequently, the output voltage rises less steeply than it would without D_1 and R_3 in the circuit. [See the output waveform in Figure 9-14(a).] When the output falls below $V_1 + V_F$, diode D_1 is reverse biased, R_3 is no longer in parallel with R_2, and the output is once again $V_i R_2 / (R_1 + R_2)$.

Similarly, during the negative half-cycle of the input, the output is $V_o = V_i R_2 / (R_1 + R_2)$ until V_o goes below $-(V_1 + V_F)$. Then, D_2 becomes forward biased, putting R_4 in parallel with R_2 and making

$$V_o \approx -V_1 - V_F - V_i \frac{R_2 \| R_4}{R_1 + R_2 \| R_4}$$

With $R_3 = R_4$, the negative half-cycle of the output is similar in shape to the positive half-cycle.

When six or more diodes are employed, all connected via resistors to different bias voltage levels, as illustrated in Figure 9-14(b), a good sine wave approximation

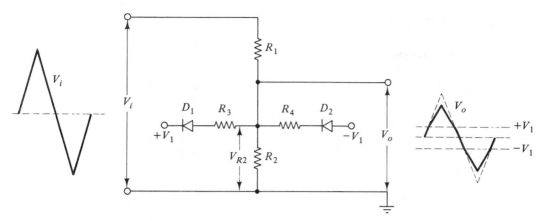

(a) Triangular wave input is attenuated and reshaped
by the potential divider and diode loading circuit

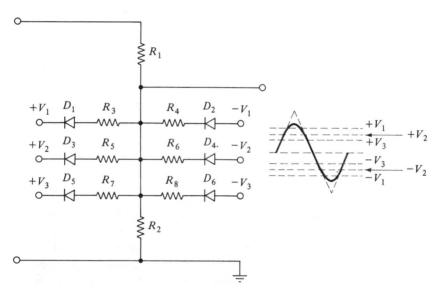

(b) Multilevel diode loading circuit can reshape a
triangular waveform into a good approximation
of a sine wave

Figure 9-14 A triangular waveform can be shaped into a good sine wave approximation by diode loading. In (a), when $V_o < V_1$, $V_o = V_i[R_2/(R_1 + R_2)]$. When $V_o > V_1$, $V_o \simeq V_1 + V_F + V_i[R_2\|R_3/(R_1 + R_2\|R_3)]$.

can be achieved. With six diodes, three positive voltage levels, and three negative voltage levels, the slope of the output wave changes shape three times during each quarter-cycle. Assuming correctly selected bias voltages and resistor values, the output waveform is as illustrated in Figure 9-14(b).

9-10 IC FUNCTION GENERATOR

Circuit Description

The 8038 is an integrated circuit function generator which can produce sine, square, triangular, sawtooth, and pulse waveforms. It may also be operated as a voltage-controlled oscillator, and as a sweep frequency generator.

The function block diagram for the 8038 is illustrated in Figure 9-15(a), and the terminal connections are shown in Figure 9-15(b). The 8038 has several similarities to a 555 timer. There are two comparators and a flip-flop, as well as buffer output stages. The comparators trigger the flip-flop from one state to the other when they detect the upper and lower limits of the voltage on (*externally connected*) capacitor C. Two constant-current sources are included, one for charging C, and the other for discharging C. The levels of these currents are determined by selection of external resistors.

Because the capacitor is charged and discharged linearly, its voltage waveform is triangular. This is buffered, as shown in Figure 9-15(a), to give a low-impedance output at terminal 3. The triangular wave is then passed to a sine converter stage (see Section 9-9) which produces a sine wave output at terminal 2. The output of the flip-flop is also buffered, so that a square wave is available at terminal 9.

The duty cycle of the square wave can be altered by charging and discharging the capacitor at different rates. Thus, instead of a square wave, a pulse waveform is generated. Similarly, different charge and discharge rates convert the triangular wave into a sawtooth.

Supply Voltage and Output Amplitude

The supply voltage for the 8038 is specified as a minimum of $+10$ V and a maximum of $+30$ V (positive at terminal 6, ground at terminal 11). Alternatively, a dual-polarity supply of ±5 V to ±15 V may be used (positive at terminal 6, negative at terminal 11). When a single-polarity supply is employed, the output waveforms are symmetrical above and below half the supply level. With a dual-polarity supply, the outputs are symmetrical above and below ground. This is not necessarily true for the pulse/square wave output, as will be explained.

The peak-to-peak amplitude of the triangular output is typically 33% of the supply voltage, while that of the sine wave is typically 22% of the supply. Thus, if a 10 V supply is employed, the triangular output should be approximately 3.3 V peak-to-peak. If a ±15 V supply is used, the triangular amplitude should be 33% of 30 V—approximately 10 V peak-to-peak, or ±5 V with respect to ground.

The output stage of the square wave buffer is an open-circuited transistor collector terminal. It must be connected via a *pull-up resistor* to a supply which is positive with respect to the voltage at terminal 11. Thus, it could be connected to $+V_{CC}$ when a single-polarity supply is used [Figure 9-16(a)], or to ground when a dual-polarity supply is employed [Figure 9-16(b)]. Alternatively, it could be connected to a separate supply

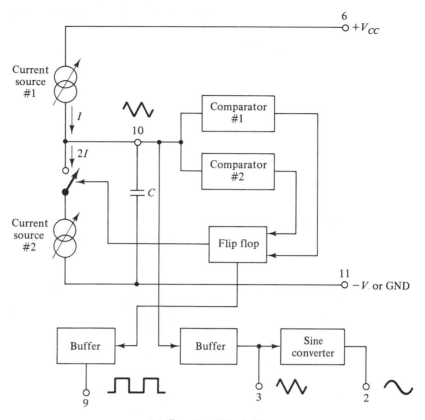

(a) Functional block diagram

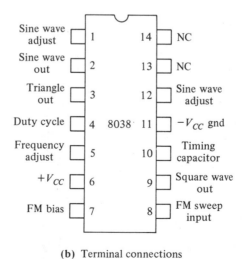

(b) Terminal connections

Figure 9-15 Function block diagram and terminal connections for the 8038 IC function generator. An externally connected capacitor is charged and discharged linearly, to produce a triangular waveform. Square and sine waveforms are also generated. (Courtesy of INTERSIL).

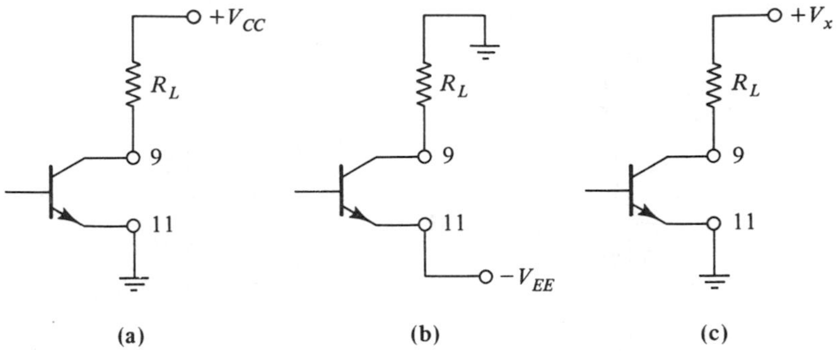

(a)   (b)   (c)

Figure 9-16 The output at terminal 9 is an open-circuited transistor collector. It must be connected via a pull-up resistor to a supply voltage more positive than the voltage at terminal 11.

V_X, as illustrated in Figure 9-16(c). This last case allows the output from terminal 9 to be used to drive any digital logic gate. (See Chapter 12.)

The amplitude of the output from terminal 9 depends upon the supply voltage to which it is connected. When low, the pulse/square wave output is the typical transistor saturation voltage of approximately 0.2 V. When high, the output is $+V_{CC}$ when connected as in Figure 9-16(a), ground when arranged as in Figure 9-16(b), and $+V_X$ when supplied as shown in Figure 9-16(c). The resistance of R_L depends upon the circuit to be connected to terminal 9. In the absence of any significant load, a current of 1 mA is normally selected to operate the output transistor. Then, R_L is simply calculated as (supply voltage to R_L)/1 mA.

Basic Waveform Generator

The basic circuit for an 8038 used as a fixed-frequency function generator is shown in Figure 9-17. The positive supply is connected to terminal 6, and the negative to terminal 11. Resistors R_A and R_B, from $+V_{CC}$ to terminals 4 and 5, set the levels of the charge and discharge currents for the capacitor. Capacitor C is connected from terminal 10 to $-V_{EE}$. The resistance of R_C, which connects terminal 12 to $-V_{EE}$, affects the harmonic distortion content in the sine wave output. If the harmonic distortion is not very important, the manufacturer recommends that R_C be 82 kΩ. Terminal 8 is directly connected to terminal 7. Design of the basic waveform generator only involves selection of R_A, R_B, and C.

One-fifth of the total supply voltage is developed across resistor R_A. So the equation relating R_A and the charging current is

$$I_A = \frac{V_T}{5 R_A} \tag{9-1}$$

where V_T is the total supply voltage—for example, 30 V for a ± 15 V supply.

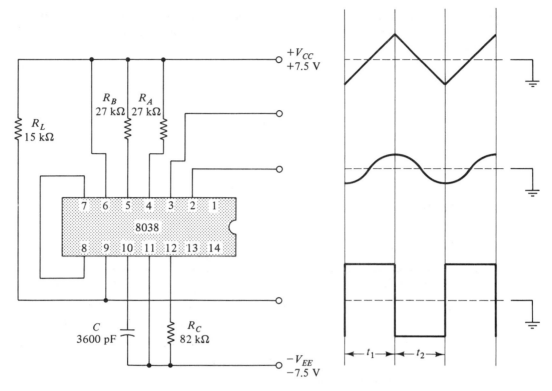

Figure 9-17 Basic waveform generator using an 8038 IC function generator. Resistors R_A and R_B determine the levels of constant charge and discharge currents for capacitor C. With $R_A = R_B = R$, $f = 0.3/(CR)$.

The device manufacturer recommends that the capacitor charging current be within the limits 10 μA to 1 mA, although the circuit can operate with current levels of 1 μA to 5 mA. High charging currents require high capacitance values. So for minimum capacitances, low current levels should be used. Capacitors employed for timing or waveform generation should, of course, have capacitance values much greater than stray or instrument capacitances. Once a current level is selected, R_A can be calculated from Eq. 9-1.

When the square waveform is to have a 50% duty cycle, the capacitor discharge current must be made equal to the charge current. Consequently, R_B should be selected equal to R_A. This may not give a precise 50% duty cycle, so some resistance adjustment may be necessary.

The capacitor equation is derived from Eq. 2-8, which relates constant charging current, voltage change, and capacitance. From Eq. 2-8,

$$t = \frac{C \, \Delta V}{I}$$

where $\Delta V = V_T/3$, for V_T the total supply voltage and $t = t_1$. (See the waveforms in Figure 9-17.) Substituting for ΔV and I (from Eq. 9-1), the equation for charging time becomes

$$t_1 = C \times \frac{V_T}{3} \times \frac{5\,R_A}{V_T}$$

or $$t_1 = \frac{C\,R_A}{0.6}$$ (9-2)

For 50% duty cycle, with $R_B = R_A = R$,

$$T = 2\,t_1 = \frac{C\,R}{0.3}$$

Since $f = 1/T$,

$$f = \frac{0.3}{C\,R}$$ (9-3)

EXAMPLE 9-11

An 8038 IC function generator is to be employed to produce a 3 kHz triangular waveform with a peak-to-peak amplitude of 5 V, symmetrical about ground level. Select an appropriate supply voltage, and calculate R_A, R_B, and R_L. Also, determine the amplitude of the square and sine wave outputs.

Solution   The circuit is as in Figure 9-17.

$$\text{Triangular } V_{o(\text{p-to-p})} \approx 33\% \text{ of } V_T$$

or $$V_T = 3\,V_o = 3 \times 5 \text{ V}$$

$$= 15 \text{ V}$$

For the output to be symmetrical about ground, use

$$V_{CC} = +7.5 \text{ V and } V_{EE} = -7.5 \text{ V}$$

Let $$I_A = I_B = I = 100 \text{ μA}$$

Then, from Eq. 9-1,

$$R_A = R_B = \frac{V_T}{5\,I} = \frac{15 \text{ V}}{5 \times 100 \text{ μA}}$$

$$= 30 \text{ kΩ (use 27 kΩ standard value)}$$

From Eq. 9-3,

$$C = \frac{0.3}{f\,R} = \frac{0.3}{3 \text{ kHz} \times 27 \text{ kΩ}}$$

$$\approx 3700 \text{ pF (use 3600 pF standard value)}$$

$$R_L = \frac{V_T}{1\ \text{mA}} = \frac{15\ \text{V}}{1\ \text{mA}}$$

$$= 15\ \text{k}\Omega\ \text{(standard value)}$$

$$R_C = 82\ \text{k}\Omega\ \text{(as recommended)}$$

Adjusting the Duty Cycle

When the duty cycle of the rectangular output waveform is other than 50%, R_A is determined from Eq. 9-1, after suitable levels of V_T and I_A have been selected. C can then be calculated from Eq. 9-2. To determine the level of I_B, first note from Eq. 2-8 that

$$I_A = \frac{C\,\Delta V}{t_1}$$

and that

$$I_B = \frac{C\,\Delta V}{t_2}$$

Therefore,

$$I_B = I_A \frac{t_1}{t_2} \qquad (9\text{-}4)$$

A knowledge of the internal circuitry of the 8038 reveals that the discharge current I_B through R_B can also be determined by the equation

$$I_B = \frac{2\,V_T}{5\,R_B} - I_A \qquad (9\text{-}5)$$

which gives

$$R_B = \frac{2\,V_T}{5\,(I_A + I_B)} \qquad (9\text{-}6)$$

The duty cycle of the waveform generator outputs can be made adjustable by replacing a portion of R_A and R_B with variable resistors, as illustrated in Figure 9-18(a). A fixed-value resistor should be retained as part of each, so that when the variable portions are reduced to zero, the current levels do not exceed the maximum specified currents.

When the resistances of R_A and R_B are very different, the rectangular output is a pulse waveform and the triangular output becomes a sawtooth waveform. [See Figure 9-18(b).] Obviously, when a sawtooth waveform is desired, it can be produced by selecting R_A and R_B for the appropriate duty cycle. In these circumstances, the sine output waveform is quite distorted.

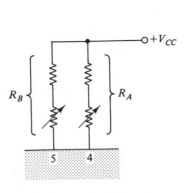

(a) Duty cycle adjustment

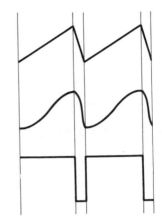

(b) Waveforms with duty cycle
     approximately 80%

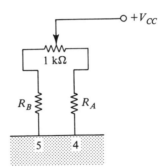

(c) With $R_A = R_B$, a small
     amount of duty cycle
     adjustment is possible —
     around 50%

Figure 9-18   When R_A and R_B in the basic waveform generator (Figure 9-17) are made
adjustable, the duty cycle (and frequency) of the output can be altered.

When a fixed-frequency output with adjustable duty cycle is required, the arrangement in Figure 9-18(c) might be appropriate. This shows a method by which the duty
cycle can be adjusted by small amounts around 50%, without significantly altering the
output frequency.

EXAMPLE 9-12

Using an 8038 IC function generator, design a pulse generator to produce a positive
pulse with PW = 200 μs and a pulse repetition frequency of 1 kHz. The output
amplitude is to be approximately 10 V.

Solution

$$t_1 = PW = 200 \ \mu s$$

$$T = \frac{1}{f} = \frac{1}{1 \ kHz}$$

$$= 1 \ ms$$

$$t_2 = T - t_1 = 1 \ ms - 200 \ \mu s$$

$$= 800 \ \mu s$$

For a positive pulse with 10 V amplitude, use $V_{CC} = 10 \ V$, let

$$I_A = I_{(max)} = 1 \ mA$$

From Eq. 9-4,

$$I_B = I_A \frac{t_1}{t_2} = 1 \ mA \times \frac{200 \ \mu s}{800 \ \mu s}$$

$$= 250 \ \mu A$$

From Eq. 9-1,

$$R_A = \frac{V_T}{5 \ I} = \frac{10 \ V}{5 \times 1 \ mA}$$

$$= 2 \ k\Omega \ (\text{use } 2.2 \ k\Omega \ \text{standard value})$$

From Eq. 9-2,

$$C = \frac{0.6 \ t_1}{R_A} = \frac{0.6 \times 200 \ \mu s}{2.2 \ k\Omega}$$

$$= 0.055 \ \mu F \ (\text{use } 0.06 \ \mu F \ \text{standard value})$$

Eq. 9-6

$$R_B = \frac{2 \ V_T}{5 \ (I_A + I_B)} = \frac{2 \times 10 \ V}{5 \ (1 \ mA + 250 \ \mu A)}$$

$$= 3.2 \ k\Omega \ (\text{use } 3.3 \ k\Omega \ \text{standard value})$$

$$R_L = \frac{V_T}{1 \ mA} = \frac{10 \ V}{1 \ mA}$$

$$= 10 \ k\Omega \ (\text{standard value})$$

$$R_C = 82 \ k\Omega \ (\text{as recommended})$$

Adjusting the Frequency

In Figure 9-19(a), R_A and R_B are replaced by a single resistor R, a portion of which may be adjustable, as illustrated. In this case, the charge and discharge currents are halved, and Eq. 9-3 becomes

$$f = \frac{0.15}{C\,R} \tag{9-7}$$

The modification in Figure 9-19(a) affords an inexpensive means of constructing a variable-frequency function generator, but only if the duty cycle of the output is not important.

When the internal circuitry of the 8038 is examined, it is found that terminal 7 is connected to a simple potential divider. The voltage from $+V_{CC}$ to terminal 7 is $V_T/5$, and this is the bias voltage connected to terminal 8 in the circuit of Figure 9-17. This bias voltage determines the voltage drop across the capacitor charge and discharge resistors R_A and R_B. Terminal 8 may be disconnected from terminal 7 and connected instead to an adjustable potential divider, as illustrated in Figure 9-19(b). This arrangement affords adjustment of the voltage across R_A and R_B, to alter the charging and discharging times, and consequently, alter the output frequency. Adjusting the moving contact of the potentiometer towards $+V_{CC}$ reduces the bias voltage V_8, thus reducing the output frequency. The resistance values illustrated in Figure 9-19(b) give almost 1000:1 adjustment in output frequency.

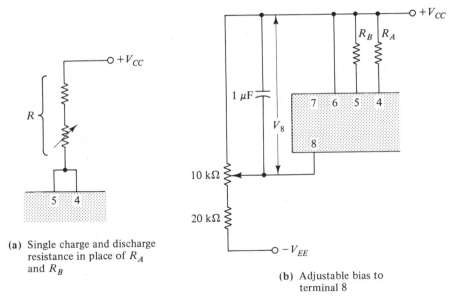

(a) Single charge and discharge resistance in place of R_A and R_B

(b) Adjustable bias to terminal 8

Figure 9-19   Modifications to the basic 8038 waveform generator circuit to provide frequency control.

Note that, in Figure 9-19(b), V_8 could be replaced by an external bias voltage to create a *voltage-controlled oscillator*. If a ramp voltage is substituted in place of V_8, the circuit becomes a *sweep frequency generator*. When terminal 8 is left connected to terminal 7, as in Figure 9-17, and a signal is capacitor coupled to terminal 8, *frequency modulation* of the output can be performed.

EXAMPLE 9-13

Modify the 8038 waveform generator designed in Example 9-11 to make the output frequency adjustable to the widest possible range while still retaining the fixed value of C. Calculate the output frequency extremes.

Solution   Modify the circuit as in Figure 9-19(a). Let $I_{(min)} = 10$ μA and $I_{(max)} = 1$ mA, as recommended. Since the charge and discharge currents are halved with the modification, Eq. 9-1 becomes

$$I = \frac{V_T}{10\,R}$$

giving

$$R = \frac{V_T}{10\,I}$$

$$R_{(max)} = \frac{V_T}{10\,I_{(min)}} = \frac{15\text{ V}}{10 \times 10\text{ μA}}$$

$$= 150\text{ k}\Omega$$

$$R_{(min)} = \frac{V_T}{10\,I_{(max)}} = \frac{15\text{ V}}{10 \times 1\text{ mA}}$$

$$= 1.5\text{ k}\Omega$$

From Eq. 9-7,

$$f_{(min)} = \frac{0.15}{C\,R_{(max)}} = \frac{0.15}{3600\text{ pF} \times 150\text{ k}\Omega}$$

$$\approx 280\text{ Hz}$$

and

$$f_{(max)} = \frac{0.15}{C\,R_{(min)}} = \frac{0.15}{3600\text{ pF} \times 1.5\text{ k}\Omega}$$

$$\approx 28\text{ kHz}$$

Output Impedances and Currents

The output impedance from the triangular waveform buffer (terminal 3) is typically 200 Ω. At the pulse/square waveform output (terminal 9), the output impedance depends upon the selected resistance for R_L. (See Figure 9-16.) Both terminal 3 and terminal 9 can sink a maximum of 25 mA. The maximum source current for terminal 9 depends upon R_L and its supply. The output impedance at the sine waveform output (terminal 2) is typically 1 kΩ. This is relatively high, and may require external buffering for some applications.

Output Waveform Accuracy

The triangular/ramp output waveform has a typical linearity of 0.1% (0.05% for the 8038BC/BM). The sine wave output has a maximum total harmonic distortion of 0.8%, which can be reduced to 0.5% with the use of additional external adjustable resistors. For the pulse/square waveform, the typical rise and fall times are 100 ns and 40 ns, respectively.

REVIEW QUESTIONS AND PROBLEMS

9-1. Sketch the circuit of a simple CR ramp generator. Briefly explain its operation and its limitations. Also, sketch the typical input and output waveforms.

9-2. Design a CR ramp generator to give an output of 3 V peak. The supply voltage is 20 V, and the load to be connected at the output is 330 kΩ. The ramp is to be triggered by a negative-going pulse with an amplitude of 4 V, PW of 3 ms, and time interval between pulses of 0.3 ms. Take the transistor $h_{FE(min)} = 70$.

9-3. Sketch the circuit of a CR ramp generator using a transistor constant-current circuit. Briefly explain how the circuit operates. Also, sketch typical input and output waveforms.

9-4. Design a transistor constant-current circuit for the CR ramp generator designed in Problem 9-2.

9-5. Redesign the constant-current circuit of problem 9-4 to make the ramp amplitude adjustable from 2 V to 4 V.

9-6. Sketch a FET constant-current circuit. Explain how the circuit operates, and discuss the necessary current and voltage levels.

9-7. A FET with the transconductance characteristics in Figure 9-4(b) is to have a constant current of 1.5 mA. Calculate the maximum and minimum resistance values for R_S, and determine the minimum drain-source voltage.

9-8. Using the FET constant-current circuit from Problem 9-7, determine a suitable supply voltage and capacitance value to generate a linear 5 V, 500 μs ramp.

9-9. Sketch the circuit of a UJT relaxation oscillator with adjustable output frequency. Sketch the output waveform, and show how the circuit can be synchronized by external pulses. Briefly explain the operation of the circuit.

9-10. Design a relaxation oscillator using a 2N3980 UJT. The supply voltage is 25 V, and the output frequency is to be 2 kHz. Calculate the amplitude of the output waveform.

9-11. Using a 2N6027 PUT, design a relaxation oscillator to operate from a supply of 20 V. The output is to be 7 V peak at a frequency of 3 kHz.

9-12. Sketch the circuit of a transistor bootstrap ramp generator. Show the waveforms, and explain the operation of the circuit.

9-13. A transistor bootstrap generator is to produce an output of 7 V, with a time period of 2.5 ms. The load resistor is to be 1.2 kΩ, and the ramp is to be linear to within 3%. Design a suitable circuit using transistors with $h_{FE(min)} = 120$ and $V_{CC} = \pm 20$ V. The input pulse has PA = -5 V, PW = 2.5 ms, and space width = 1 ms.

9-14. Sketch the circuit of a bootstrap ramp generator using an IC operational amplifier. Briefly explain the operation of the circuit, drawing a comparison between the IC and transistor bootstrap circuits.

9-15. Design a bootstrap generator using a 741 operational amplifier. The circuit specification is the same as for the circuit in problem 9-13.

9-16. Sketch the circuit of a free-running bootstrap ramp generator. Show the waveforms and carefully explain the operation of the circuit. Also, show how the input frequency and amplitude may be controlled.

9-17. Design a free-running bootstrap ramp generator using 741 IC operational amplifiers. The output ramp is to have an amplitude of ± 3 V and a frequency of 2 kHz. Use a supply voltage of ± 12 V.

9-18. Sketch a Miller integrator circuit connected to operate as a ramp generator. Show the input and output waveforms, and explain the circuit operation.

9-19. Design a Miller integrator ramp generator to produce an output waveform with a peak amplitude of 3 V. The input is a -8 V pulse with PW = 1.2 ms and space width = 100 μs. Select a suitable IC operational amplifier and supply voltage. Specify the FET.

9-20. Sketch the circuit of a free-running triangular waveform generator using IC operational amplifiers. Show all the waveforms in the circuit, and carefully explain the overall circuit operation. Also, show how the output amplitude and frequency may be controlled.

9-21. Design a free-running triangular waveform generator to have an output of ± 2.5 V at a frequency of 500 Hz. Select suitable operational amplifiers, and use a supply of ± 12 V.

9-22. The circuit designed for problem 9-21 is to have its output amplitude and frequency adjustable by $\pm 20\%$. Make the necessary design modifications.

9-23. Design a free-running triangular waveform generator to have $V_o = (\pm 2 \text{ V to } \pm 6 \text{ V})$ and $f = 300$ Hz. Select suitable IC operational amplifiers and supply voltage. Discuss frequency compensation of the operational amplifiers.

9-24. Sketch a pulse generator circuit using operational amplifiers and/or voltage comparators. The circuit should have adjustments for pulse amplitude, pulse frequency, pulse width, and dc offset. Explain the circuit operation, and discuss op-amp and comparator selection and op-amp compensation.

9-25. Design the square wave circuit for a pulse generator (as in Figure 9-12) which is to have an output frequency of 100 PPS to 10 000 PPS. Select a suitable operational amplifier or comparator. Use $V_{CC} = \pm 15$ V.

9-26. Design the monostable circuit for a pulse generator (as in Figure 9-12) which is to have an output pulse width of 100 μs to 1 ms. Select a suitable operational amplifier or comparator. Use $V_{CC} = \pm 15$ V.

9-27. Design the attenuator and dc level control for a pulse generator (as in Figure 9-12) which is to have an output amplitude ranging from 300 mV to 3 V and a dc level controllable over ±3 V. Select a suitable operational amplifier. Use $V_{CC} = \pm 15$ V.

9-28. Sketch the circuit of a function generator using operational amplifiers. Explain the circuit operation.

9-29. A function generator, as in Question 9-28, is to be designed to produce a triangular output with an amplitude of 3 V peak-to-peak. The rectangular output wave is to have pulse width range = space width range = (400 μs to 1 ms), and pulse amplitude ≈ ±9 V. Select a suitable supply voltage and operational amplifiers, and design the circuit.

9-30. Sketch a diode loading circuit for converting triangular waveforms into sine waves. Explain the circuit operation.

9-31. A two-level diode loading circuit, as in Figure 9-14(a), is to attenuate and reshape a triangular waveform which has an amplitude of ±5 V. The input is to be initially attenuated by a factor of approximately 0.8, and then by approximately 0.5 when the output exceeds 2 V. Starting with $R_2 = 10$ kΩ, determine suitable values for R_1, R_3, R_4, $+V_1$, and $-V_1$.

9-32. Sketch the functional block diagram of an 8038 IC function generator. Briefly explain the circuit operation and output waveforms. Discuss the supply voltage options and the output voltage amplitudes.

9-33. Sketch the circuit and outputs of a basic waveform generator using an 8038 IC function generator. Briefly discuss the circuit operation.

9-34. An 8038 waveform generator, as in Question 9-33, is to have a 6 V peak-to-peak triangular output with a frequency of 5 kHz. Select a suitable supply voltage, calculate all component values, and determine the amplitude of the sine and square wave outputs.

9-35. The circuit designed for Problem 9-34 is to have a duty cycle adjustment of ±15%. Show the necessary modifications, and calculate the new component values.

9-36. A sawtooth waveform generator is to be designed to use an 8038 IC function generator. The sawtooth is to have a positive-going amplitude of +3 V with a time period of 750 μs. The negative-going time is to be 100 μs. Sketch the circuit, select a suitable supply voltage, and determine all component values.

9-37. Sketch circuits to show two methods of adjusting the output frequency of an 8038 function generator. Briefly explain. Also, explain how the circuit can be made to function as a voltage-controlled oscillator, and as a sweep frequency generator.

9-38. The circuit designed for Problem 9-34 is to be modified to have a single resistor frequency control, as in Figure 9-19(a). Calculate the new capacitance of C, and the maximum and minimum resistance values for R if the output frequency range is to be 1 kHz to 12 kHz.

9-39. Instead of using a single adjustable resistor, as in Problem 9-38, the circuit designed for Problem 9-34 is to have its frequency adjusted by using variable resistors for R_A and R_B, as in Figure 9-18(a). Determine the new capacitance value, and suitable maximum and minimum resistances for R_A and R_B. Also, calculate the resultant maximum and minimum pulse and space widths in the rectangular output waveform.

Chapter 10

Basic Logic Gates and Logic Functions

INTRODUCTION

A *logic gate* is an electronic circuit which monitors several voltage levels to determine such conditions as: are all of the inputs to the circuit *high*; or, perhaps, is only one of them *high*. An output voltage is produced by the gate only when the desired input condition is obtained.

A logic system may involve a great many gates, and each gate may have many components. Because of this, it would be confusing to show a logic system as an electronic circuit complete with all components. Instead, each type of gate is represented by a particular graphic symbol. The two basic gates are the **AND** gate and the **OR** gate. Derived from these two are the **NAND** gate and the **NOR** gate.

10-1 DIODE *AND* GATE

The circuit of a diode *AND* gate with three input terminals is shown in Figure 10-1. If one or more of the input terminals (i.e., diode cathodes) are grounded, then the diode (or diodes) are forward biased. Consequently, the output voltage V_o is equal to the diode forward voltage drop V_F. Suppose that the supply voltage is $V_{CC} = 5$ V, and that an input of 5 V is applied to terminal A, while terminals B and C are grounded. In this case, diode D_1 is reverse biased, while D_2 and D_3 remain forward biased, and $V_o = V_F$. If levels of 5 V are applied to all three inputs, no current flows through R_1, and $V_o = V_{CC} = 5$ V. Thus,

> a *high* output voltage is obtained from the *AND* gate only when *high* input voltages are present at input A, **and** at input B, **and** at input C.

Hence the name *AND* gate.

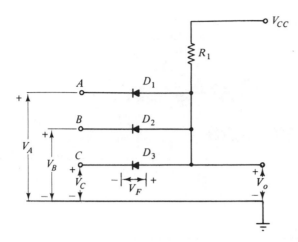

Figure 10-1 Circuit of a three-input diode AND gate. A *high* output is obtained when *high* inputs are present at terminal A, ***and*** at terminal B, ***and*** at terminal C.

An *AND* gate may have as few as two input terminals, or it may have a great many input terminals. In all cases an output is obtained only when the correct input voltage levels are provided at every input terminal.

For all logic gates, the level of input and output voltages are usually described as either *high* or *low*. Depending upon the particular gate circuit, a *high* level might be between 3 V and 6 V (or even higher), while a *low* voltage level might be less than 1 V. The high level is usually designated *1*, and the low level is designated *0*.

EXAMPLE 10-1

An *AND* gate has three input terminals which are connected to the collectors of saturated transistors, as illustrated in Figure 10-2. The transistors can each take an *additional* collector current of 0.5 mA. Design a suitable gate circuit, and determine the low and high output levels from the gate. Use $V_{CC} = 5$ V, and diodes with $V_F = 0.7$ V.

Solution   From Figure 10-2, the maximum additional collector current I_1 flows through Q_1 when Q_1 is *on* and Q_2 and Q_3 are *off*. So,

$$I_{R1} = I_{1(max)}$$

$$V_{CC} = (I_{R1} R_1) + V_F + V_{CE(sat)}$$

$$R_1 = \frac{V_{CC} - V_F - V_{CE(sat)}}{I_{R1}}$$

$$= \frac{5\text{ V} - 0.7\text{ V} - 0.2\text{ V}}{0.5\text{ mA}}$$

$$= 8.2\text{ k}\Omega \qquad \text{(standard value)}$$

This is the minimum value for R_1 to limit the transistor additional collector current to 0.5 mA. R_1 could be made larger than 8.2 kΩ, in which case the current would be smaller.

Output voltages are

$$V_{o(\text{low})} = V_{CE(\text{sat})} + V_F$$

$$= 0.2 \text{ V} + 0.7 \text{ V} = 0.9 \text{ V}$$

$$V_{o(\text{high})} = V_{CC} = 5 \text{ V}$$

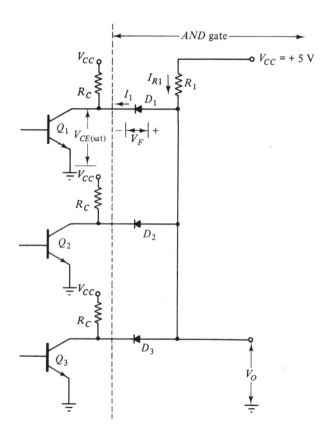

Figure 10-2 Diode *AND* gate with inputs controlled by transistors. See Example 10-1.

10-2 DIODE *OR* GATE

A three-input diode *OR* gate is illustrated in Figure 10-3. It is obvious from the gate circuit that the output is zero when all three inputs are at ground level. If a 5 V input is applied to terminal A, D_1 is forward biased and V_o becomes (5 V $- V_F$). If terminals

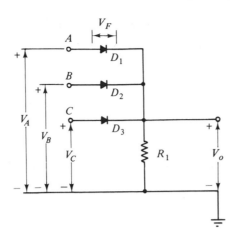

Figure 10-3 Circuit of a three-input diode *OR* gate. A *high* output is obtained when a *high* input is present at terminal *A*, *or* at terminal *B*, *or* at terminal *C*.

B and *C* are grounded at this time, diodes D_2 and D_3 are reverse biased. Instead of being applied to terminal *A*, the positive input might be applied to terminal *B* or *C* to obtain a positive output voltage.

> A *high* output voltage is obtained from the *OR* gate when a *high* input is applied to terminal *A*, *or* to terminal *B*, *or* to terminal *C*.

Hence the name *OR* gate.

As in the case of the *AND* gate, an *OR* gate may have only two input terminals, or many input terminals.

EXAMPLE 10-2

The *OR* gate shown in Figure 10-4 has two input terminals, each of which is supplied from flip-flops having $R_C = 3.3$ kΩ. The supply voltage to the flip-flops is $V_{CC} = 5$ V. The gate output voltage is to be at least 3.5 V for a 1 logic level. Design a suitable circuit, using diodes with $V_F = 0.7$ V.

Solution   Refer to the circuit shown in Figure 10-4. When one of the gate inputs is *high*,

$$V_o = V_{CC} - (I_1 R_C) - V_F$$

and

$$I_{R1} = I_1 = \frac{V_{CC} - V_F - V_o}{R_C}$$

Thus,

$$I_R = \frac{5 \text{ V} - 0.7 \text{ V} - 3.5 \text{ V}}{3.3 \text{ k}\Omega} = 0.24 \text{ mA}$$

Also,

$$R_1 = \frac{V_o}{I_{R1}}$$

$$= \frac{3.5 \text{ V}}{0.24 \text{ mA}}$$

$$\approx 14.6 \text{ k}\Omega \qquad \text{(use 15 k}\Omega\text{)}$$

14.6 kΩ is a minimum value for R_1, to maintain the output voltage at a minimum of 3.5 V. R_1 could be made larger, in which case V_o would be larger.

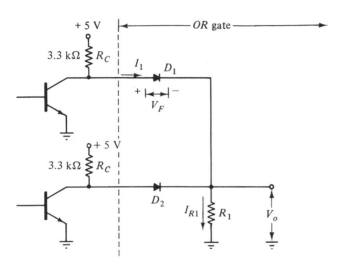

Figure 10-4 Diode *OR* gate with inputs controlled by transistors. See Example 10-2.

10-3 *NOT, NAND,* AND *NOR* GATES

NOT and *NAND* Gates

As already explained, a positive logic diode *AND* gate has a *low* voltage output when one or more of its inputs are *low*, and a *high* output when all inputs are *high*. If a transistor inverter is connected at the output of the *AND* gate, the inverter output is *high* when one or more of the *AND* inputs are *low*, and *low* when all *AND* gate inputs are *high*. Used in this fashion, the inverter is termed a *NOT* gate. The combination of the *NOT* gate and the *AND* gate is then referred to as a *NOT-AND* gate, or a *NAND* gate.

Figure 10-5 shows an integrated circuit DTL (diode transistor logic) *NAND* gate composed of a diode *AND* gate and an inverter. R_1, D_1, D_2, and D_3 constitute the *AND* gate. The inverter is formed by transistor Q_1 with connector resistor R_C and bias resistor R_B. When all input terminals are at ground level, the voltage at point X is the voltage drop across the input diodes (i.e., $V_X = V_F$). If diodes D_4 and D_5 were not

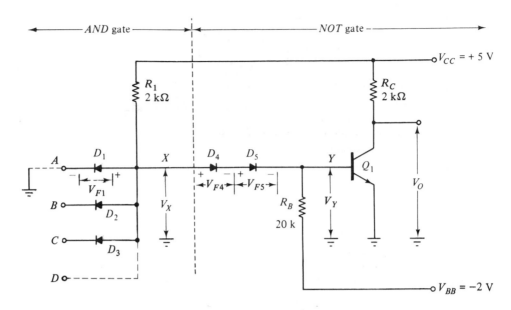

Figure 10-5 Diode-transistor *NAND* gate, consisting of an *AND* gate and an inverter stage, or *NOT* gate. A *low* output is produced only when *high* inputs are present at terminal *A*, *and* at terminal *B*, *and* at terminal *C*.

present, V_X would be sufficient to forward-bias the base-emitter junction of Q_1. The negative supply $-V_{BB}$ keeps diodes D_4 and D_5 forward biased, so that when the inputs are at 0 V the transistor base voltage is

$$V_Y = V_X - (V_{F4} + V_{F5})$$

$$= V_F - V_{F4} - V_{F5}$$

For silicon devices,

$$V_Y \approx 0.7 \text{ V} - 0.7 \text{ V} - 0.7 \text{ V}$$

$$= -0.7 \text{ V}$$

Therefore, when any one input to the *NAND* gate is at 0 V, Q_1 is biased *off*, and the output voltage is V_{CC}.

    Suppose all inputs to the *NAND* gate are made sufficiently positive to reverse-bias D_1, D_2, and D_3. Now, V_Y depends upon the values of R_1 and R_B, and upon the levels of V_{CC} and $-V_{BB}$. If these quantities are all correctly selected, V_Y is positive at this time, Q_1 is driven into saturation, and the output voltage goes to $V_{CE(\text{sat})}$. When any one input to the *NAND* gate is at logic *0*, the gate output is at *1*. When input *A* *and* input *B* *and* input *C* are at *1*, the output of the *NAND* gate is level *0*.

    Input terminal *D* in Figure 10-5 is a direct connection to the diode anodes. Thus, it provides for the connection of additional diodes to increase the number of input terminals. With this facility the gate is said to be *expandable*.

NOR Gate

A transistor inverter (or *NOT* gate) connected at the output of a positive logic *OR* gate produces a *low* output when any one of the inputs is *high*. The complete circuit is termed a *NOT-OR* gate, or *NOR* gate. A diode-transistor *NOR* gate circuit is shown in Figure 10-6. When all of the inputs are *low* (at *logic 0*), transistor Q_1 is *off*, and the output is *high* (*logic 1*). A *high* input at terminal A, *or* at terminal B, *or* at terminal C, biases the transistor *on*, and produces a *low* output.

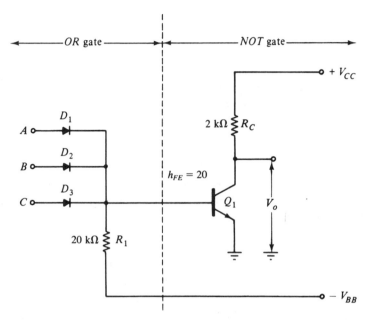

Figure 10-6   Diode-transistor *NOR* gate, consisting of an *OR* gate and an inverter stage, or *NOT* gate. A *low* output is produced when *high* inputs are present at terminal A, *or* at terminal B, *or* at terminal C.

10-4 THE *AND* OPERATION

Logic Symbol and Truth Table

The circuit of an *AND* gate is explained in Section 10-1. Recall that, to obtain a *logic 1* output from a three-input terminal *AND* gate, input levels of *1* must be present at input terminal A, *and* at terminal B, *and* at terminal C. An *AND* gate with any number of inputs produces a *1* output only when a *1* is present at every input terminal.

The logic symbol for a two-input *AND* gate is shown in Figure 10-7(a), and its *truth table* is illustrated in Figure 10-7(b). The terms *true* and *false* are sometimes employed instead of *high* and *low* for logic levels. The truth table lists the *true* and

A	B	$X = A \cdot B$
0	0	0
0	1	0
1	0	0
1	1	1

(a) Logic symbol for two-input AND gate. A *1* output is produced when *1* inputs are present at terminal *A*, **and** at terminal *B*

(b) Truth table for two-input AND gate

Figure 10-7 Logic symbol and truth table for a two-input *AND* gate. The output of the gate is $X = A \cdot B$, where *A* and *B* are inputs, and the dot represents Boolean multiplication.

false conditions for a logic gate with various inputs. The left column of the truth table lists all possible combinations of *A* and *B* inputs. The right side shows the outputs that result from each input combination.

The output of the *AND* gate is defined as

$$X = A \cdot B \qquad (10\text{-}1)$$

which is stated as $X = A$ *and* B.

Equation 10-1 is the *Boolean algebra*[*] expression for the output of a two-input *AND* gate. The dot represents Boolean multiplication, which, as will be seen, is similar to ordinary arithmetic multiplication. Sometimes the dot is omitted, and the *AND* gate output expression becomes

$$X = AB$$

Consider the truth table in Figure 10-7(b). Its right (output) column is headed $X = A \cdot B$. When both inputs are *1*, the gate output is

$$X = A \cdot B$$
$$= 1 \times 1 = 1$$

When one input is *0*, and the other is *1*,

$$X = A \cdot B$$
$$= 0 \times 1 = 0$$

Finally, if both inputs are *0*,

$$X = A \cdot B$$
$$= 0 \times 0 = 0$$

It is clear that to have a *1* output from an *AND* gate, every input must be equal to *1*. When any input is at *0*, the output is *0*, regardless of the levels at other input terminals.

[*] Developed by English mathematican George Boole (1815–1864).

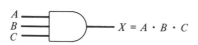

A	B	C	X = A · B · C
0	0	0	0
0	0	1	0
0	1	0	0
0	1	1	0
1	0	0	0
1	0	1	0
1	1	0	0
1	1	1	1

(a) Logic symbol for three-input AND gate. A *1* output is produced when a *1* input is present at terminal *A*, **and** at terminal *B*, **and** at terminal *C*

(b) Truth table for three-input AND gate

Figure 10-8 Logic symbol and truth table for a three-input *AND* gate. The output of the gate is $X = A \cdot B \cdot C$, where *A*, *B*, and *C* are inputs, and the dot once again represents Boolean multiplication.

Although only a two-input *AND* gate has been discussed, the same rules apply to all *AND* gates, whatever the number of input terminals. For example, for a three-input *AND* gate, the Boolean expression for the output is

$$X = A \cdot B \cdot C \tag{10-2}$$

The logic symbol and truth table for a three-input *AND* gate are illustrated in Figure 10-8.

Input and Output Waveforms

EXAMPLE 10-3

The three-input *AND* gate in Figure 10-9 has input waveforms *A*, *B*, and *C*, as illustrated. Plot the output waveform.

Solution   Referring to the input waveforms, it is seen that all three inputs are *high* (at *logic 1*) only during the times between points 2 and 3, and between points 10 and 11, on the time scale. At all other times, one or more of the inputs is *low*. Therefore, a *1* output level is obtained only during the time intervals 2 to 3 and 10 to 11, giving the output waveform illustrated.

AND Gate Application

A simple application of an *AND* gate is illustrated in Figure 10-10. The logic system shown is a safety arrangement for an elevator. A push-button switch for starting the elevator motor is connected to one input of the *AND* gate (input *A*), while door switches are connected to all of the other inputs. When all the door switches are closed, *1* levels are applied to each input of the gate, with the exception of input *A*. Closing the

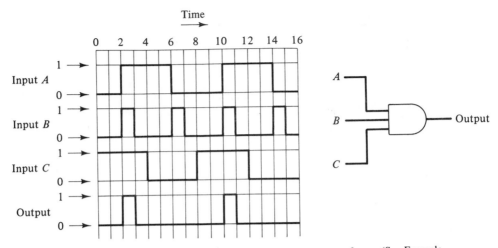

Figure 10-9 Three-input *AND* gate, with input and output waveforms. (See Example 10-3.)

push-button switch now provides a *1* level at *A*. Thus, the *AND* gate output goes from *0* to *1*, and the motor controller is energized to start the elevator motor. If one or more of the door switches are open, the *AND* gate output remains at *0* when the push-button is pressed, and the elevator motor cannot be started.

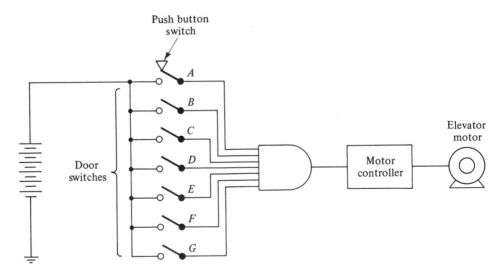

Figure 10-10 Simple logic system for control of an elevator motor. The push-button switch starts the elevator motor only when all of the door switches are closed. When one or more door switches are open, *1* levels are not present at every input of the *AND* gate, and the motor cannot be started.

Number of Input Combinations

Consider the various input combinations for a two-input gate once again. As shown in the truth table in Figure 10-7(b), there are *four* possible combinations: *00*, *01*, *10*, and *11*. For a three-input gate, *eight* different input combinations are possible. (See Figure 10-8). The number of different input combinations for any gate is 2^n, where n is the number of inputs. Thus, for a two-input gate there are $2^2 = 4$ possible input combinations, for a three-input gate $2^3 = 8$ possible input combinations, and for a four-input gate, $2^4 = 16$ possible input combinations.

10-5 THE *OR* OPERATION

Logic Symbol and Truth Table

Recall from Section 10-2 that an *OR* gate with three input terminals produces a *1* output level when a *1* level is present at input terminal *A*, *or* at terminal *B*, *or* at terminal *C*. Similarly, an *OR* gate with any number of input terminals produces a *1* output when a *1* is present at *any* input terminal.

Figure 10-11(a) shows the logic symbol for an *OR* gate with an output *X* and two inputs identified as *A* and *B*. The truth table for the two-input *OR* gate is shown in Figure 10-11(b). The left columns of the truth table list all possible combinations of *A* and *B* inputs. The right side of the table shows the outputs that result from each combination of inputs. The table shows that when both *A* and *B* inputs are *0*, the output of the *OR* gate is *0*. When $A = 0$ and $B = 1$, the output is *1*. Inputs of $A = 1$ and $B = 0$ also give a *1* output, as do inputs of $A = 1$ and $B = 1$.

The right column of the truth table is headed by the Boolean algebra expression for the output of the two-input *OR* gate,

$$X = A + B \qquad\qquad\qquad (10\text{-}3)$$

A	B	X = A + B
0	0	0
0	1	1
1	0	1
1	1	1

(a) Logic symbol for two-input OR gate. A *1* output is produced when a *1* input is present at terminal *A*, *or* at terminal *B*

(b) Truth table for two-input OR gate

Figure 10-11 Logic symbol and truth table for a two-input *OR* gate. The output of the gate is $X = A + B$, where *A* and *B* are inputs and the plus sign represents Boolean addition.

In this case the plus sign represents *Boolean addition*; *it does not represent ordinary arithmetic addition* Boolean addition is different from ordinary addition in which $1 + 1 = 2$. In digital logic, the output can only be *1* or *0* (*on* or *off*, *high* or *low*, *true* or *false*). Consequently, output (and input) levels can never be greater than *1*.

Equation 10-3 can be stated as *X equals A or B*, or *X equals A plus B*.

Using Boolean algebra, when both *A* and *B* are equal to *1*,

$$X = A + B$$

$$= 1 + 1 = 1$$

Also, when one gate input is *0* and the other is *1*,

$$X = A + B$$

$$= 0 + 1 = 1$$

alternatively, $$X = 1 + 0 = 1$$

It is clear that the result of Boolean addition is *1* when *any* input has a level of *1*. Also, the result is *0* only when all inputs are at level *0*. In general, these rules apply to all *OR* gates regardless of the number of input terminals.

For a three-input *OR* gate, the Boolean algebra expression for the output is

$$X = A + B + C \qquad (10\text{-}4)$$

The logic symbol and truth table for a three-input *OR* gate are illustrated in Figure 10-12.

The *OR* gates just described are sometimes termed *inclusive OR gates* to distinguish them from the *exclusive OR gate*, which is described in Section 11-6. Where the term *OR gate* is used, it is understood to refer to an inclusive *OR* gate, unless otherwise specified.

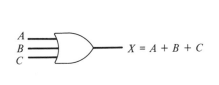

A	B	C	$X = A + B + C$
0	0	0	0
0	0	1	1
0	1	0	1
0	1	1	1
1	0	0	1
1	0	1	1
1	1	0	1
1	1	1	1

(a) Logic symbol for three-input OR gate. A *1* output is produced when a *1* input is present at terminal *A*, *or* at terminal *B*, *or* at terminal *C*

(b) Truth table for three-input OR gate

Figure 10-12   Logic symbol and truth table for a three-input *OR* gate. The output of the gate is $X = A + B + C$, where *A*, *B*, and *C* are inputs, and the plus signs once again represents Boolean addition.

EXAMPLE 10-4

An *OR* gate with four input terminals has the following inputs: *0100, 0111, 0000, 1000, 1100*. Prepare a truth table showing the outputs obtained for each input.

Solution

A B C D	$X = A + B + C + D$
0 1 0 0	1
0 1 1 1	1
0 0 0 0	0
1 0 0 0	1
1 1 0 0	1

OR Gate Application

As an example of one application of an *OR* gate, consider the simple logic system in Figure 10-13. A boiler, used in a hot-water space heating system, is to be controlled by two thermostats. When the air temperature drops below a predetermined level, the air thermostat provides a *high* output, which calls for the boiler to be switched *on*. Similarly, when the water temperature falls below a preset level, a *high* output is produced by the water thermostat, to switch the boiler *on*.

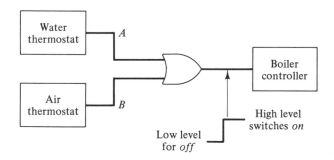

Figure 10-13 Logic system for controlling a boiler used in a hot-water space heating system. The boiler is switched *on* when the air temperature falls below a predetermined level *or* when the water temperature falls below a preset level.

The desired operation is achieved by the use of an *OR* gate, as illustrated, and a controller which switches the boiler *on* when the *OR* gate output is *high*.

10-6 THE *NOT* OPERATION

The circuit of a *NOT* gate is explained in Section 10-3. Recall that the *NOT* gate is simply an inverter. When a *high* level (*1*) is applied as input, the output of the *NOT* gate is *low* (*0*). Conversely, a *0* input produces a *1* output. So, a *NOT* gate inverts or *complements* the input.

(a) Logic symbol for NOT gate   (b) Truth table for NOT gate

Figure 10-14 Logic symbol and truth table for a *NOT* gate, or inverter. The small circle, or *bubble*, at the output denotes inversion. The output level is always the complement of the input.

The logic symbol and truth table for the *NOT* gate are illustrated in Figure 10-14. The symbol is seen to be an amplifier triangle with a small circle, or *bubble*, at its output. The bubble represents inversion, so it could appear at the input instead of the output.

The Boolean algebra expression for the *NOT* gate output is shown at the top of the truth table as

$$X = \overline{A} \tag{10-5}$$

The bar above the *A* denotes inversion. Sometimes a *prime* (*A'*) is used instead of a bar. The equation is variously read as *X equals NOT A*, *X equals the complement of A*, or *X equals the inverse of A*.

It is clear that when the input of a *NOT* gate is *0*, the output is *NOT 0*, or

$$X = \overline{A} = 1$$

When the input is *1*, the output is *NOT 1*, or

$$X = \overline{A} = 0$$

A *NOT* gate has only one input (and one output). There are no multi-input *NOT* gates.

If two *NOT* gates are cascaded, as illustrated in Figure 10-15, the output of the first gate is

$$X = NOT\ A$$

or

$$X = \overline{A}$$

The output of the second gate is

$$X = NOT\ NOT\ A$$

or

$$X = \overline{\overline{A}} = A$$

Thus, two cascaded *NOT* gates cancel each other out.

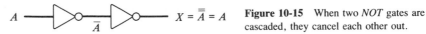

Figure 10-15 When two *NOT* gates are cascaded, they cancel each other out.

10-7 THE *NAND* OPERATION

As already discussed in Section 10-3, the *NAND* gate is simply an *AND* gate followed by a *NOT* gate. Recall that the name *NAND* gate comes from the combination *NOT-AND*. To explore the *NAND* operation, the logic symbols of cascaded *AND* and *NOT* gates are shown in Figure 10-16(a). The normally used *NAND* gate symbol is an *AND* gate symbol with a bubble at its output to indicate that inversion occurs. [See Figure 10-16(b).]

The output of a *NAND* gate can be expressed as

$$X = \overline{A \cdot B} \tag{10-6}$$

The truth table in Figure 10-16(c) shows that the *NAND* gate output is *0* when all inputs are *1*. For all other input conditions, the output is *1*. Comparing the truth table of the *NAND* gate to the truth table for an *AND* gate [Figure 10-7(b)], it is seen that the output of the *NAND* gate is always the inverse of the *AND* gate output, for the same input conditions.

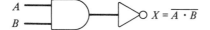

(a) AND and NOT gates combined   (b) Usual logic symbol for a NAND gate
    to produce the NAND function

A	B	$X = \overline{A \cdot B}$
0	0	1
0	1	1
1	0	1
1	1	0

(c) Truth table for two-input NAND gate

Figure 10-16 A *NAND* gate is made up of an *AND* gate followed by a *NOT* gate. For any combination of inputs, the *NAND* gate output is always the inverse of the output from an *AND* gate.

EXAMPLE 10-5

Determine the output waveform, and write the corresponding truth table, for a three-input *NAND* gate with the waveforms shown in Figure 10-17.

Solution   The simplest approach to determining the output waveform of the *NAND* gate is to first draw the waveform that would be obtained from the output of an *AND* gate with the given inputs. This can then be inverted to produce the *NAND* gate output waveform.

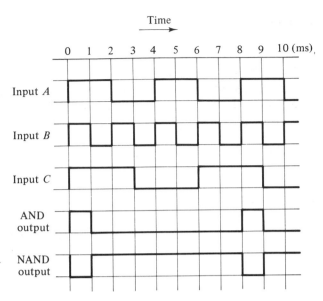

(a) Input and output waveforms

Time	A	B	C	$X = \overline{ABC}$
0–1	1	1	1	0
1–2	1	0	1	1
2–3	0	1	1	1
3–4	0	0	0	1
4–5	1	1	0	1
5–6	1	0	0	1
6–7	0	1	1	1
7–8	0	0	1	1
8–9	1	1	1	0
9–10	1	0	0	1

(b) Truth table

Figure 10-17 Output waveform and truth table determination for a *NAND* gate with the given input waveforms.

Referring to the input waveforms, inputs A, B, and C are all *1* during the time interval between 0 and 1 ms. Therefore, as illustrated, the output of an *AND* gate would be *1* during this time.

From 1 ms to 8 ms, at least one input is always at *0* when the other two are at *1*. Consequently, the *AND* gate output remains *0* for this time interval.

From 8 ms to 9 ms, all three inputs are again equal to *1*. This, once again, gives an *AND* gate output of *1*.

Between 9 ms and 10 ms, only input *A* is at *1*. So the *AND* gate output remains at *0*.

The *NAND* gate output waveform is the inverse of the *AND* gate output, as illustrated.

To write the truth table, inputs *A*, *B*, and *C* are identified as either *0* or *1* during each 1 ms time interval. Then, for each combination of inputs, the output of the *NAND* gate is simply the inverse of the output that would be obtained with an *AND* gate. From the waveforms, *A*, *B*, and *C* all have *1* inputs during the time 0 to 1 ms. The output from an *AND* gate with these inputs would be *1*. Using a *NAND* gate, the output becomes *0*.

10-8 THE *NOR* OPERATION

The *NOR* gate, as discussed in Section 10-3, is simply an *OR* gate followed by a *NOT* gate. Recall that the name comes from *NOT-OR*. Figure 10-18(a) shows *OR* gate and *NOT* gate logic symbols combined to produce the *NOR* operation. Figure 10-18(b) gives the usual *NOR* gate logic symbol, consisting of an *OR* gate symbol with a bubble at its output, to show once again that output inversion occurs.

The Boolean algebra expression for the *NOR* gate output is

$$X = \overline{A + B} \qquad (10\text{-}7)$$

The truth table for the *NOR* gate in Figure 10-18(c) shows that the gate output is *0* when any input is *1*. When all inputs are *0*, the *NOR* gate output is *1*. Comparison

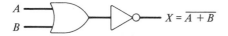

(a) OR and NOT gates combined
   to produce the NOR function

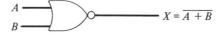

(b) Usual logic symbol for a NOR gate

A	B	$X = \overline{A + B}$
0	0	1
0	1	0
1	0	0
1	1	0

(c) Truth table for a two-input NOR gate

Figure 10-18 A *NOR* gate is composed of an *OR* gate and a *NOT* gate. The *NOR* gate output is always the inverse of the output obtained from an *OR* gate with the same inputs.

of the *NOR* gate truth table to the truth table of the *OR* gate [Figure 10-11(b)] shows that the *NOR* gate output is always the inverse of the *OR* gate output.

As in the case of the *OR* gate, the term *inclusive* is sometimes used in reference to the *NOR* gate. This is to distinguish it from a gate known as the *exclusive NOR* gate, which is discussed in Section 11-6.

EXAMPLE 10-6

A *NOR* gate with five input terminals has the following input combinations: *11001*, *00111*, *10111*, *00000*, *11111*. Write a truth table showing the outputs that would be obtained with these inputs.

Solution

A B C D E	$X = \overline{A + B + C + D + E}$
1 1 0 0 1	0
0 0 1 1 1	0
1 0 1 1 1	0
0 0 0 0 0	1
1 1 1 1 1	0

REVIEW QUESTIONS AND PROBLEMS

10-1. Sketch the circuit of a diode *AND* gate with three input terminals. Briefly explain the operation of the circuit.

10-2. Design a four-input diode *AND* gate using a 9 V supply. The gate inputs are to be controlled from the collector terminals of saturated transistors, each of which can pass an additional collector current of 1 mA. Determine the *low* and *high* output levels for the gate.

10-3. Sketch the circuit of a diode *OR* gate with three input terminals. Briefly explain the operation of the circuit.

10-4. A diode *OR* gate is to have an output which switches from a *low* level of 0 V to a *high* of at least 2 V. The inputs to the *OR* gate are connected to transistors which have collector resistors of $R_C = 4.7$ kΩ and a supply of $V_{CC} = 9$ V. Design a suitable circuit.

10-5. Sketch the circuit of a DTL *NAND* gate. Carefully explain the operation of the circuit, and discuss the function of each component.

10-6. Sketch the circuit of a DTL *NOR* gate. Carefully explain the operation of the circuit, and discuss the function of each component.

10-7. Sketch the logic symbol for a two-input *AND* gate. Write the Boolean equation for the gate and prepare the truth table.

10-8. An *AND* gate with four input terminals has the following inputs: *0011*, *0100*, *1100*, *1111*. Prepare a truth table, showing the outputs obtained for each input.

10-9. A three-input *AND* gate has the input waveforms (*A*, *B*, and *C*) shown in Figure 10-17, except that waveform C is inverted. Determine the output waveform and prepare the corresponding truth table.

10-10. An audio alarm is to sound in an automobile if the ignition is switched *on* and any one of four doors is not properly closed. Sketch an appropriate logic circuit and briefly explain.

10-11. An *npn* transistor is to be switched *on* when all three input signals are at logic *1*. Sketch a suitable circuit and briefly explain.

10-12. Determine the number of different input signal combinations possible for three-input, six-input, and nine-input *AND* gates.

10-13. Sketch the logic symbol for a two-input *OR* gate. Write the Boolean equation for the gate, and prepare the truth table.

10-14. An *OR* gate with four input terminals has the following inputs: *0011, 0101, 1101, 1111*. Prepare a truth table, showing the outputs obtained for each input.

10-15. A three-input *OR* gate has the input waveforms A, B, and C shown in Figure 10-9, with waveform C inverted. Determine the output waveform and prepare the corresponding truth table.

10-16. A home burglar alarm is to be activated when any one of six windows or two doors is opened. Sketch an appropriate logic circuit and explain its operation.

10-17. A light-emitting diode is to be switched *on* when any one of five input signals is at logic *1*. Sketch the logic circuit and briefly explain.

10-18. Sketch the logic symbol for a *NOT* gate. Write the Boolean equation for the gate and prepare the truth table.

10-19. Sketch the logic symbol for a three-input *NAND* gate. Show the composition of the gate and briefly explain. Write the Boolean equation for the *NAND* gate and prepare the truth table.

10-20. A *NAND* gate with three input terminals has the input waveforms A, B, and C shown in Figure 10-9, except that waveform A is inverted. Determine the output waveform and prepare the corresponding truth table.

10-21. A *NAND* gate with four input terminals has the following inputs: *0000, 0010, 1101, 1111*. Prepare a truth table, showing the outputs obtained for each input.

10-22. Sketch the logic symbol for a four-input *NOR* gate. Show the composition of the gate and briefly explain. Write the Boolean equation for the *NOR* gate and prepare the truth table.

10-23. A *NOR* gate with three input terminals has the input waveforms A, B, and C shown in Figure 10-9, except that waveform B is inverted. Determine the output waveform and prepare the corresponding truth table.

10-24. A *NOR* gate with five input terminals has the following inputs: *00000, 00101, 11010, 11110*. Prepare a truth table, showing the outputs obtained for each input.

10-25. A *NOR* gate with three input terminals has the input waveforms A, B, and C shown in Figure 10-17. Determine the output waveform and prepare the corresponding truth table.

Chapter 11

Logic Circuits

INTRODUCTION

A *logic circuit* may be made up of several different types of logic gates. Every logic circuit can be analyzed by writing the Boolean equation for the circuit and then using the equation to prepare a truth table. A logic circuit may be designed to perform a desired operation by first writing the Boolean equation and then selecting appropriate gates to fulfill each part of the equation. Complicated Boolean equations can be simplified by the use of certain Boolean algebra rules. The simplified equation results in a simpler logic circuit. *NAND* gates and *NOR* gates can each simulate the function of any other type of gate. Complete logic circuits can be constructed using only *NAND logic* or *NOR logic*.

11-1 BASIC LOGIC GATE COMBINATIONS

The logic circuit examples in Chapter 10 each involved only a single logic gate. Most practical logic circuits use many interconnected gates. To commence the study of multi-gate logic circuits, first consider some simple examples of two-gate circuits.

EXAMPLE 11-1

The elevator control circuit in Figure 10-10 is to be modified to permit a maintenance technician to start the motor regardless of the condition of the other switches. Determine the necessary modification.

Solution Referring to Figure 10-10, the operation of the circuit can be described by the following statement:

The motor starts when a *high* input is provided at terminals *A and B and C and D and E and F and G*

Obviously, this statement shows that an *AND* gate is required for the original circuit.

When an additional switch is included, an additional input terminal *H* must be provided. Now, to meet the original requirement, and also to provide for the maintenance technician's switch, the circuit statement must be modified as follows:

The motor starts when a *high* input is provided at terminals (*A and B and C and D and E and F and G*) *or H*.

This statement shows that an *AND* gate and an *OR* gate are required.

The modified circuit is shown in Figure 11-1. Closing switch *H* provides a *high* input to the *OR* gate, resulting in a *high* output to the motor controller and the subsequent starting of the motor. The condition of the other switches has no effect on this operation. Also, with switch *H* left open, all of the other switches must be closed before the motor can start.

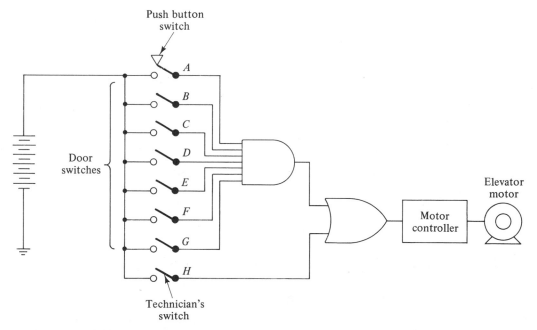

Figure 11-1   Modification to the elevator control circuit in Figure 10-10, to provide a maintenance technician's switch *H* which starts the motor regardless of the condition of the other switches.

EXAMPLE 11-2

The boiler control circuit in Figure 10-13 is to be modified to provide safety functions. The boiler is to switch *off* if the water temperature exceeds a prescribed maximum. A second water thermostat is included to detect the upper temperature limit. This produces a *high* output level while the water temperature is below the maximum. Also, the boiler is to switch *off* if the quantity of water is below a minimum safe level. A water level transducer is included, and this produces a *high* output while the water level remains above the minimum. Make the necessary circuit modifications.

Solution   The statement which describes the operation of the circuit in Figure 10-13 is:

The boiler starts when a *high* input is provided at terminals *A or B*

When the two additional inputs are included, the circuit statement must be modified as follows:

The boiler starts when a *high* input is provided at terminals (*A or B*) *and at C and D.*

It is clear that an *AND* gate and an *OR* gate are required.

The modified circuit is shown in Figure 11-2. Inputs *A* (from the low water temperature detector) or *B* (from the air thermostat) switch *on* the boiler only when inputs *C* and *D* are also present. If the water temperature exceeds the maximum, input *C* will go *low*, causing the *AND* gate output to fall, and consequently switching the boiler *off*. Similarly, if the water level falls below the prescribed minimum, a *low* output is produced by the water level detector (input *D*). This again results in a *low* output from the *AND* gate, and in the boiler being switched *off*.

Examples 11-1 and 11-2 demonstrate relatively simple logic circuits, each involving two different types of gates. For more complex circuits, the Boolean equation for the complete logic circuit must first be derived. The circuit is then designed to suit the equation.

11-2 BOOLEAN EQUATIONS FOR BASIC GATE COMBINATIONS

Combination of *AND* and *OR* Gates

Consider Figure 11-3, which shows the output of an *AND* gate connected to one input of an *OR* gate. The output from the *AND* gate is $B \cdot C$. Thus, the inputs to the *OR* gate are *A* and $B \cdot C$. The final output from the *OR* gate is

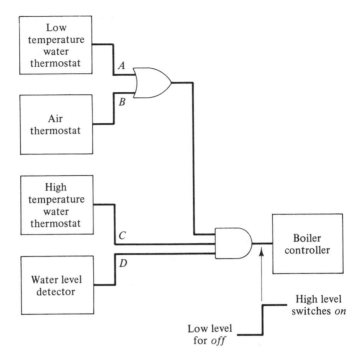

Figure 11-2 Modification to the boiler control circuit in Figure 10-13, to provide safety functions. The boiler can be switched *on* only if the water temperature remains below a prescribed maximum and the water level remains above a specified minimum.

$$X = A + B \cdot C \qquad (11\text{-}1)$$

Equation 11-1 describes the function of the complete logic diagram in Figure 11-3. In a similar manner, equations may be derived for all logic diagrams, however complex.

The value of X for any given set of inputs may be determined from Eq. 11-1 by substituting *1* and *0* levels into the equation. Thus, if $A = 1$, $B = 0$ and $C = 1$,

$$X = 1 + 0 \cdot 1$$

$$= 1$$

It is important to remember that the normal algebraic procedures apply when solving Boolean equations, i.e., multiplication is performed before addition. When this procedure is not intended, parentheses must be employed as necessary.

The accuracy of the evaluation of the circuit output from the Boolean equation

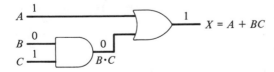

Figure 11-3 Logic system using *AND* and *OR* gate combination. The output of the *AND* gate is $B \cdot C$, giving inputs to the *OR* gate of A and $B \cdot C$. This gives $X = A + BC$.

may be rapidly checked by writing the specified *1* and *0* inputs on the logic circuit diagram. In Figure 11-3, a *1* has been inserted at the *A* input terminal. Similarly, *0* and *1* have been placed at the *B* and *C* inputs, respectively. The output of the *AND* gate is now clearly *0*, since *B* and *C* would both have to be *1* to give a *1* output. The inputs to the *OR* gate are *A* = *1* and *BC* = *0*. The *OR* gate gives a *1* output if any of its inputs are *1*. Therefore, the final output is *X* = *1*, which coincides with the evaluation of Eq. 11-1.

A truth table can be written for the circuit in Figure 11-3 to reveal the output levels for every possible combination of inputs:

A	B	C	X = A + BC
0	0	0	0
0	0	1	0
0	1	0	0
0	1	1	1
1	0	0	1
1	0	1	1
1	1	0	1
1	1	1	1

Another *AND-OR* gate combination is illustrated in Figure 11-4. In this case, the output of an *OR* gate is applied to one input terminal of an *AND* gate. The *OR* gate output is *B* + *C*, and the final output of the *AND* gate is

$$X = A \cdot (B + C) \tag{11-2}$$

Note that, if parentheses were not present in Eq. 11-2, the output would be evaluated as,

$$X = (A \cdot B) + C$$

which would be incorrect.

Putting *A* = 1, *B* = 0, and *C* = 1 into Eq. 11-2,

$$X = 1 \cdot (0 + 1)$$

$$= 1$$

Evaluating the output directly by inserting *A* = 1, *B* = 0, and *C* = *1* into the logic circuit diagram (see Figure 11-4), confirms that the output of the *OR* gate is *1*. This gives *A* = *1* and *B* + *C* = *1* as inputs to the *AND* gate, resulting in *X* = *1* as the final output, which corresponds with the result obtained from the equation.

Figure 11-4 Logic system consisting of an *OR* gate and an *AND* gate. The output of the *OR* gate is *B* + *C*, giving inputs to the *AND* gate of *A* and *B* + *C*. This gives *X* = *A* · (*B* + *C*).

AND, OR, and *NOT* Gate Combinations

Figure 11-5(a) shows an *AND* gate with a *NOT* gate at one of its inputs. The output of the *NOT* gate is $\overline{A}$ when its input is A. This gives $\overline{A}$ and B as inputs to the *AND* gate. Consequently, the final output is

$$X = \overline{A} \cdot B \qquad (11\text{-}3)$$

Suppose that inputs of $A = 1$ and $B = 1$ are applied to the circuit in Figure 11-5(a). Then the two inputs to the *AND* gate are $\overline{A} = 0$ and $B = 1$. Thus, the output is

$$X = \overline{A}B = 0 \cdot 1 = 0$$

When $A = 0$ and $B = 1$, $\overline{A} = 1$. Thus, the *AND* gate inputs are now $\overline{A} = 1$ and $B = 1$, resulting in

$$X = 1 \cdot 1 = 1$$

A ———$\triangleright\!\!\circ$—$\overline{A}$—$\boxed{\text{AND}}$——— $X = \overline{A} \cdot B$
B ———————————

(a) *AND* and *NOT* gate combination

A ———$\triangleright\!\!\circ$—$\overline{A}$—$\boxed{\text{OR}}$——— $X = \overline{A} + B$
B ———————————

(b) *NOR* and *NOT* combination

Figure 11-5 *AND* gate and *OR* gate, with *NOT* gates inverting one input of each. The inverted input is changed from A to $\overline{A}$.

In Figure 11-5(b), and *OR* gate with a *NOT* gate inverting one of its inputs is illustrated. The *OR* gate inputs are seen to be $\overline{A}$ and B. The output is

$$X = \overline{A} + B \qquad (11\text{-}4)$$

The truth table for Eq. 11-4 is

A	B	$X = \overline{A} + B$
0	0	1
0	1	1
1	0	0
1	1	1

11-3 ANALYSIS OF LOGIC CIRCUITS

All logic circuit diagrams, however complex, are analyzed in the same step-by-step way discussed in Section 11-2. Starting at the input terminals, the output equation for each gate is evaluated to determine the inputs to subsequent gates. The procedure is repeated for each stage, until the equation for the final output is obtained. Using the equation for the complete circuit, the output level can be determined for any given set of inputs. The accuracy of this evaluation may be checked by inserting the specified *1* and *0* levels at the inputs on the logic circuit diagram. The resultant output of each gate is then determined, and the appropriate *1* and *0* levels are inserted, until the final output level is obtained. As well as being employed to evaluate the output for a given set of inputs, the Boolean equation for the logic circuit may be used to write a truth table, to show the circuit output for all possible input combinations.

EXAMPLE 11-3

Write the Boolean equation for the output from the logic circuit shown in Figure 11-6(a). Then evaluate X when $A = 1$, $B = 0$, $C = 1$, and $D = 1$. Also, determine the output of the system by inserting appropriate *1* and *0* levels throughout the diagram.

Solution

Deriving the circuit equation

Refer to Figure 11-6(b)

Output of *NOT* gate: $\overline{A}$
Output of *AND* gate: $B \cdot C \cdot D$
Output of *OR* gate: $X = \overline{A} + B \cdot C \cdot D$

For the specified inputs

$$X = \overline{A} + B \cdot C \cdot D$$
$$= \overline{1} + 0 \cdot 1 \cdot 1$$
$$= 0 + 0 = 0$$

Using 1 and 0 levels throughout the diagram

Refer to Figure 11-6(c)

Output of *NOT* gate: $\overline{A} = \overline{1} = 0$
Output of *AND* gate: $B \cdot C \cdot D = 0 \cdot 1 \cdot 1 = 0$
Output of *OR* gate: $X = 0 + 0 = 0$

EXAMPLE 11-4

Write the Boolean equation for the logic diagram in Figure 11-7. Also, write a truth table for the circuit.

Solution

Deriving equation

Output of gate 1: $\overline{BC}$
Output of gate 2: $\overline{B + C}$
Output of gate 3: $A + \overline{BC}$
Output of gate 4: $X = (A + \overline{BC}) \cdot (\overline{B + C})$

Truth table

A	B	C	$X = (A + \overline{BC}) \cdot (\overline{B} + C)$
0	0	0	1
0	0	1	0
0	1	0	0
0	1	1	0
1	0	0	1
1	0	1	0
1	1	0	0
1	1	1	0

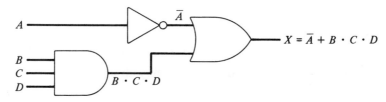

(a) Logic circuit

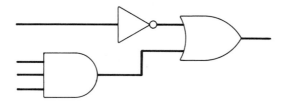

$$X = \overline{A} + B \cdot C \cdot D$$

(b) Inputs and outputs identified

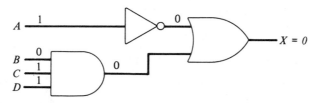

$$X = 0$$

(c) *1* and *0* levels inserted

Figure 11-6 Logic diagram for Example 11-3.

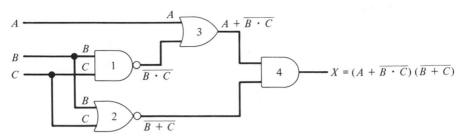

Figure 11-7 Logic diagram for Example 11-4.

11-4 DESIGNING LOGIC CIRCUITS

Logic circuit designers may be presented with a Boolean equation for which a logic circuit is to be designed. Alternatively, a problem may be presented, requiring a truth table to be prepared and an equation to be derived before the circuit design can commence.

As demonstrated in Section 11-3, logic circuit analysis is performed by working from the input to the output. For logic circuit design, the reverse is true: first, the output stage is determined from the Boolean equation, and then, working from output to inputs, gates for each preceding stage are selected.

EXAMPLE 11-5

Design a logic circuit to give an output $X = (\overline{A\,B} + \overline{A}\,C) \cdot (\overline{A\,D + C})$

Solution   The output stage is an *AND* gate with inputs

$$(\overline{A\,B} + \overline{A}\,C) \text{ and } (\overline{A\,D + C}) \qquad\qquad \text{[Figure 11-8(a)]}$$

$(\overline{A\,B} + \overline{A}\,C)$ is obtained from an *OR* gate with inputs

$$\overline{A\,B} \text{ and } \overline{A}\,C \qquad\qquad \text{[Figure 11-8(b)]}$$

$(\overline{A\,D + C})$ is obtained from a *NOR* gate with inputs

$$A\,D \text{ and } C \qquad\qquad \text{[Figure 11-8(b)]}$$

$\overline{A\,B}$ is obtained from a *NAND* gate with inputs

$$A \text{ and } B \qquad\qquad \text{[Figure 11-8(c)]}$$

$\overline{A}\,C$ is obtained from an *AND* gate with inputs

$$\overline{A} \text{ and } C \qquad\qquad \text{[Figure 11-8(c)]}$$

$A\,D$ is obtained from an *AND* gate with inputs

$$A \text{ and } D \qquad\qquad \text{[Figure 11-8(c)]}$$

$\overline{A}$ is obtained from A by use of a *NOT* gate   [Figure 11-8(c)]
The complete logic circuit is now as illustrated in Figure 11-8(c).

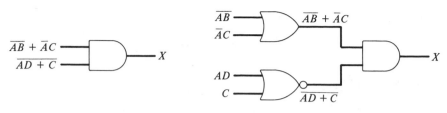

(a) Output gate   (b) Output and next-to-output gates

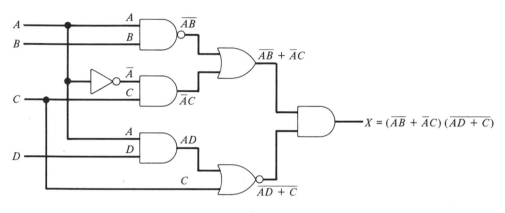

(c) Complete logic circuit

Figure 11-8 Development of a logic circuit for the equation $X = (\overline{AB} + \overline{A}C) \cdot (\overline{AD + C})$. (Illustration for Example 11-5.)

EXAMPLE 11-6

An electric power generating station supplies current to three loads. While any one load is switched *on*, only a single generator is required to operate. If more than one load is *on*, an auxiliary generator must be started. Design a logic circuit to provide an output signal to start the auxiliary generator.

Solution

Boolean equation

The three loads are A, B, and C. When a *1* is received from any two or all three of the loads, a *1* output must be produced by the logic circuit.
Therefore,

$$X = 1 \text{ when } A\,B\,C = 1$$

giving $X = A\,B\,C$

Also, $X = 1 \text{ when } A\,B = 1 \text{ and } C = 0$

giving $X = A\,B\,\overline{C}$

and $X = A\,\overline{B}\,C$

and $\quad\quad\quad\quad\quad\quad X = \overline{A}\,B\,C$

giving $\quad\quad\quad\quad\quad X = A\,B\,C + \overline{A}\,B\,C + A\,\overline{B}\,C + A\,B\,\overline{C}$

Logic circuit

The output stage is an *OR* gate with inputs of

$$A\,B\,C, \overline{A}\,B\,C, A\,\overline{B}\,C, A\,B\,\overline{C} \quad\quad\quad\text{[Figure 11-9(a)]}$$

$A\,B\,C$ is obtained from an *AND* gate with inputs

$$A, B, \text{ and } C \quad\quad\quad\text{[Figure 11-9(b)]}$$

$\overline{A}\,B\,C$ is obtained from an *AND* gate with inputs

$$\overline{A}, B, \text{ and } C \quad\quad\quad\text{[Figure 11-9(b)]}$$

$A\,\overline{B}\,C$ is obtained from an *AND* gate with inputs

$$A, \overline{B}, \text{ and } C \quad\quad\quad\text{[Figure 11-9(b)]}$$

$A\,B\,\overline{C}$ is obtained from an *AND* gate with inputs

$$A, B, \text{ and } \overline{C} \quad\quad\quad\text{(Figure 11-9(b))}$$

$\overline{A}, \overline{B},$ and $\overline{C}$ are obtained from $A, B,$ and C by the use of *NOT* gates
$$\text{[Figure 11-9(c)]}$$

The complete logic circuit is now as illustrated in Figure 11-9(c).

Example 11-6 is sometimes described as *the three judges*, requiring an output signal when a majority of the judges agree upon a decision. The circuit in Figure 11-9 can be further simplified, as discussed in Section 11-5.

EXAMPLE 11-7

An automatic door-opener for a garage is to operate on receipt of a radio signal with a four-bit code. The decoding circuit is to be programmable, so that it can be set to give an output for any predetermined combination of inputs. Design a suitable logic circuit.

Solution The equations for the circuit are $X = A\,B\,C\,D$, $X = \overline{A\,B\,C\,D}$, and any combination of inputs between these two extremes.

The output stage of the circuit has to be an *AND* gate with four inputs, selectable from

$$A \text{ or } \overline{A}, B \text{ or } \overline{B}, C \text{ or } \overline{C}, \text{ and } D \text{ or } \overline{D} \quad\quad\text{(Figure 11-10)}$$

To make A or $\overline{A}$ selectable, a *NOT* gate and switch are required, as illustrated in Figure 11-10. Similarly, a *NOT* gate and switch are required for each of the other input stages, so that B or $\overline{B}$, C or $\overline{C}$, and D or $\overline{D}$, may be selected.

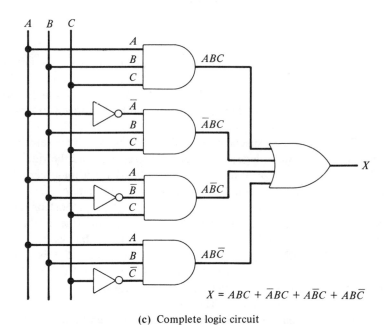

(a) Output gate

(b) Output and next-to-output gates

(c) Complete logic circuit

$$X = ABC + \overline{A}BC + A\overline{B}C + AB\overline{C}$$

Figure 11-9 Development of a logic circuit to suit the equation $X = ABC + \overline{A}BC + A\overline{B}C + AB\overline{C}$. When any two, or all three, of A, B, and C are I, a I output is produced. (Illustration for Example 11-6.)

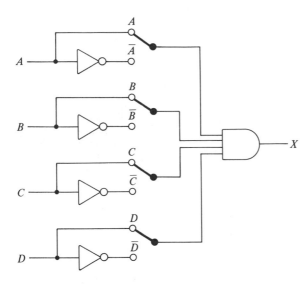

Figure 11-10 Decoding circuit for an automatic garage door-opener. The circuit can be programmed to respond to a predetermined code, selectable from any combination of inputs. (Logic circuit for Example 11-7.)

11-5 SIMPLIFYING LOGIC CIRCUIT DESIGNS

Rules for Boolean Algebra

The basic rules of ordinary algebra also apply to Boolean algebra. For example, the expression $A\ B\ C$ can be written as $B\ A\ C$, or as $C\ B\ A$. Similarly, $A + B + C$ can be written in any order. Also, for $A(B + C)$, the A can be multiplied through to give $A\ B + A\ C$. Alternatively, the latter expression can be factorized to give $A(B + C)$, if this form is found to be most convenient.

There are several special rules for Boolean algebra, however, that are quite different from those of ordinary algebra. Together, these rules and the rules of ordinary algebra can be applied to simplify the Boolean equations for a logic circuit. Simplification of the equation results in a simpler, less expensive logic circuit.

Consider the twelve rules listed in Figure 11-11. Most of these are justified by the logic gates shown alongside each rule. Thus, for Rule 1, the corresponding gate shows that when A and 0 are ANDed together, the output is 0. Also, A and 1 ANDed together give A (Rule 2). Rule 3 states that an output of A is also obtained when A and A are ANDed together. Look at the AND gate for Rule 3. If A is *high*, the two inputs to the gate are *high*, and a *high* output results. If A is *low*, the gate inputs are both *low*, giving a *low* output. Thus $A\ A = A$.

It is seen that, in any Boolean equation, $A\ 0$ can be replaced with 0, while A can be substituted for both $A\ 1$ and $A\ A$.

Rules 4 through 8 in Fig. 11-11 are all justified by the logic gates shown alongside them. Rule 9 is proved by the use of Rule 6, while a truth table is employed to prove Rule 10. Rules 11 and 12 are known as *De Morgan's Theorem*, the truth of which is demonstrated in Figures 11-12 and 11-13.

(1) $A0 = 0$

(2) $A1 = A$

(3) $AA = A$

(4) $A\bar{A} = 0$

(5) $A + 0 = A$

(6) $A + 1 = 1$

(7) $A + A = A$

(8) $A + \bar{A} = 1$

(9) $A + AB = A$
$A + AB = A(1 + B)$
$\quad = A(1)$, see (6) above
$\quad = A$

(10) $A + \bar{A}B = A + B$

A	B	$A + B$	$A + \bar{A}B$
0	0	0	0
0	1	1	1
1	0	1	1
1	1	1	1

(11) $\overline{AB} = \bar{A} + \bar{B}$ $\left.\rule{0pt}{2.5em}\right\}$ DeMorgan's theorem

(12) $\overline{A\bar{B}} = \bar{A} + B$

Figure 11-11   Rules for Boolean algebra. These may be applied to simplify the Boolean equation for a logic circuit, resulting in a simplification of the circuit.

Figure 11-12(a) shows a two-input *NAND* gate and its truth table. A two-input *OR* gate with its inputs inverted (signified by the input bubbles) is shown, together with its truth table, in Figure 11-12(b). The truth table for the *NAND* gate has already been discussed (Section 10-8). The *OR* gate with its inputs inverted is new. If it were an ordinary *OR* gate, its output would be $X = A + B$. But because its inputs are inverted, its output equation is $X = \bar{A} + \bar{B}$. When A and B are *0*, $\bar{A}$ and $\bar{B}$ both equal

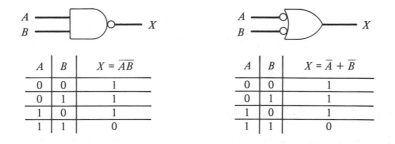

A	B	$X = \overline{AB}$
0	0	1
0	1	1
1	0	1
1	1	0

A	B	$X = \bar{A} + \bar{B}$
0	0	1
0	1	1
1	0	1
1	1	0

(a) Gate and truth table for $X = \overline{AB}$ **(b)** Gate and truth table for $X = \bar{A} + \bar{B}$

Figure 11-12   Proof of De Morgan's theorem that $\overline{AB} = \bar{A} + \bar{B}$. The identical truth tables for $X = \overline{AB}$ and $X = \bar{A} + \bar{B}$ demonstrate the equality.

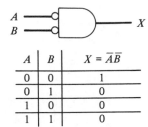

(a) Gate and truth table for $X = \overline{A}\overline{B}$ (b) Gate and truth table for $X = \overline{A + B}$

Figure 11-13 Proof of De Morgan's theorem that $\overline{A}\,\overline{B} = \overline{A + B}$. Once again, the identical truth tables for $X = \overline{A}\,\overline{B}$ and $X = \overline{A + B}$ demonstrate the equality.

1, and $X = 1$, as illustrated in the truth table. When $A = 1$ and $B = 1$, $\overline{A} = 0$ and $\overline{B} = 0$, giving $X = 0$. Finally, when either A or B is 0, its inverse is 1, and consequently, the *OR* gate output is $X = 1$.

Now compare the two truth tables in Figure 11-12. They are identical. Therefore, $X = \overline{A}\overline{B}$ and $X = \overline{A} + \overline{B}$ can be substituted for each other. So $\overline{A}\overline{B} = \overline{A} + \overline{B}$, as stated by Rule 11.

Two more logic gates and their truth tables are illustrated in Figure 11-13. Figure 11-13(a) shows an *AND* gate with its inputs inverted, together with its truth table. In Figure 11-13(b), a *NOR* gate and its truth table are shown. Once again, the two truth tables are identical, showing that $\overline{A}\,\overline{B} = \overline{A + B}$, as stated by Rule 12.

Although gates with only two input terminals have been considered in the proof of De Morgan's theorem, it can be demonstrated that Rules 11 and 12 can be extended to cover gates with more than two inputs. Thus, by extension

$$\overline{A}\,\overline{B}\,\overline{C} = \overline{A + B + C}$$

and

$$\overline{A}\,\overline{B}\,\overline{C} = \overline{A} + \overline{B} + \overline{C}$$

Application of Rules 11 and 12 to the Boolean equation for a logic circuit is sometimes termed *De Morganizing*. Two simple memory aids for De Morganizing are

break the bar and change the sign

and

join the bars and change the sign.

For Rule 11, breaking the bar above $\overline{A}\overline{B}$ and changing the sign gives $\overline{A} + \overline{B}$. Alternatively, joining the bars above $\overline{A} + \overline{B}$ and changing the sign produces $\overline{A}\overline{B}$. Rule 12 can be explained in a similar way.

Logic Circuit Simplification

EXAMPLE 11-8

Simplify the Boolean equation $X = (A + \overline{B}\,\overline{C}) \cdot (\overline{B + C})$ from Example 11-4. Design a new logic circuit from the simplified equation.

Solution

$$X = (A + \overline{B\,C})(\overline{B + C})$$

From Rule 11 in Figure 11-11,

$$\overline{B\,C} = \overline{B} + \overline{C}$$

and from Rule 12, $\overline{B + C} = \overline{B}\,\overline{C}$

Substituting into (1) yield:

$$X = (A + \overline{B} + \overline{C})(\overline{B}\,\overline{C})$$
$$= A\,\overline{B}\,\overline{C} + \overline{B}\,\overline{B}\,\overline{C} + \overline{B}\,\overline{C}\,\overline{C}$$

From Rule 3 in Fig. 11-11,

$$\overline{B}\,\overline{B} = \overline{B}$$

and $\overline{C}\,\overline{C} = \overline{C}$

Therefore, $X = A\,\overline{B}\,\overline{C} + \overline{B}\,\overline{C} + \overline{B}\,\overline{C}$

From Rule 7, $\overline{B}\,\overline{C} + \overline{B}\,\overline{C} = \overline{B}\,\overline{C}$

giving $X = A\,\overline{B}\,\overline{C} + \overline{B}\,\overline{C}$
$$= \overline{B}\,\overline{C}(A + 1)$$

From Rule 6, $A + 1 = 1$

Therefore, $X = \overline{B}\,\overline{C}$

From Rule 12, $\overline{B}\,\overline{C} = \overline{B + C}$

Therefore, $X = \overline{B + C}$

This result suggests that the output of the logic circuit for Example 11-4 (the circuit in Figure 11-7) is the same as that from a *NOR* gate with inputs of B and C, and that the condition of A is of no consequence. To check on this, write the truth table for $X = \overline{B + C}$, and compare it to the truth table for $X = (A + \overline{B\,C})$ $(\overline{B + C})$ from Example 11-4.

Truth tables

A	B	C	$X = (A + \overline{B\,C}) \cdot (\overline{B + C})$	$\overline{B}$	$\overline{C}$	$X = \overline{B + C}$
0	0	0	1	1	1	1
0	0	1	0	1	0	0
0	1	0	0	0	1	0
0	1	1	0	0	0	0
1	0	0	1	1	1	1
1	0	1	0	1	0	0
1	1	0	0	0	1	0
1	1	1	0	0	0	0

The simplified logic circuit, drawn for the Boolean equation $X = \overline{B + C}$, is shown in Figure 11-14. This can be substituted for the circuit in Figure 11-7.

EXAMPLE 11-9

Simplify the Boolean equation $X = A B C + \overline{A} B C + A \overline{B} C + A B \overline{C}$ from Example 11-6. Design a new logic circuit from the simplified equation.

Solution

$$X = A B C + \overline{A} B C + A \overline{B} C + A B \overline{C} \qquad (1)$$

From Rule 7 in Figure 11-11,

$$A + A = A$$
or $$A B C + A B C = A B C$$
and $$A B C + A B C + A B C = A B C$$

Substituting for $A B C$ in Eq. 1, and rearranging the equation,

$$X = (A B C + \overline{A} B C) + (A B C + A \overline{B} C) + (A B C + A B \overline{C})$$
$$= B C(A + \overline{A}) + A C(B + \overline{B}) + A B(C + \overline{C})$$

From Rule 8 in Figure 11-11,

$$A + \overline{A} = 1, B + \overline{B} = 1, \text{ and } C + \overline{C} = 1$$

Therefore $$X = B C + A C + A B$$
or $$X = A B + A C + B C$$

Write the truth tables for $X = A B C + \overline{A} B C + A \overline{B} C + A B \overline{C}$ and $X = A B + A C + B C$ to check that they both give the same result for any combination of inputs.

Truth tables

A	B	C	$X = ABC + \overline{A}BC + A\overline{B}C + AB\overline{C}$	$X = AB + AC + BC$
0	0	0	0	0
0	0	1	0	0
0	1	0	0	0
0	1	1	1	1
1	0	0	0	0
1	0	1	1	1
1	1	0	1	1
1	1	1	1	1

A new, simpler logic circuit can now be designed from the simplified equation. Instead of three *NOT* gates, four three-input *AND* gates and a four-input *OR* gate, as in Figure 11-9, only three two-input *AND* gates and a three-input *OR* gate are required. (See Figure 11-15.) The simplified logic circuit is seen to be less expensive than the original circuit.

Figure 11-14   The logic circuit for Example 11-8, drawn for the Boolean equation $X = \overline{B + C}$ is simply a *NOR* gate. This can be substituted for the circuit in Figure 11-7.

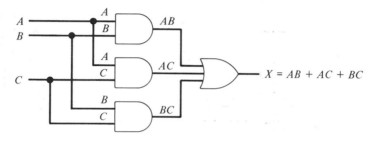

Figure 11-15   Logic circuit for Example 11-9, designed for the Boolean equation $X = A B + A C + B C$. This simplified logic circuit may be substituted for the circuit in Figure 11-9.

11-6 EXCLUSIVE *OR* AND EXCLUSIVE *NOR* GATES

Operation of *XOR* and *XNOR* Gates

It is mentioned in Sections 10-5 and 10-8 that *OR* and *NOR* gates are sometimes referred to as *inclusive OR* and *inclusive NOR* gates, respectively. To discover the origin of these terms, examine the truth tables for *OR* and *NOR* gates reproduced in Figure 11-16(a). Note that a *1* output is generated by the *OR* gate when both inputs are *1*, as well as in the cases when one of the inputs is *1* and the other is *0*. The *OR* gate operation is said to *include* the condition of both inputs being equal to *1*. Similarly, the *NOR* gate produces a *0* output (inverse of the *OR*) when both inputs are *1*, the same as when one input is equal to *1* and the other is equal to *0*. Thus, the *NOR* gate operation *includes* the condition of both inputs being equal to *1*. Hence the use of the term *inclusive*.

Now consider the gates and truth tables shown in Figure 11-16(b). The *exclusive OR* gate (*XOR* gate) responds to inputs just like an *OR* gate, with one exception: the input condition of A = *1* and B = *1*. In this case, the *XOR* gate produces a *0* output, while the *OR* gate gives a *1* output. Therefore, the condition A = *1* and B = *1* is said

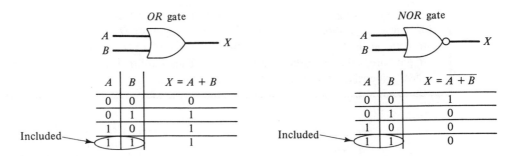

(a) *Inclusive OR* and *inclusive NOR* gates and truth tables

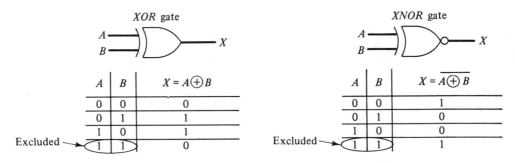

(b) *Exclusive OR* and *exclusive NOR* gates and truth tables

Figure 11-16 Logic symbols and truth tables for *inclusive* and *exclusive OR* and *NOR* gates. The exclusive *OR* (*XOR*) gate and exclusive *NOR* (*XNOR*) gate *exclude* the condition $A = 1$ and $B = 1$.

to be *excluded* by the operation of the *XOR* gate. Similarly, the *XNOR* gate responds to inputs like a *NOR* gate, except when $A = 1$ and $B = 1$. Here again, the output is the same as when $A = 0$ and $B = 0$. Thus, the condition $A = 1$ and $B = 1$ is *excluded* by the operation of the *XNOR* gate. Hence the use of the term *exclusive*.

Note the graphic symbols for the *XOR* and *XNOR* gates in Figure 11-16(b). Note also that *XOR* and *XNOR* gates are always two-input logic gates: there are no multi-input *XOR* and *XNOR* gates.

As illustrated in Fig. 11-16(b), the output equation of the *XOR* gate is

$$X = A \oplus B \tag{11-5}$$

where the circled plus sign denotes the exclusive-type operation of the gate. For the *XNOR* gate, the output equation is

$$X = \overline{A \oplus B} \tag{11-6}$$

Logic Circuits of *XOR* and *XNOR* Gates

To operate as discussed above, an *XOR* gate has to be composed of several other gates. Using the techniques explored in Section 11-5, the Boolean equation and logic circuit for an *XOR* gate can be readily determined.

EXAMPLE 11-10

Write the Boolean equation for an *XOR* gate, and design a suitable logic circuit.

Solution   The operation of the *XOR* gate may be described as

$$X \text{ equals } (A \text{ or } B), \text{ and not } (A \text{ and } B)$$

which gives $X = (A + B)(\overline{A\,B})$ (11-7)

Logic circuit for Eq. 11-7

Output stage	An *AND* gate with inputs $(A + B)$ and $(\overline{A\,B})$
$(A + B)$	Obtained from an *OR* gate with inputs A and B
$\overline{A\,B}$	Obtained from a *NAND* gate with inputs A and B

The circuit for $X = (A + B)(\overline{A\,B})$ is now as illustrated in Figure 11-17(a). Equation 11-7 can be simplified by the use of some of the Boolean algebra rules listed in Figure 11-11:

$$X = (A + B)(\overline{A\,B})$$
$$= A\,\overline{A\,B} + B\,\overline{A\,B}$$
$$= A(\overline{A} + \overline{B}) + B(\overline{A} + \overline{B})$$
$$= A\,\overline{A} + A\,\overline{B} + B\,\overline{A} + B\,\overline{B}$$
$$= 0 + A\,\overline{B} + B\,\overline{A} + 0$$

or $X = A\,\overline{B} + B\,\overline{A}$ (11-8)

Logic circuit for Eq. 11-8

Output stage	An *OR* gate with inputs $A\,\overline{B}$ and $B\,\overline{A}$
$A\,\overline{B}$	Obtained from an *AND* gate with inputs A and $\overline{B}$
$B\,\overline{A}$	Obtained from an *AND* gate with inputs B and $\overline{A}$
$\overline{A}$ and $\overline{B}$	Obtained from A and B by the use of *NOT* gates

The circuit for $X = A\,\overline{B} + B\,\overline{A}$ is illustrated in Figure 11-17(b).

Equation 11-8 is usually offered to define the operation of an *XOR* gate. The logic circuit designed for this equation [Figure 11-17(b)] is also usually shown as the circuit of the *XOR* gate.

Boolean equations for the *XNOR* gate can be derived from the equations for the

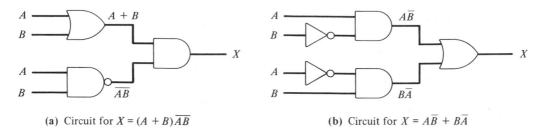

(a) Circuit for $X = (A + B)\overline{AB}$ (b) Circuit for $X = A\overline{B} + B\overline{A}$

Figure 11-17 Two possible logic circuits for an *XOR* gate. (Illustration for Example 11-10.)

XOR gate, and two possible logic circuits can also be designed. The equation usually offered to define the *XNOR* gate is

$$X = \overline{A}\,\overline{B} + A\,B \qquad (11\text{-}9)$$

Applications of *XOR* and *XNOR* Gates

One application of an *XOR* gate is illustrated in Figure 11-18. Input *A* is an input signal to be controlled, while input *B* is a control level. With *B* held at *0*, the output is always $X = A$. If $A = 1$, then $X = 1$, and if $A = 0$, $X = 0$. [See the *XOR* truth table in Figure 11-16(b).] With *B* held at *1*, the output is inverted to $X = \overline{A}$. (See the table again.) The circuit is known as a *controlled inverter* or *controlled complementer*.

An *XNOR* gate is commonly applied as a *digital comparator* or *coincidence gate*. Referring to the *XNOR* gate truth table in Figure 11-16(b), note that a *1* output is obtained when $A = B$, regardless of whether *A* and *B* are *0* or *1*.

Now look at Figure 11-19, which shows two binary numbers, $A\,B\,C$ and $D\,E\,F$, applied as inputs to a logic circuit. The logic circuit consists of three *XNOR* gates and one *AND* gate. When $A\,B\,C$ and $D\,E\,F$ are equal, as illustrated, each *XNOR* gate has equal inputs at its two terminals. Consequently, each *XNOR* gate produces as *1* output. With all three *XNOR* gates having *1* outputs, the *AND* gate output is $X = 1$. When the binary numbers $A\,B\,C$ and $D\,E\,F$ are not equal, at least one of the *XNOR* gates has unequal input levels. This results in an *XNOR* gate output of *0*, and in a final *AND* gate output of $X = 0$. Thus, this logic circuit compares two binary numbers, producing a *1* output when the numbers are equal, and a *0* output when the numbers are unequal.

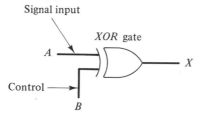

Figure 11-18 Use of an *XOR* gate as a *controlled inverter*. When $B = 0$, $X = A$; when $B = 1$, $X = \overline{A}$.

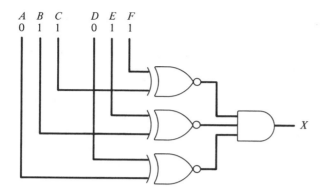

A B C   D E F
0 1 1   0 1 1

Figure 11-19 Use of *XNOR* gates in a digital comparator circuit. A *1* output is produced when $A B C = D E F$. When the input numbers are unequal, the output is 0.

X

11-7 GATE CONVERSIONS: UNIVERSAL BUILDING BLOCKS

Reducing and Expanding Gate Inputs

For some applications, an available gate may have more input terminals than are required. This is easily taken care of in the case of an *OR* gate by grounding the unwanted inputs, as illustrated in Figure 11-20(a). The *OR* gate would function correctly if the spare input terminals were simply left unconnected; however, this might facilitate pickup of noise spikes. For an *AND* gate to function correctly with more input terminals than

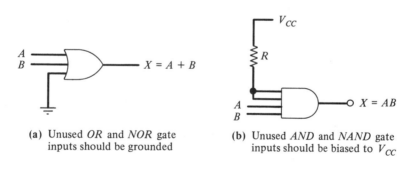

(a) Unused *OR* and *NOR* gate inputs should be grounded

(b) Unused *AND* and *NAND* gate inputs should be biased to V_{CC}

$X = A + B$

V_{CC}

R

$X = AB$

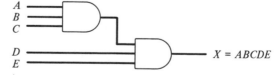

(c) Several *AND* gates (or several *OR* gates) may be connected together to function as a single gate

$X = ABCDE$

Figure 11-20 Unused input terminals on an *OR* gate should be grounded to prevent pickup of noise spikes. Spare input terminals on an *AND* gate must be connected to V_{CC}. Two or more gates may be combined to function as a single gate with an increased number of input terminals.

are required, the unused terminals must have a *1* input. As shown in Figure 11-20(b), this is done by connecting the terminals to the logic circuit supply voltage via a resistor. The value of the resistor depends upon the type of logic gates used, i.e., the actual hardware. (See Chapter 12.) *NOR* and *NAND* gates with spare input terminals should be treated in the same way as *OR* and *AND* gates, respectively.

When the available gates have too few terminals for a particular application, several gates may be combined to operate as a single gate. Depending upon the type of logic gates used (see Chapter 12), it might be possible to connect the output terminals together. Otherwise, *AND* and *OR* gates may be connected as illustrated in Figure 11-20(c). Two three-input gates are shown connected to function as a single gate with five input terminals.

NAND Gate Conversions

NAND gates can be connected to function as any other kind of gate. Consider the two arrangements shown in Figure 11-21(a). In one case, a two-input *NAND* gate has its input terminals connected together, and an input signal *A* is applied. When *A* is *1*,

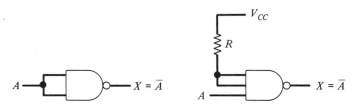

(a) A *NAND* gate can be made to function as a *NOT* gate by connecting all inputs together, or by biasing all but one input to V_{CC}

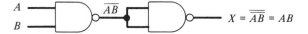

(b) Two *NAND* gates may be connected to function as an *AND* gate

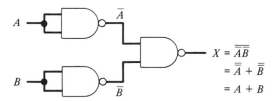

(c) Three *NAND* gates connected to function as an *OR* gate

Figure 11-21 *NAND* gates may be connected to function as *NOT*, *AND*, and *OR* gates, allowing logic circuits to be constructed entirely of *NAND* gates.

both input terminals are at *1*, and the output is *0*. When *A* is *0*, the output is *1*. Obviously, the output is always $X = \overline{A}$. Thus, the *NAND* gate is connected to function as a *NOT* gate. The second part of Figure 11-21(a) shows a *NAND* gate with all but one of its input terminals biased to V_{CC}. A level *1* is applied constantly at these inputs. The remaining input terminal is treated as the *NOT* gate input.

Figure 11-21(b) shows one *NAND* gate followed by another one which is connected to function as a *NOT* gate. The output of the first *NAND* gate is inverted by the second, so that the combination operates as an *AND* gate.

In Figure 11-21(c), two *NAND* inverters are connected at the inputs of another *NAND* gate. The inputs to the third gate are $\overline{A}$ and $\overline{B}$, giving an output of $X = A + B$, as illustrated. It is seen that this combination of three *NAND* gates functions as an *OR* gate.

The advantage of being able to use *NAND* gates to perform any logic function, is that a company in the business of developing or manufacturing logic circuits need stock only one type of logic gate. Thus, *NAND* gates are known as *universal building blocks*, and circuits constructed entirely of *NAND* gates are said to use *NAND logic*.

Logic circuits are first designed in the usual way, employing all types of gates necessary. Then, the conversion to *NAND* logic is made by replacing each gate with its *NAND* equivalent. When this is done, several occurrences of two cascaded *NOT* gates are usually found. Since successive *NOT* gates cancel, they should both be deleted from the circuit. This, of course, results in greater simplification and economy.

EXAMPLE 11-11

Convert the circuit of Figure 11-15 to *NAND* logic.

Solution   Replace all three *AND* gates with their *NAND* equivalent [Figure 11-21(b).] Also, replace the *OR* gate with its *NAND* equivalent [Figure 11-21(c).] The result is shown in Figure 11-22(a).

In Figure 11-22(a), there are three pairs of cascaded *NOT* gates which cancel each other out. These can be deleted to give the final *NAND* logic circuit shown in Figure 11-22(b).

To prove that the circuit in Figure 11-22(b) performs like the circuit in Figure 11-15, the Boolean equation is written and then simplified as illustrated.

NOR Gate Conversions

Like *NAND* gates, *NOR* gates can be connected to function as other types of gates, and *NOR* logic can be used instead of *NAND* logic as *universal building blocks*. Figure 11-23 shows *NOR* gates connected to function as *NOT* gates, *OR* gates, and *AND* gates.

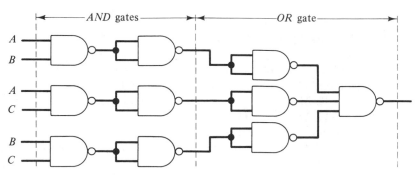

(a) Conversion of Fig. 11-15 to *NAND logic* by substituting the *NAND* equivalent circuit for each gate

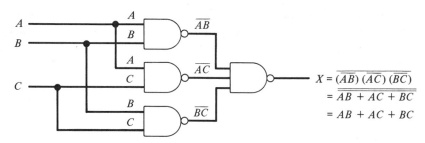

$$X = \overline{(\overline{AB})\,(\overline{AC})\,(\overline{BC})}$$
$$= \overline{\overline{AB} + \overline{AC} + \overline{BC}}$$
$$= AB + AC + BC$$

(b) Deletion of cascaded *NOT* gates further simplifies the logic circuit

Figure 11-22 Conversion of *AND-OR* logic into *NAND* logic. (Illustration for Example 11-11.)

11-8 POSITIVE LOGIC AND NEGATIVE LOGIC

For all of the gates discussed so far, a *high* input level has been treated as the presence of a signal, and a *low* input level has been regarded as the absence of a signal. Gates that operate with such input levels can be termed *positive logic gates*, because they require positive inputs to produce positive outputs. Another way of stating this is: the gates have an *active high input* and an *active high output*.

Consider the *NAND* and *NOR* gates once again, as reproduced in Figure 11-24(a). Recall that the bubble at the output terminal of each symbol indicates that a *NOT* gate is included, to invert the output. When the required *active high* input levels are present, the output is *low*. When the inputs are absent, the output is *high*. Thus, *NAND* and *NOR* gates have *active low* output levels. This is identified by the presence of the bubble.

Suppose that bubbles are placed at the inputs of *AND* and *OR* gates, as illustrated in Figure 11-24(b). This, of course, indicates that the inputs are inverted. Consequently,

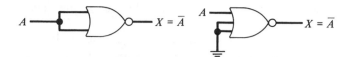

(a) A *NOR* gate can be made to function as a *NOT* gate by connecting all inputs together, or by biasing all but one input to ground

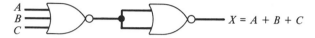

(b) Two *NOR* gates may be connected to function as an *OR* gate

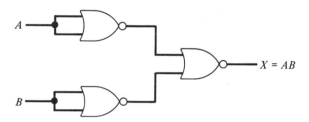

(c) Three *NOR* gates connected to function as an *AND* gate

Figure 11-23   *NOR* gates may be connected to function as *NOT*, *OR*, and *AND* gates, allowing logic circuits to be constructed entirely of *NOR* gates.

high input levels enter the regular portion of the gates (i.e., after the bubbles) as *low* levels. To obtain a *high* output from the *AND* gate illustrated, a *low* input level must be present *both* at terminal *A and* at terminal *B*. In the case of the *OR* gate in Figure 11-24(b), a *high* output is produced when a *low* input is present *either* at terminal *A or*

(a) *NAND* and *NOR* gates have *active high* inputs and *active low* outputs

(b) Gates with *active low* inputs and *active high* outputs

Figure 11-24   The presence of bubbles at the input or output terminals of a logic gate indicates *active low* input or output levels.

at terminal B. These gates may be described as having *active low* inputs and *active high* outputs.

Now consider the logic gate symbols illustrated in Figure 11-25. Figure 11-25(a) shows an *AND* gate with *active low* inputs and *active low* outputs. Because negative-going (from *high* to *low*) inputs and outputs are involved, the gate is termed a *negative-logic AND* gate. The operation of the gate can be described by the statement; a *low* output is produced when a *low* level is present at input A *and* at input B *and* at input C. If a *high* level is present at input A *or* at input B *or* at input C, a *high* output is obtained. Reread the last sentence; it describes a positive-logic *OR* gate [Figure 11-25(b)]. Thus, a positive-logic *OR* gate can behave as a negative-logic *AND* gate. In fact, the two gates are identical, and when a negative-logic *AND* gate is required, a positive-logic *OR* gate is employed.

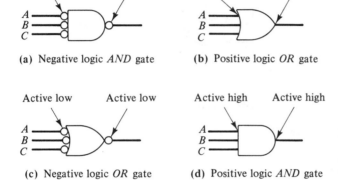

Active low   Active low   Active high   Active high

(a) Negative logic *AND* gate   (b) Positive logic *OR* gate

Active low   Active low   Active high   Active high

(c) Negative logic *OR* gate   (d) Positive logic *AND* gate

Figure 11-25 A negative-logic *AND* gate is identical to a positive-logic *OR* gate. A negative-logic *OR* gate is identical to a positive-logic *AND* gate.

A similar relationship exists between a negative-logic *OR* gate and a positive-logic *AND* gate [Figure 11-25(c) and (d)]. The negative logic *OR* gate (bubbles at inputs and output) produces a *low* output when a *low* level is present at input A *or* at input B *or* at input C. A *high* output is obtained from a negative logic *OR* gate only when a *high* level is present at input A *and* at input B *and* at input C. It is seen that a positive-logic *AND* gate can function as a negative-logic *OR* gate.

REVIEW QUESTIONS AND PROBLEMS

11-1. Modify the logic circuit in Figure 11-3 to give $X = AB + BC$. Briefly explain the modification.

11-2. The logic circuit in Figure 11-4 is to be modified to have the equation $X = A(B + C) + D$. Make the necessary modification and briefly explain.

11-3. Waveforms A, B, and C in Figure 10-9 are applied as inputs to the circuit in Figure 11-4. Determine the output waveform.

11-4. Waveforms A, B, and C in Figure 10-17 are applied as inputs to the circuit in Figure 11-3. Determine the output waveform.

11-5. An audio (beeper) signal is to be activated when the key has been removed from the ignition of an automobile and the headlights are left on. The signal is also to be activated if the key is in the ignition lock and the driver's door is opened. A 1 level is produced by the headlight switch when it is *on*. A 1 is also output from the ignition lock when the key is in the lock, and a 1 is available from the driver's door when it is open. Write the Boolean equation and truth table for this problem.

11-6. Design a logic circuit for Problem 11-5. Insert the specified 1 levels at the circuit inputs, and show that they produce a 1 output.

11-7. A logic circuit is to be designed to produce an output when input E is *low*, and when A and B are *high*, or C and D are *high*. Write the Boolean equation and truth table, and design the circuit.

11-8. The logic circuit in Figure 11-7 is modified by inserting a *NOT* gate at input A and replacing the *AND* gate with an *OR* gate. Write the Boolean equation, simplify it as far as possible, and write the truth table for the new circuit.

11-9. In the logic circuit in Figure 11-8, *OR* gates are substituted in place of the three *AND* gates, and the single *OR* gate is replaced by a *NAND* gate. Write the Boolean equation, simplify it as far as possible, and write the truth table for the new circuit.

11-10. Design a logic circuit to have the Boolean equation $X = (A\,B\,C + B\,C)\,(\overline{B + D})$.

11-11. Design a logic circuit to have the Boolean equation $X = (A\,\overline{B} + B\,\overline{C} + C\,\overline{D})\,(\overline{A}\,B + D\,\overline{C})$.

11-12. An automobile has two seat switches to detect the presence of passengers, and two seat belt switches to detect fastened seat belts. Each switch produces a 1 output when activated. A signal light is to flash when the ignition is switched *on* and any passenger is present without his or her seat belt fastened. Design a suitable logic circuit.

11-13. Write the equations for De Morgan's theorem. Sketch the appropriate gates and write the truth tables to prove the equations.

11-14. Simplify the equation of Problem 11-8 and redesign the circuit. Write the new truth table to check that the performance of the circuit is the same as it was for the original equation.

11-15. Simplify the equation of Problem 11-10 and redesign the circuit.

11-16. Simplify the equation of Problem 11-11 and redesign the circuit.

11-17. Design a logic circuit to have the equation $X = A\,B\,C + A\,\overline{B}\,\overline{C} + \overline{A}\,\overline{B}\,C$. Then, simplify the equation and redesign the circuit.

11-18. Sketch the logic symbol for the *XOR* gate and write the truth table. Explain the difference between the *XOR* and *OR* gates.

11-19. Sketch the logic symbol for the *XNOR* gate and write the truth table. Explain the difference between the *XNOR* and *NOR* gates.

11-20. Sketch logic circuits to show applications of *XOR* and *XNOR* gates. Explain.

11-21. A light emitting diode is to be switched *on* or *off* from two locations. Each time the input at either location changes state from 0 to 1 or from 1 to 0, the light changes state. Write the truth table and design a suitable logic circuit.

11-22. Draw sketches to show how unused gate input terminals should be connected for each type of gate. Also, show how the number of *AND* or *OR* gate input terminals may be increased.

11-23. Show how *NOT*, *AND*, and *OR* gates may be simulated using *NAND* logic. Explain.

11-24. Show how *NOT*, *AND*, and *OR* gates may be simulated using *NOR* logic. Explain.

11-25. Convert the circuit in Figure 11-8(c) to *NAND logic*, and simplify it as far as possible.

11-26. Convert the circuit in Figure 11-8(c) to *NOR* logic, and simplify it as far as possible.

11-27. Convert the circuit in Figure 11-4 to *NAND* logic and to *NOR logic*. Select the most suitable of the two.

11-28. Using illustrations, explain *positive logic* and *negative logic*.

Chapter 12

Integrated Circuit Logic Gates

INTRODUCTION

Logic gates may be constructed of diodes and transistors, combinations of bipolar transistors, combinations of field effect transistors, etc. Two major integrated circuit logic types are *transistor transistor logic* (TTL), which is subdivided into several families, and *complementary MOSFET logic* (CMOS). Other important logic types are: *diode transistor logic* (DTL), *emitter-coupled logic* (ECL), and *integrated injection logic* (I^2L). The most important characteristics of TTL and CMOS are that TTL gates switch very fast, while CMOS has very low power dissipation. Other characteristics of logic gates are supply voltage, input and output voltage levels, input and output current levels, noise immunity, and the number of gate inputs that can be supplied by one gate output. Usually, a single type of logic gate is used throughout a logic system, but sometimes different types of logic have to be interfaced.

12-1 LOGIC GATE PERFORMANCE FACTORS

Supply Voltage

Most integrated circuit logic gates are designed to operate with a supply of 5 V. Some types of gates can use a higher supply voltage, and one type (CMOS) can operate with a supply as low as 1 V. Where a V_{CC} of 5 V is specified, the actual voltage must usually be within ± 0.25 V of 5 V for reliable gate operation.

Input Voltages

Each type of IC logic gate has a *minimum high-input voltage*, regarded as the minimum voltage required to represent a logic *1* input level. This voltage is designated $V_{IH(\text{min})}$. If an input of 2 V or higher will cause the output of a given type of gate to change state, but any level less than 2 V may not change the output, then $V_{IH(\text{min})} = 2$ V.

Figure 12-1(a) shows the circuit of a *resistor transistor logic* (RTL) *NOR* gate. The minimum input necessary to switch Q_1 from *off* into saturation, and thus cause the output to go from *high* to *low*, is

$$V_{IH(\text{min})} = V_{BE} + I_{IH}\, R_B$$

where input current I_{IH} is large enough to drive the transistor into saturation.

Each type of gate also has a *maximum low-input voltage*, $V_{IL(\text{max})}$, which is the highest level acceptable as a logic *0* input. The maximum *low*-input level for the circuit of Figure 12-1(a) must be much less than the V_{BE} level that biases the transistor *on*. In this case, with silicon transistors, $V_{IL(\text{max})}$ might be around 0.2 V.

Input and Output Currents

The level of current at a gate input terminal normally depends upon whether the gate input voltage is *high* or *low*. The *high-level input current* is designated I_{IH}, and the *low-level input current* is I_{IL}. The actual input current levels obviously depend upon the gate circuitry.

For the RTL *NOR* gate in Figure 12-1(a), the minimum input current when the input voltage is *high* (I_{IH}), must be high enough to cause the transistor to saturate. This might typically be 1 mA or more. In the case of the diode *AND* gate in Figure 12-1(b), I_{IH} is the leakage current of the reverse-biased diode, which is likely to be less than 1 μA.

The low-level input current I_{IL} in Figure 12-1(a) occurs when Q_2 is *off*. Thus, it is a junction leakage current with a level less than 1 μA. For the gate in Figure 12-1(b), I_{IL} is the current through a forward-biased diode, so it might typically be 1 mA. This current actually flows out of the input terminal, as illustrated. On a logic gate data sheet, it would be listed as a negative quantity.

The output current from a gate is specified as a *high-level output current* I_{OH} or a *low-level output current* I_{OL}, depending upon whether the gate output voltage is *high* or *low*. The high-level output current is usually a positive quantity: the current flows out of the gate output terminal. [See Figure 12-2(a).] In this case, the gate is said to *source* the output current, and the gate output may be termed a *current source*. The low-level output current is normally a negative quantity; current flows into the output terminal. [See Figure 12-2(b).] The gate is said to *sink* this current, and the gate output is referred to as a *current sink*. I_{OH} and I_{OL} must be carefully considered when several gates inputs are connected to one gate output. (See the discussion on *fan-out* below.)

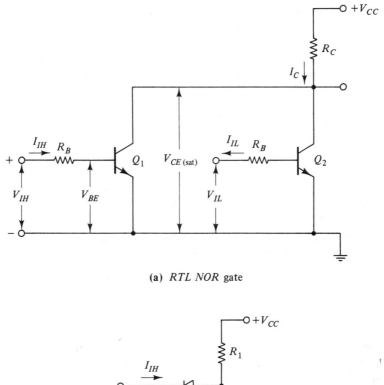

(a) *RTL NOR* gate

(b) Diode *AND* gate

Figure 12-1 Logic gate input voltage and current levels. $V_{IH(min)}$ is the minimum input voltage that can represent logic *1*. $V_{IL(max)}$ is the maximum input voltage for representing logic *0*. I_{IH} and I_{IL} are the *high*-level and *low*-level input currents, respectively.

Output Voltages

The *minimum high-output voltage* from a gate, or a logic *1* output level, is designated $V_{OH(min)}$. In the case of a gate with a transistor output, as illustrated in Figure 12-2(a), the *high* output level would be

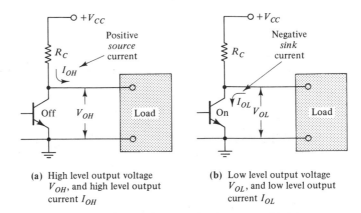

(a) High level output voltage V_{OH}, and high level output current I_{OH}

(b) Low level output voltage V_{OL}, and low level output current I_{OL}

Figure 12-2   Logic gate output voltage and current levels. $V_{OH(min)}$ is the minimum output voltage that represents a logic *1*. $V_{OL(max)}$ is the maximum output voltage for representing logic *0*. I_{OH} and I_{OL} are the *high*-level and *low*-level output currents, respectively.

$$V_{OH} = V_{CC} - I_{OH} R_C$$

where I_{OH} is the output current. With the output of most gates driving the inputs of similar gates, it is important that the minimum logic *1* output voltage be equal to or greater than the minimum logic *1* input voltage. $V_{OH(min)}$ must be larger than $V_{IH(min)}$ (see *Noise Immunity* below).

The logic *0* output level for a gate is specified in terms of a *maximum low-output voltage* $V_{OL(max)}$. For the gate output stage illustrated in Figure 12-2(b), $V_{OL(max)}$ would equal the transistor saturation voltage, which is typically around 0.2 V. For any given type of logic gate driving another similar gate, the maximum logic *0* output voltage must be equal to or less than the maximum logic *0* input voltage. $V_{OL(max)}$ must be less than $V_{IL(max)}$ (see *Noise Immunity*).

Fan-out, or Loading Factor

Because logic gates are connected in complex combinations, the output of every gate must be capable of driving the inputs of several other similar gates. The maximum number of similar gate inputs that any one gate output can drive is termed the *dc fan-out*, or *loading factor*, of the gate. If $I_{OL(max)}$ is the maximum gate output current, and $I_{IL(max)}$ is the maximum gate input current, or *unit load*, then

$$\text{fan-out} = \frac{I_{OL(max)}}{I_{IL(max)}} \tag{12-1}$$

If more gate inputs than specified by the fan-out are connected to the output of a gate, the gate may not function correctly. Indeed, even when the fan-out is not exceeded, each additional load at a gate output tends to increase the switching time of the gate. For high-speed operation, IC manufacturers usually recommend a *maximum loading factor* which is less than the dc fan-out capability of the gate.

EXAMPLE 12-1

Determine the fan-out for the DTL *NAND* gate shown in Figure 12-3. Assume that transistors Q_1 and Q_2 have $h_{FE(min)} = 20$, and that all diodes and transistors are silicon.

Solution

$$I_2 = \frac{V_{BE1}}{R_2} = \frac{0.7 \text{ V}}{5 \text{ k}\Omega}$$

$$= 140 \text{ } \mu\text{A}$$

$$V_A = V_{F4} + V_{F5} + V_{BE1}$$

$$= 0.7 \text{ V} + 0.7 \text{ V} + 0.7 \text{ V}$$

$$= 2.1 \text{ V}$$

$$I_1 = \frac{V_{CC} - V_A}{R_1} = \frac{5 \text{ V} - 2.1 \text{ V}}{2 \text{ k}\Omega}$$

$$= 1.45 \text{ mA}$$

$$I_B = I_1 - I_2 = 1.45 \text{ mA} - 140 \text{ } \mu\text{A}$$

$$= 1.31 \text{ mA}$$

$$I_{C1} = h_{FE} I_B = 20 \times 1.31 \text{ mA}$$

$$= 26.2 \text{ mA}$$

$$I_3 = \frac{V_{CC} - V_{CE(sat)}}{R_3} = \frac{5 \text{ V} - 0.2 \text{ V}}{6 \text{ k}\Omega}$$

$$= 0.8 \text{ mA}$$

The maximum low-level output current is

$$I_{OL} = I_{C1} - I_3 = 26.2 \text{ mA} - 0.8 \text{ mA}$$

$$= 25.4 \text{ mA}$$

With Q_2 *off*, the unit load is

$$I_{IL} = \frac{V_{CC} - V_{F6}}{R_4} = \frac{5 \text{ V} - 0.7 \text{ V}}{2 \text{ k}\Omega}$$

$$= 2.15 \text{ mA}$$

$$\frac{I_{OL}}{I_{IL}} = \frac{25.4 \text{ mA}}{2.15 \text{ mA}} = 11.8$$

$$\text{Fan-out} = 11$$

Propagation Delay Time

The switching speed of a logic gate is defined in terms of its *propagation delay time*. This is the time required for the gate to switch from its *low* output state to its *high* output state, or *vice versa*. The quantity varies with load conditions, and is dependent upon the type of gate circuit; for example, the transistors may have to be switched out of saturation, or they may be unsaturated. In the DTL gates in Figure 12-3, the transistors are saturated when *on*. Typical propagation delay time for integrated circuit DTL is 25 to 30 ns. Some other types of logic gates have propagation delay times of 2 ns or less.

The method of measuring the propagation delay time is illustrated in Figure 12-4. The input waveform e_{in} is shown as a solid line, and the output e_{out} is shown broken. Time t_{PLH} is the time for the output to go from *low* to *high*, and t_{PHL} is the time to switch from *high* to *low*. These times, which are not necessarily equal, are measured between the 50% levels of input and output, as illustrated. The propagation

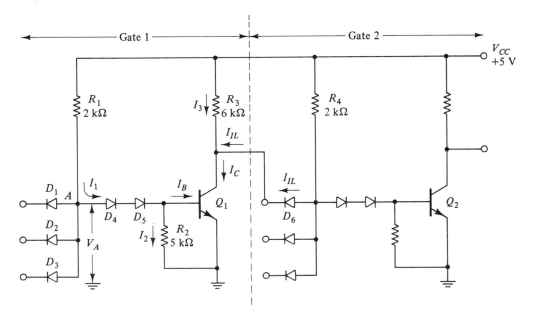

Figure 12-3 Circuit for Example 12-1. One *diode transistor logic* (DTL) *NAND* gate driving another similar gate. The gate fan-out may be calculated from a knowledge of the low-level input and output currents, I_{IL} and I_{OL}.

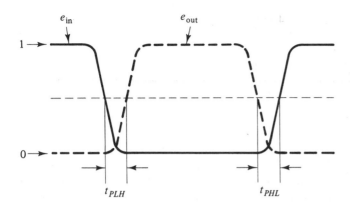

Figure 12-4 Method of measuring propagation delay time for a logic gate. t_{PLH} is the time for the output to go from *low* to *high*, and t_{PHL} is the time to switch from *high* to *low*.

delay time specified on the logic gate data sheets is usually the average of t_{PLH} and t_{PHL}.

Noise Immunity

As previously explained, each gate has a maximum *low*-input voltage $V_{IL(max)}$ which is the highest level acceptable as a logic *0* input, and a minimum *high*-input voltage, $V_{IH(min)}$, which is the lowest level acceptable as a logic *1* input. These two input levels are illustrated in Figure 12-5, with the region between them identified as an *indeterminate range*. If any voltage between these two levels is applied as an input to a gate, the output of the gate is unpredictable.

Each gate also has a maximum *low*-output voltage $V_{OL(max)}$ which is the maximum output level that represents a logic *0* output, and a minimum *high*-output voltage $V_{OH(min)}$ to represent a logic *1* output. These two levels are shown in Figure 12-5.

When one gate is connected to drive another similar gate, the maximum logic *0* output must be equal to or less than the maximum logic *0* input level. Also, the minimum logic *1* output level must be equal to or greater than the minimum logic *1* input level. The difference between $V_{OL(max)}$ and $V_{IL(max)}$ is termed the *low-state noise margin* for the gate, and the difference between $V_{OH(min)}$ and $V_{IH(min)}$ is referred to as the *high-state noise margin*. Noise spikes with amplitudes greater than either noise margin may drive a gate into the indeterminate range, possibly producing unwanted triggering.

The noise margins of a gate afford a method of comparing each type of gate for noise susceptibility. But the *noise immunity* of a logic gate does not depend solely upon noise margins. If a circuit has a low input impedance, noise spikes are potentially divided, and consequently, are less likely to cause unwanted switching. Similarly, a gate that is driven from a low impedance source (the low output impedance of another gate) will be more immune to noise voltages than one with a high source impedance. Also, a gate that switches slowly is less sensitive to fast noise spikes than one that has a very short propagation delay time.

Clearly, many factors are involved in the noise immunity for a given type of logic gate. Instead of trying to rate the noise immunity of each type of gate in terms

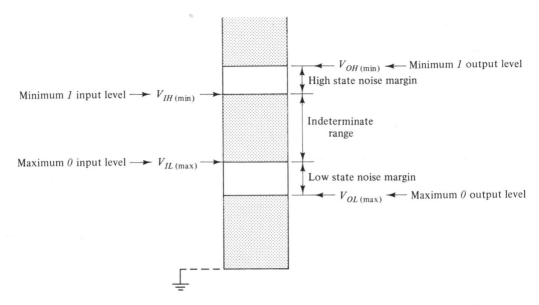

Figure 12-5　Each logic gate has a maximum logic *0* input voltage and a minimum logic *1* input voltage. Any input voltage which falls within the *indeterminate range* between these two levels might produce unwanted gate triggering.

of voltage levels, switching time, and impedance, noise immunity is usually described as *poor*, *fair*, *good*, or *excellent*.

Power Dissipation

The power dissipated by each gate in a logic circuit can be a very important consideration. Where a logic circuit is to operate with a battery power supply, it must use gates with the lowest possible power dissipation to minimize the current drain on the battery. The current drain is less important for a logic circuit with an ac supply, but even in this case the heat generated by a large number of gates can present a problem. In general, high power dissipation is accepted with gates that must switch very fast. Typical power dissipations range from 10 nW to 25 mW per gate, depending upon the gate circuitry.

12-2 DTL, HTL, and RTL

Diode Transistor Logic (DTL)

Diode transistor logic is discussed in Section 10-3, and another DTL gate is shown in Figure 12-3. As already explained, DTL circuits are essentially diode gates followed by transistor inverter stages. Integrated circuit DTL operates from a 5 V supply, has a fan-out of 8, propagation delay time of 30 ns, and power dissipation of 15 mW per gate.

High-Threshold Logic (HTL)

High-threshold logic gate circuits are similar to DTL circuits, except that diode D_5 in the circuit of Figure 12-3 is replaced with a 6.8 V Zener diode. This gives the gate a minimum high-input voltage of approximately 7.5 V and, consequently, makes HTL much more immune to noise spikes than DTL. Because of the presence of the Zener diode, the supply voltage for HTL has to be 15 V instead of the 5 V for DTL. HTL typically has a fan-out of 10, propagation delay time of 120 ns, and power dissipation of 50 mW per gate.

Resistor Transistor Logic (RTL)

The *resistor transistor logic* circuit shown in Figure 12-1(a) produces a *high* output when both input terminals are at ground level, and a *low* output when one of the inputs has a high level. Thus, the circuit is a *NOR* gate. Typical integrated circuit RTL uses a V_{CC} of 3 V, has a fan-out of 5, propagation delay time of 12 ns, and power dissipation of 20 mW per gate.

12-3 STANDARD TTL

TTL Gate Circuit

In *transistor transistor logic* (TTL or T^{2}L), the input signals are applied directly to transistor terminals. Consider the basic TTL circuit shown in Figure 12-6(a). The output transistor Q_2 is controlled by the voltage at the collector terminal of transistor Q_1. When the input terminal (the Q_1 emitter) is grounded, sufficient base current I_B flows to keep Q_1 in saturation. The collector voltage of Q_1 is $V_{CE(sat)}$ above ground. Typically, $V_{CE(sat)}$ is 0.2 V, which is not high enough to bias Q_2 *on*. Therefore, when the input voltage is low, Q_2 is *off* and the output level is *high*.

If a positive voltage is applied to the input terminal, Q_1 remains in saturation (I_B is still large enough) and Q_1 collector voltage goes to $V_i + V_{CE(sat)}$. Depending upon the actual level of input voltage, sufficient base current can be supplied to Q_2 to drive it into saturation, causing the output to switch *low*. Figure 12-6(b) shows Q_1 replaced by diodes representing the base-emitter and collector-base junctions. The arrangement is similar to that of a DTL circuit. It is clear that the input voltage could easily be made large enough to reverse-bias the base-emitter junction of Q_1. When this occurs, the collector-base junction of Q_1 remains forward biased, and current I_1 flows to saturate the output transistor.

Figure 12-6(c) shows a basic three-input TTL circuit. Q_1 is seen to be a transistor with three emitter terminals. This is fabricated easily in integrated circuit form. The

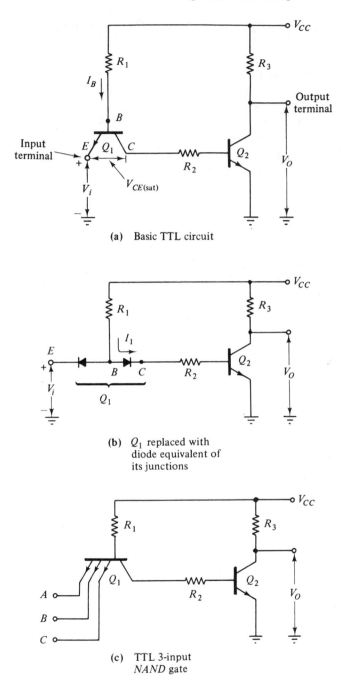

(a)   Basic TTL circuit

(b) Q_1 replaced with diode equivalent of its junctions

(c)   TTL 3-input NAND gate

Figure 12-6 Basic circuitry of a TTL gate. Signal inputs are applied to the emitter terminals of a saturated transistor. The junctions of this transistor behave like two diodes in a DTL gate. A multiemitter transistor is employed to provide several input terminals.

three emitters are the input terminals to the gate. For Q_1 collector to rise above $V_{CE(sat)}$, input *A and* input *B and* input *C* must be *high* positive levels. Because of this, and because the output voltage level goes from *high* to *low*, the circuit is a *NAND* gate.

7400 TTL

Standard TTL is usually referred to as *74* or *7400* series, although the individual circuits may have a *54* number instead of a *74* number. Circuits with a *54* number can operate over a temperature range of $-55°C$ to $+125°C$, and the supply voltage limits are 4.5 V to 5.5 V. For *74* number circuits, the temperature range is 0°C to 70°C, and the supply voltage must be between 4.75 V and 5.25 V.

The circuit of a three-input *7400 series* integrated circuit TTL *NAND* gate is shown in Figure 12-7. The diodes connected from ground to each input terminal become forward biased only when the input voltage goes negative. The function of the diodes is to limit the amplitude of negative spikes appearing at the gate inputs. The arrangement of the output transistors (Q_2, Q_3, and Q_4) is referred to as *totem pole*. When Q_2 is *off*, R_3 biases Q_4 *off*, and R_2 biases Q_3 *on*. Thus, Q_3 provides *active pull-up* (or low output impedance) when the gate output voltage is *high*. When Q_2 is *on* (in saturation), base current supplied to Q_4 drives Q_4 into saturation. Consequently, the output voltage is pulled down, and Q_4 offers a low output impedance when the gate output is in its *low* state. At this time, Q_3 is biased *off* by the voltage drop across R_2. This is assisted by the presence of diode D_4.

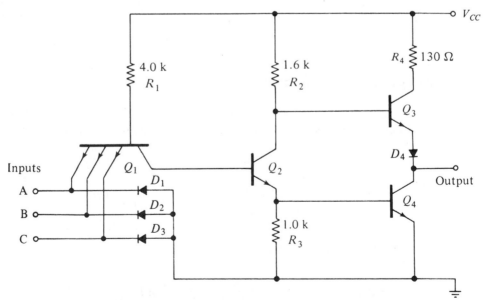

Figure 12-7   7400/5400 series integrated circuit three-input TTL *NAND* gate. Diodes D_1, D_2, and D_3 protect the circuit from excessive negative input voltages. The output stage is termed a *totem pole* circuit.

The multi-emitter input transistor used with TTL is normally in saturation (i.e., its collector-base junction is forward biased), even when all the input voltages are *high*. An exception to this occurs at the instant that one input terminal goes *low*. With Q_2 still *on*, Q_1 collector-base junction becomes reverse biased, and a large current flows *from* Q_2 base into Q_1 collector. The effect of this is to cause Q_2 to switch *off* very rapidly. Q_2 also switches *on* very rapidly when Q_1 inputs go high. This is because Q_1 collector-base junction remains forward biased during the *on* switching time for Q_2. These two effects make TTL one of the fastest of all integrated circuit logic types.

Input and Output Voltages

The logic *0* and logic *1* input voltage levels for standard TTL are specified in Appendix 1-18 as

$$V_{IL(\text{max})} = 0.8 \text{ V and } V_{IH(\text{min})} = 2 \text{ V}$$

The output levels are listed as

$$V_{OL(\text{max})} = 0.4 \text{ V and } V_{OH(\text{min})} = 2.4 \text{ V}$$

Consider the (usual) case of one TTL gate driving another. When the first gate output is at logic *0*, its maximum output voltage $V_{OL(\text{max})}$ is 0.4 V lower than the maximum $V_{IL(\text{max})}$ that the second gate can accept as a *0* input. Thus, as explained in Section 12-1, there is a noise margin of 0.4 V. Noise spikes would have to exceed this voltage before the gate input would be driven into its indeterminate range. Similarly, at the logic *1* level, the minimum high output $V_{OH(\text{min})}$ is 0.4 V greater than the minimum level $V_{IH(\text{min})}$ that the driven gate can accept as a *1* input. Here again, the noise margin is 0.4 V.

Input and Output Currents

When each input terminal of a TTL gate is *high*, only a very low (emitter-base leakage) input current flows. When the input voltage is *low*, a (emitter) current flows *out* of the input terminal. The maximum *high-level input current* I_{IH} is listed on the data sheet as 40 μA. (See Appendix 1-18.) The *low-level input current* I_{IL} is stated as -1.6 mA maximum. The minus sign indicates that current flows out of the input terminal. This value of -1.6 mA is the *unit load* for this type of logic circuitry.

When the output terminal of a TTL gate is *low*, it can *sink* the low-level input current I_{IL} from the input terminals of several TTL gates. The maximum *low-level output current* I_{OL} is specified on the data sheet as 16 mA. Therefore, the maximum dc fan-out is

$$\frac{I_{OL}}{I_{IL}} = \frac{16 \text{ mA}}{1.6 \text{ mA}} = 10$$

The *high*-level output current I_{OH} is listed as 400 μA, so the gate is not designed to drive any significant load in its *high*-output state.

Performance

Standard (74/54 series) TTL has a fan-out of 10, propagation delay time of 10 ns, and power dissipation of approximately 10 mW per gate. Because the gate input terminals are transistor emitters, they have a low-input impedance, which gives good noise immunity.

12-4 OPEN-COLLECTOR TTL

An *open-collector TTL* logic gate has an output stage which consists of a single transistor with its collector terminal unconnected. [See Figure 12-8(a).] The complete gate circuit is similar to that in Figure 12-7, with components Q_3, R_4, and D_4 omitted. An external *pull-up resistor* R_C must be included, as illustrated in Figure 12-8(a), to provide a path for the transistor collector current.

Figure 12-8(b) shows that a single collector resistor can serve the outputs of several open-collector gates. The gate outputs are connected together in an arrangement termed *wired-AND* configuration. Because each gate has an open-collector output, it is not loaded by the other gates with which it is connected in common, regardless of the condition of the gates. (This would not be the case with totem pole output stages.) If the output of any one gate is *low*, the *wired-AND* output is *low*. The *wired-AND* output is *high* only when a *high* level is present at terminal A *and* at terminal B *and* at terminal C. [See Figure 12-8(b).] It is seen that this wiring arrangement functions as an *AND* gate, hence the name *wired-AND*. The inputs of other gates may be connected to the *wired-AND* output.

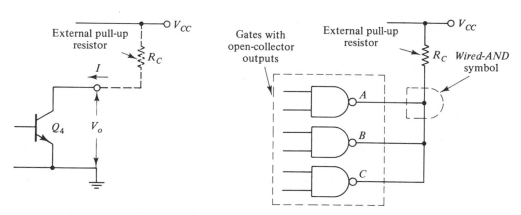

(a) Open-collector TTL output stage

(b) Three open-collector TTL gates connected in *wired-AND* configuration

Figure 12-8   Open-collector TTL gates have transistors with open-circuited collector terminals as their output stages. This permits gate output terminals to be connected together to function as an *AND* gate.

The advantage of the *wired-AND* arrangement available with open-collector TTL is that it can be substituted in place of a logic gate. Consider the digital comparator circuit in Figure 11-19. The *AND* gate can be eliminated if open-collector TTL *XNOR* gates are available.

EXAMPLE 12-2

A *wired-AND* circuit, as in Figure 12-8(b), is to drive the inputs of five TTL gates. Determine a suitable resistance for R_C.

Solution   When the output is *low*, the maximum input current to each driven gate is

$$I_{IL(max)} = 1.6 \text{ mA}$$

Total load (input) current for the five gates is

$$I_L = 5 \times I_{IL(max)} = 5 \times 1.6 \text{ mA}$$

$$= 8 \text{ mA}$$

Assuming only one *wired-AND* gate is *on*,

$$I_{RC} = I_{OL(max)} - I_L$$

$$= 16 \text{ mA} - 8 \text{ mA}$$

$$= 8 \text{ mA}$$

$$V_{RC} = V_{CC} - V_{OL(max)}$$

$$= 5 \text{ V} - 0.4 \text{ V}$$

$$= 4.6 \text{ V}$$

$$R_C = \frac{V_{RC}}{I_{RC}} = \frac{4.6 \text{ V}}{8 \text{ mA}}$$

$$= 575 \text{ }\Omega \qquad \text{(use 680 }\Omega\text{ standard value)}$$

Note that a lower current level could be used for I_{RC} if a slower switching time is acceptable. The highest possible current changes junction and stray capacitances fastest, and thus gives the shortest switching times.

12-5 TRI-STATE TTL (TSL)

A *tri-state TTL* (or *TSL*) logic gate has a *control input* as well as the usual input and output terminals. Figure 12-9 shows the circuit arrangement and logic symbol for a TSL *NAND* gate. Note that the control input terminal goes to an inverter. The output

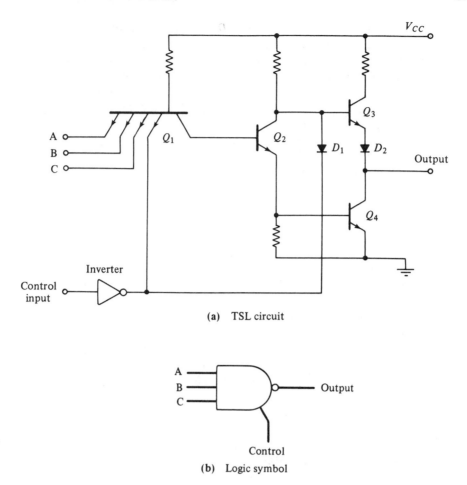

(a)   TSL circuit

(b)   Logic symbol

Figure 12-9 *Tri-state TTL (TSL)* circuits have a third output state: in addition to being *high* or *low*, the output may be placed in a *high-impedance* state. This is accomplished by setting transistors Q_3 and Q_4 in an *off* condition.

of the inverter is connected to one emitter on transistor Q_1, and to the base of Q_3 via diode D_1.

When the control input is *low*, the inverter output is *high*. This reverse biases D_1 and provides a *high* input to the connected emitter of Q_1. In this condition the *NAND* gate functions normally; when all the gate inputs are *high*, the output is *low*; when one or more inputs are *low*, the output is *high*.

With a *high* input applied to the control terminal, the inverter output goes *low*, forward biasing D_1 and the connected emitter of Q_1. Now Q_1 is held in a *low* state, regardless of the level of the other gate input terminals. Thus, Q_2 and Q_4 are *off*. In addition, the base of Q_3 is held in a *low* state by (forward-biased) diode D_1. Consequently, Q_3 is *off*. Both output transistors Q_3 and Q_4 are *off*, and the output terminal offers a high impedance to all circuits that are connected to it. This condition is the *third state* of the TSL circuit: the output of a TSL gate may be *high* or *low*, or have a high output impedance.

TSL gates are used in logic systems where the outputs of several gates are connected in parallel to a single input of another circuit. All gates are usually maintained in the high output impedance state and are *sampled*, or switched *on* briefly one at a time, by the control signals applied in sequence. This avoids the possibility of the output of one gate short-circuiting another gate output.

Another aspect of the TSL gate is that the circuit *input* impedance also becomes high when the gate is placed in its high *output* impedance state.

12-6 OTHER TTL TYPES

As well as the standard *54/74* series TTL, the data sheet in Appendix 1-18 lists four other types of TTL gates: *54H/74H*, *54L/74L*, *54S/74S*, and *54LS/74LS*.

High-speed TTL (*54H/74H*)

The circuit speed is increased by reducing the resistance of the resistors, and by including an additional emitter-follower transistor to drive one of the output transistors. (See the *H00-H30* circuit diagram on the data sheet.) Because of the reduced resistor values, the supply current is approximately double that for standard TTL, resulting in an average per-gate power dissipation of 22.5 mW. The typical propagation delay time for *54H/74H* TTL is 6 ns.

Low-power TTL (*54L/74L*)

In this case the resistor values are increased above those normally employed in standard TTL. (See the *L00-L30* circuit and table of component values on the data sheet.) The result is lower supply currents and an average power dissipation per gate of only 1 mW. Reduced switching speed is another consequence of the increased resistance values. The typical propagation delay time is 35 ns for low-power TTL. The major applications of this logic series are found in portable battery-operated equipment, where supply currents must be minimized.

Schottky TTL (*54S/74S*)

This logic family employs *Schottky transistors* to further increase the circuit switching speed. A *Schottky transistor* is a bipolar transistor with a *Schottky diode* connected between its collector and base terminals, as illustrated in Figure 12-10(a). A *Schottky diode* has a junction of silicon and metal. Like other diodes, it is a one-way device, but its major characteristics are that it switches very fast and that its forward drop is

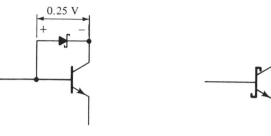

(a) Transistor with
    Schottky diode

(b) Schottky
    transistor
    symbol

Figure 12-10 A Schottky transistor is a bipolar transistor with a Schottky diode connected from base to collector. This prevents the transistor from saturating, and thus increases its switching speed.

typically 0.25 V. The presence of the Schottky diode prevents the transistor from going into saturation, and consequently the transistor switching speed is minimized.

The Schottky transistor circuit symbol is illustrated in Figure 12-10(b), and an S00-S133 Schottky TTL *NAND* gate circuit is shown on the data sheet in Appendix 1-18.

The typical propagation delay time for Schottky TTL is 3 ns, and the average power dissipation per gate is around 20 mW. An improved version known as *advanced Schottky TTL* (54/74 AS) boasts a 1.5 ns typical gate delay time, with 20 mW per gate power dissipation. Obviously, this type of logic circuit should be used where high speed is the most important consideration.

Low-power Schottky TTL (*54LS/74LS*)

In this family, as in the 54L/74L family, the resistor values are increased in order to minimize power dissipation. However, because Schottky transistors are used, the typical propagation delay time is relatively small, at 9 ns. Power dissipation per gate is around 2 mW. The circuit for low-power Schottky TTL in Appendix 1-18 (LS00-LS30) shows that this type of TTL uses diode inputs instead of the usual multi-emitter input transistor. *Advanced low-power Schottky TTL* has a typical gate delay time of 4 ns with 1 mW per gate power dissipation.

12-7 EMITTER-COUPLED LOGIC (ECL)

One major limitation on the switching speed of logic circuits is the *storage time* of saturated transistors. The storage time is the time required to drive a transistor out of saturation, that is, to reverse the forward bias on the collector-base junction. In *emitter-coupled logic* (ECL), also termed *current mode logic*, the transistors are maintained in an unsaturated condition. This eliminates the transistor storage time and results in logic gates which switch very fast indeed.

The schematic diagram of a typical integrated circuit ECL gate is shown in Figure 12-11. The circuit uses a negative supply $-V_{EE}$, and the positive supply terminal V_{CC}

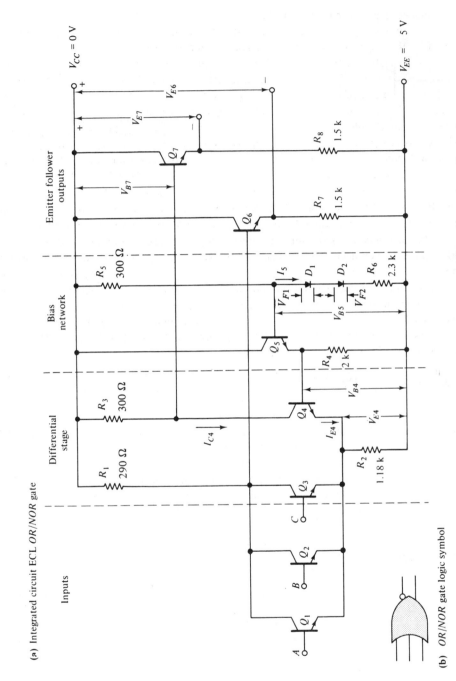

(a) Integrated circuit ECL *OR/NOR* gate

(b) *OR/NOR* gate logic symbol

Figure 12-11 Circuit and logic symbol for ECL *OR/NOR* gate. The circuit arrangement keeps the transistors from saturating, so that they can be switched *off* very fast.

is grounded. Transistor Q_5 has its base bias voltage provided by the potential divider composed of R_5, R_6, D_1, and D_2. The diodes provide temperature compensation for changes in the V_{BE} of Q_5. Q_5 operates as an emitter follower to provide a low impedance bias to the base of transistor Q_4. With a constant bias voltage at Q_4 base, the voltage drop across emitter resistor R_2 is also maintained constant, as long as the input voltages are low enough to keep transistors Q_1, Q_2, and Q_3 in the *off* state. In this circumstance, the emitter current and collector current of Q_4 are held constant, and the transistor is maintained in an unsaturated condition. With Q_4 *on*, the output voltage via emitter follower Q_7 is *low*, and that via emitter follower Q_6 is *high*. When a positive voltage is applied to terminal *A or* terminal *B or* terminal *C*, the emitter voltage of Q_4 is pulled up above its base level. Consequently, Q_4 switches *off* as Q_1, Q_2, or Q_3 switches *on*. When this occurs, the voltage at the base of Q_6 falls and that at Q_7 base rises.

It is seen that when the input voltages at terminals *A*, *B*, and *C* are *low*, the output voltage at Q_7 emitter is also *low*. Q_7 output becomes *high* when a *high* input is applied to terminal *A or* terminal *B or* terminal *C*. Thus, the gate functions as an *or* gate when the output is derived from Q_7 emitter. Q_6 emitter voltage is *high* when the inputs are *low*, and *low* when any of terminal *A*, *B*, or *C* inputs are *high*. Therefore, with output taken from Q_6 emitter, the circuit functions as a *NOR* gate. The *OR/NOR* logic symbol is shown in Figure 12-11(b).

EXAMPLE 12-3

The *OR/NOR* gate circuit in Figure 12-11(a) has supply voltages of -5 V and ground. Determine the output voltages when inputs *A*, *B*, and *C* are *low*.

Solution   With inputs *A*, *B*, and *C* low,

$$I_5 = \frac{(0 - V_{EE}) - V_{F1} - V_{F2}}{R_5 + R_6}$$

$$= \frac{0 - (-5 \text{ V}) - 0.7 \text{ V} - 0.7 \text{ V}}{300 \ \Omega + 2.3 \text{ k}\Omega} \approx 1.4 \text{ mA}$$

$$V_{B5} = (I_5 R_6) + V_{F1} + V_{F2}$$

$$= (1.4 \text{ mA} \times 2.3 \text{ k}\Omega) + 0.7 \text{ V} + 0.7 \text{ V}$$

$$\approx 4.6 \text{ V}$$

$$V_{B4} = V_{B5} - V_{BE5}$$

$$= 4.6 \text{ V} - 0.7 \text{ V} = 3.9 \text{ V}$$

$$V_{E4} = V_{B4} - V_{BE4}$$

$$= 3.9 \text{ V} - 0.7 \text{ V}$$

$$= 3.2 \text{ V}$$

$$I_{E4} = \frac{V_{E4}}{R_2}$$

$$= \frac{3.2 \text{ V}}{1.18 \text{ k}\Omega} \approx 2.7 \text{ mA}$$

$$I_{C4} \approx I_{E4} = 2.7 \text{ mA}$$

$$V_{B7} \approx V_{CC} - I_{C4}R_3 \qquad (\text{neglect } I_{B7})$$

$$= 0 \text{ V} - (2.7 \text{ mA} \times 300 \text{ }\Omega)$$

$$= -0.81 \text{ V}$$

$$V_{E7} = V_{B7} - V_{BE7}$$

$$= -0.81 \text{ V} - 0.7 \text{ V} \approx -1.5 \text{ V}$$

This is the *low* state of the output at Q_7 emitter.

   With Q_1, Q_2, and Q_3 biased *off*, only I_{B6} flows through R_1. Consider ($I_{B6} \times R_1$) as negligible. Then

$$V_{E6} \approx V_{CC} - V_{BE6}$$

$$= 0 - 0.7 \text{ V}$$

$$= -0.7 \text{ V}$$

This is the *high* state of the output at Q_6 emitter.

   From Example 12-3, the *high* output level for the ECL gate is -0.7 V, and the *low* output level is -1.5 V. When applied to the input of another gate, these *high* and *low* levels must be capable of switching the gate from one state to another. Consider the circuit in Figure 12-11(a), and assume that input terminal C is connected to the output of another similar gate. When the *low* output level (-1.5 V) is applied to terminal C, V_{B3} is 3.5 V above V_{EE}. In Example 12-3, V_{E4} was found to be 3.2 V above V_{EE}, and this is also the voltage at the emitter of Q_3. Since $V_{BE3} = (3.5 \text{ V} - 3.2 \text{ V}) = 0.3$ V, Q_3 base-emitter is actually forward biased by 0.3 V. This is not sufficient to bias a silicon transistor into conduction, so Q_3 remains *off*. However, an increase of approximately 250 mV at the base of Q_3 (e.g., a noise spike) could cause the transistor to at least partially switch *on*. A similar analysis of the circuit conditions when Q_3 is *on* and Q_4 is *off* shows that switching could again occur with a -250 mV spike.

   The principal drawback of integrated circuit ECL compared to other IC logic families is now evident. That drawback is its sensitivity to low-level noise on the order of ±250 mV. The high input resistance and very fast switching speed of ECL also contribute to its low noise immunity. However, the low output resistance of ECL improves the noise immunity at the input of another gate that is being driven. Another aspect of the noise sensitivity of logic gates is that most types of logic circuits generate noise spikes when transistors are switched into or out of saturation. This is not the case with ECL, because each time one transistor is switched *off* another is switched *on*. Thus, the current drawn from the supply remains approximately constant.

Another disadvantage of ECL is its relatively high power dissipation, approximately 25 mW per gate. The major advantage of ECL over other types of logic undoubtedly is the very fast switching speed. Because of the nonsaturated condition of the *on* transistors, the propagation delay time can be 2 ns or less.

Appendix 1-19 shows the data sheet for *MC306.MC307 3-input ECL gates* manufactured by Motorola. In the schematic diagram on the data sheet, the three transistors with their bases connected to terminals *6*, *7*, and *8* correspond to Q_1, Q_2, and Q_3 in Figure 12-11(a). Also, the transistor with its base connected to terminal *1* corresponds to Q_4 in the figure. The remaining two transistors in the *MC306.MC307* circuit are the emitter follower outputs. No bias network is provided in this IC gate. Instead, an external bias driver must be connected to terminal *1*.

The listed electrical characteristics of the *MC306.MC307* show that the low output voltage (*NOR* logic 0) is -1.750 V. The high output voltage (*NOR* logic 1) is -0.795 V. This gives an output voltage change of 0.955 V. The shortest propagation delay time for the *MC306.MC307* is listed as 5.5 ns.

12-8 P-MOS AND N-MOS LOGIC GATES

As discussed in Section 4-8, MOSFET switches have an extremely high input resistance, very small drain-to-source voltage drop, and very low power dissipation. The *n*-channel *enhancement mode* MOSFET is normally-*off* when its gate is at the same potential as its substrate. When the gate is made positive with respect to the substrate, an *n-type* channel is created from drain to source, and drain current flows. Similarly, the *p*-channel device has no drain current while its gate and substrate are at the same potential. The *p-type* channel appears when the gate is made negative with respect to the substrate.

P-MOS logic gates are made up of *p*-channel MOSFET transistors. No resistors or capacitors are involved. N-MOS gates are composed only of *n*-channel MOSFETs. N-MOS circuits are very similar to P-MOS circuits, with the important exception that all voltage polarities and current directions are reversed. One other important difference between P-MOS and N-MOS is that N-MOS is the faster of the two types of logic. This is due to the fact that charge carriers in *n*-channel devices are electrons while those in *p*-channel FETs are holes, and electrons have *greater mobility* than holes (i.e., they move faster).

The circuits of N-MOS *NAND* and *NOR* gates are shown in Figure 12-12. Note that Q_1 has a channel resistance (or $R_{D(on)}$ value) around 100 kΩ, while the $R_{D(on)}$ value for each of Q_2 and Q_3 is on the order of 1 kΩ. Also, in both cases, the gate of Q_1 is biased to its drain terminal. When the source terminal of Q_1 is less than V_{DD}, the gate is positive with respect to the source. This is the condition necessary to bias Q_1 *on*. Consequently, Q_1 is always in the *on* condition, and its $R_{D(on)}$ acts as a load resistor for Q_2 and Q_3.

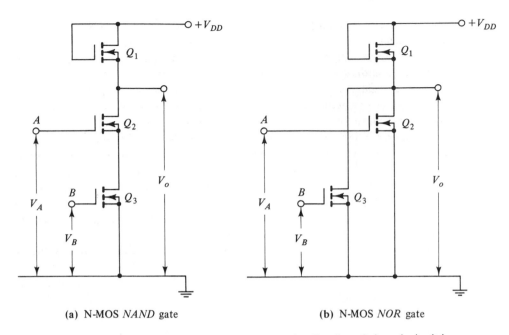

(a) N-MOS *NAND* gate   (b) N-MOS *NOR* gate

Figure 12-12   N-MOS *NAND* and *NOR* gate circuits. Transistor Q_1 in each circuit is biased *on*. For the *NAND* gate, Q_2 and Q_3 must both be switched *on* to produce a *low* output. For the *NOR* circuit, switching either Q_2 or Q_3 *on* causes the output to go *low*.

Consider the circuit in Figure 12-12(a). When input *A* and input *B* are *low* (near ground), transistors Q_2 and Q_3 are both *off*. No drain current flows, and there is no voltage drop across Q_1. The output voltage at this time is a *high* level close to $+V_{DD}$. When a *high* (positive) input is applied to the gate of Q_3, Q_3 tends to switch *on*. However, with the gate of Q_2 still held near ground, Q_2 remains an open circuit and the output remains at its *high* level. When *high* inputs are applied to the gates of Q_2 *and* Q_3, both transistors are switched *on* and current flows through the channels of all three transistors. The total $R_{D(on)}$ of Q_2 and Q_3 adds up to about 2 kΩ, while that of Q_1 is around 100 kΩ. Therefore, the voltage drop across Q_2 and Q_3 is much smaller than that across Q_1, and the output voltage is now at a *low* level.

It is seen that the circuit performs as a *NAND* gate. When any one of the inputs is *low*, the output is *high*. When input *A* *and* input *B* are high, the output voltage is *low*. As already stated, a P-MOS *NAND* gate is exactly similar to the circuit in Figure 12-12(a), except that V_{DD} must be negative, and *p*-channel devices are used.

An N-MOS *NOR* gate circuit is shown in Figure 12-12(b). When both input levels are *low*, Q_2 and Q_3 are *off*. At this time the voltage drop across Q_1 is almost zero and the output level is *high*, close to V_{DD}. When a *high* (positive) input is applied to terminal *A* *or* terminal *B*, Q_2 or Q_3 switch *on*, causing current to flow through Q_1. The voltage drop across either Q_2 or Q_3 (or both) is much smaller than that across Q_1, since the $R_{D(on)}$ of Q_1 is around 100 kΩ, while $R_{D(on)}$ for Q_2 and Q_3 is approximately 1 kΩ.

Therefore, when a *high* input is applied to terminal A *or* terminal B, the output voltage goes to a *low* level.

A P-MOS *NOR* gate is exactly similar to the circuit of Figure 12-12(b), except that V_{DD} must be a negative quantity.

P-MOS and N-MOS logic gates typically use a supply of 10 V, but can operate with lower or higher supply voltages. Using a 5 V supply, power dissipation is approximately 0.25 mW per gate, and noise margin is 1.5 V. Because there is virtually no input current to MOSFET devices, there should be no limit to the dc fan-out for P-MOS and N-MOS logic gates. But gate inputs do have capacitance, and each additional gate input connected at a gate output terminal slows down the switching speed of the gate. A fan-out of 50 is considered a normal maximum. Propagation delay time is around 50 ns for N-MOS, and 100 ns for P-MOS. The relatively large switching time is due to the high output resistance, approximately 2 kΩ, which is twenty times the typical R_O of 100 Ω for TTL. For a given load capacitance (typically, 15 pF is used when testing for t_r), an N-MOS gate will be twenty times slower than TTL.

12-9 CMOS LOGIC GATES

CMOS was introduced in Section 4-8, and the operation of the CMOS inverter was explained in that section. Although the integrated circuit fabrication process for CMOS is more complicated than that for P-MOS or N-MOS, CMOS has the very important advantage that its power dissipation per gate is much less than that for any other logic family. (Integrated injection logic can be an exception to this—see Section 12-10.) Other CMOS advantages are (1) operation from supply voltages as low as 1 V, (2) fan-out in excess of 50, and (3) excellent noise immunity.

Consider the CMOS *NAND* gate shown in Figure 12-13(a). The parallel-connected transistors Q_1 and Q_2 are *p*-channel MOSFETs, and the series-connected devices Q_3 and Q_4 are *n*-channel MOSFETs. When input terminals A and B are grounded, the gates of Q_1 and Q_2 are negative with respect to the source terminals. Therefore, Q_1 and Q_2 are biased *on*. Also, the gates of Q_3 and Q_4 are at the same potential as the device source terminals, and consequently Q_3 and Q_4 are *off*. Depending upon the actual load current and the values of $R_{D(on)}$, there will be a small voltage drop along the channels of Q_1 and Q_2. Thus, the output voltage V_O is close to the level of the supply voltage V_{DD}. When both A and B are grounded, V_O is approximately equal to V_{DD}.

When a *high* positive input voltage (equal to 0.7 V_{DD} or greater) is applied to terminal B, Q_4 is biased *on* and Q_2 is biased *off*. However, with terminal A still grounded, Q_3 remains *off*, Q_1 is still *on*, and the output voltage remains at $V_O \approx V_{DD}$. When *high* inputs are applied to terminal A *and* terminal B, both *p*-channel devices (Q_1 and Q_2) are biased *off*, and both *n*-channel FETs (Q_3 and Q_4) are biased *on*. The output now goes to $V_O \approx 0$ V.

The circuit of a CMOS *NOR* gate is shown in Figure 12-13(b). Once again, two *p*-channel devices (Q_1 and Q_2) and two *n*-channel transistors (Q_3 and Q_4) are employed. When both inputs are at ground level, Q_3 and Q_4 are biased *off*, and Q_1 and Q_2 are *on*.

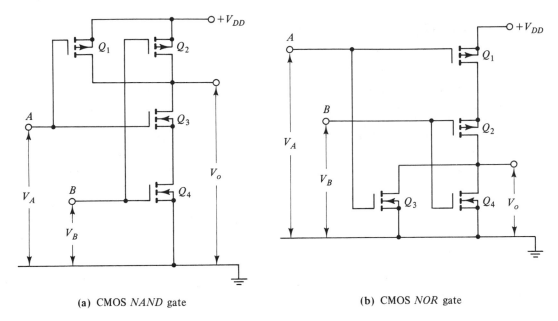

(a) CMOS *NAND* gate   (b) CMOS *NOR* gate

Figure 12-13 CMOS *NAND* and *NOR* gate circuits. Transistors Q_1 and Q_2 in each circuit are *p*-channel devices, while Q_3 and Q_4 are *n*-channel transistors. A *high* input to terminals *A* or *B* switches *on* the *n*-channel devices and switches the *p*-channel transistors *off*.

In this condition there is about a 10 mV drop from drain to source terminals in the *p*-channel transistors, and V_O is very close to V_{DD}. When terminal *A* has a *high* positive input, Q_1 switches *off* and Q_3 switches *on*. The series combination of Q_1 and Q_2 is now open circuited, and the output is shorted to ground via Q_3. Similarly, if terminal *A* remains grounded and terminal *B* has a *high* input applied, Q_2 switches *off* and Q_4 switches *on*. Again, the output goes to ground level.

The major advantage of integrated circuit CMOS logic over all other logic systems is its extremely low power dissipation. At a maximum of 10 nW per gate, the low dissipation allows greater circuit density within a given size of IC package. The resultant low supply current demand also makes CMOS ideal for battery-operated instruments. Typical supply voltages employed for CMOS are 5 V to 10 V; however, operation with a supply of 1 V to 18 V is possible. The circuitry is immune to noise levels as high as 30% of the supply voltage. The extremely high input resistance of MOSFETs gives CMOS gates typical input resistances of 10^9 Ω, and this makes it possible to have fan-outs greater than 50. Typical propagation delay time for CMOS is 25 ns. As in the case of N-MOS and P-MOS, the relatively slow switching time is due to the high output resistance.

The logic *0* and logic *1* input levels for CMOS are typically 30% and 70% of V_{DD}, respectively. With a 5 V supply, this gives $V_{IH(min)} = 3.5$ V and $V_{IL(max)} = 1.5$ V. CMOS gates draw virtually zero input current. Therefore, even with a large number of gate inputs connected to one output, the output voltages are $V_{OH(min)} \approx V_{DD}$, and $V_{OL(max)} \approx 0$.

It is very important to note that CMOS devices can be destroyed by static electricity. Sufficient static can be present in a person handling the devices to break down the gate-channel insulation. To protect against static, CMOS devices are normally shipped in some type of conductive container. Removal from the container and installation on a circuit board should preferably take place at an antistatic workbench. This should have a grounded conductive bench surface and floor, a soldering iron with a grounded tip, and provision for grounding the wrists of individuals working at the bench.

12-10 INTEGRATED INJECTION LOGIC (I^2L)

As already discussed, integrated circuit logic systems are compared in terms of switching speed, power dissipation, fan-out, and noise margin. Two other very important factors are physical size and cost of manufacture. As will be explained, individual I^2L gates require a fraction of the area of other logic types, i.e., the circuit density is much greater. Also, power dissipation per gate can be comparable with CMOS logic, very fast switching is possible, and fabrication techniques are simple and inexpensive. These improvements are due to two factors: elimination of resistors, and what is termed *merging* of transistors.

To understand the operation of I^2L, consider Figure 12-14(a), which shows a simple (normally-*on*) resistor-transistor inverter. *Npn* transistor Q_2 is supplied with base current I_1 via resistor R_1, and the collector current I_2 causes a voltage drop across R_2 which makes the transistor saturated. When the inverter input is at ground level, I_1 is diverted away from the transistor base. Thus, Q_2 is *off*, and the output is *high*. With the input *high* once again, Q_2 goes *on*, and the output is *low*.

In Figure 12-14(b), R_1 and R_2 are replaced by *pnp* transistors Q_1 and Q_3, respectively. When sufficient emitter current I_1 is supplied to Q_1, Q_1 performs exactly the same function as R_1 in Figure 12-14(a), i.e., it supplies base current to Q_2. Similarly, with an adequate level of I_2, Q_3 passes collector current to Q_2, as does R_2 in Figure 12-14(a). Here again, when the input is at ground level, I_1 is diverted and Q_2 goes *off*. When the input is *high*, Q_2 is *on*, and its collector voltage is *low*.

Integrated circuit resistors can easily occupy ten times the area of a transistor. Consequently, by replacing the resistors with transistors, a big reduction in gate area is achieved. Putting it another way, twenty (or more) purely transistor inverters might be fabricated on the area normally occupied by one resistor-transistor inverter. Furthermore, the large area occupied by each resistor results in much unwanted capacitance. Eliminating the resistors reduces the capacitance and improves the switching speed of each inverter.

Now consider Figure 12-14(c), which shows two I^2L inverters, one of which has its output connected to the input of the other. It is seen that each inverter consists of only two transistors, and that the collector load for each is the input stage of the next inverter. Thus, Q_3 is the collector load for Q_2. Note that Q_2 and Q_4 have several separate collectors, all isolated from each other, so that they may be connected to the inputs of

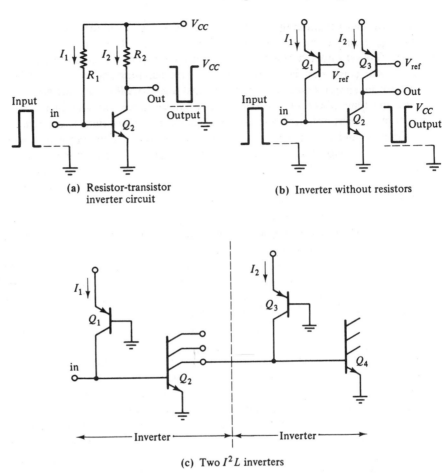

(a) Resistor-transistor
inverter circuit

(b) Inverter without resistors

(c) Two I^2L inverters

Figure 12-14 *Integrated injection logic* (I^2L) uses bipolar transistors in place of base and collector resistors. Each I^2L inverter consists of only two transistors, Q_1 supplying base current to Q_2. The input of the next gate constitutes the collector load.

several different gates. When Q_2 is *off*, these collectors could be at different (*high* or *low*) levels depending upon the state of the other gates connected to them.

In Figure 12-14(c), the base driver transistors Q_1 and Q_3 have their own bases grounded. Each of the *pnp* transistor emitters is typically 750 mV above the level of the base. When Q_2 is open circuited, Q_3 is in saturation, and its collector voltage should be about 600 mV to 700 mV above ground. This is sufficient to bias Q_4 into saturation. With Q_2 *on*, its saturation voltage is perhaps 50 mV to 100 mV, which pulls the base of Q_4 low enough to switch it *off*. The actual supply voltage at the emitter of each *pnp* transistor is typically

$$V_{CC} = V_{BE} + V_{CE(\text{sat})} = 750 \text{ mV} + 100 \text{ mV} = 850 \text{ mV}$$

Instead of thinking in terms of voltages, I^2L operation is more easily described by considering currents. In Figure 12-14(c), when Q_2 is *off*, I_2 flows into the base of Q_4 to drive it into saturation. With Q_2 *on*, I_2 is diverted through Q_2, and Q_4 is *off*.

The current supplied to the emitters of the *pnp* transistors must be regulated by the power supply. This is easily done by using one external resistor to supply current to many transistors. The charge carriers constituting the current are said to be *injected* into the transistor emitters, hence the name *integrated injection logic*. If the supply current is kept low, the power dissipation per gate is obviously minimized. However, a disadvantage of low current is that the switching time of transistors is increased. A choice must be made between fast switching and low power dissipation.

I^2L *NAND* and *NOR* gates are shown in Figures 12-15(a) and (b), respectively. The *NAND* gate is simply an inverter stage with several input terminals connected in common. If all of the collectors of (previous-stage) transistors connected to terminals A, B, and C are open circuited, then all inputs can be described as *high*. The result is

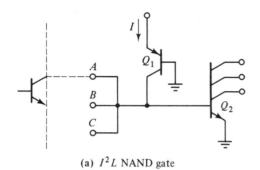

(a) I^2L NAND gate

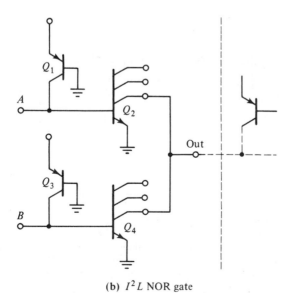

(b) I^2L NOR gate

Figure 12-15   An I^2L *NAND* gate is simply an inverter with several input terminals connected in common. A *NOR* gate uses two or more inverter stages with their outputs connected in common.

that Q_2 is *on*, and its outputs are *low*. When any one of the transistors connected to A, B, or C, is *on*, the input is *low*, Q_2 is *off*, and the gate output is *high*.

The *NOR* gate uses two inverters with their output terminals in parallel. With both inputs *low*, Q_2 and Q_4 are *off* and the output is *high*. A *high* input to either A or B causes one of the output transistors to switch *on*, pulling the gate output level *low*.

The fabrication advantages of I^2L are illustrated in Figure 12-16. The basic inverter circuit is reproduced in Figure 12-16(a), with the regions of each transistor identified as p_1, n_1, and p_2 for Q_1; and n_1, p_2, n_2, n_3, and n_4 for Q_2. Because the base of Q_1 and the emitter of Q_2 are both grounded, and because they are both *n*-type material, a

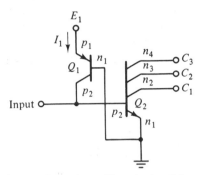

(a) *p* and *n* regions of inverter transistors

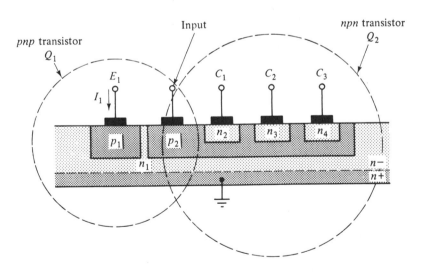

(b) Cross-section of inverter transistors

Figure 12-16   Merged transistor construction of I^2L. The base of Q_1 and the emitter of Q_2 are both *n*-type material and are connected together. So, a common *n* region is used for both. Also, a common *p* region is used for the collector of Q_1 and the base of Q_2.

single identification n_1 is employed. In fact, one single n-type region of semiconductor can be used for both. Similarly, the collector of Q_1 and the base of Q_2 are connected together and are both p-type material. Consequently, they are both identified as p_2, and a single p-type region of semiconductor material can be used for the two.

The cross-section of such an I^2L inverter is shown in Figure 12-16(b). A single bed of n-type forms region n_1. Regions p_1 and p_2 are diffused into n_1, and n_2, n_3, and n_4 are each diffused into p_2. Transistor Q_1 consists of p_1, n_1, and p_2, while Q_2 is made up of n_1, p_2, n_2, n_3, and n_4. The common n_1 region is grounded via the low resistive n^+ plane; current I_1 is injected into p_1; the common p_2 terminal is the input; the outputs are C_1, C_2, and C_3. Since they share common regions of n-type and p-type material, transistors Q_1 and Q_2 are said to be *merged*. This gives I^2L its other name: *merged transistor logic* (MTL). Many additional inverters can be fabricated using the same n-type bed, since they, too, have their n_1 regions grounded.

To fully appreciate the advantages of the I^2L merged transistor technique, it must be realized that other forms of integrated circuit logic require transistors to be isolated from each other. This involves a much more expensive fabrication process, and almost always results in unwanted junctions which must be kept in a reverse-biased state. The unwanted junctions also add speed-reducing capacitances. With I^2L, the need to isolate transistors is eliminated and there are no unwanted junctions.

Integrated injection logic can operate from low- or high-level supply voltages. Typical switching times range from 10 ns up to 250 ns, depending upon the level of injection current; however, switching times less than 1 ns are possible. Power dissipation per gate can be anywhere from 6 nW to 70 μW, again depending upon the injection current. Input and output voltage swings are approximately 700 mV. A single output of an I^2L gate can *sink* (i.e., take in) a current of 20 mA.

12-11 COMPARISON OF MAJOR IC LOGIC TYPES

The major integrated circuit logic families are compared in Table 12-1. TTL, ECL, and CMOS are the most widely used logic systems today. They are the only types that should be seriously considered for any major application. However, the large amount of hardware already in the field does not just disappear when something new is developed. So a knowledge of all currently used circuitry is important to anyone studying logic circuits.

In situations where speed is important, any of the TTL families, except for 74L, may be suitable. Where very high speed is desirable, ECL or Schottky TTL (74S) must be chosen. ECL offers a fan-out of 25 but only *fair* noise immunity, compared to Schottky TTL with a fan-out of 10 but *good* noise immunity. Schottky TTL has slightly lower power dissipation at 20 mW per gate, while the P_D for ECL is 25 mW per gate.

If switching speed is not the paramount consideration, then either CMOS or low-power TTL (74L) might be appropiate. CMOS is the faster of the two and has by far the lowest power dissipation per gate. CMOS has *excellent* noise immunity, and low-power TTL is said to have *good* noise immunity. CMOS also has a fan-out in excess

TABLE 12-1 COMPARISON OF IC LOGIC TYPES

	DTL	RTL	HTL	TTL							ECL	N-MOS and P-MOS	CMOS	I²L
				74	74H	74L	74S	74AS	74LS	74ALS				
Propagation delay time (ns)	30	12	119	10	6	33	3	1.5	9	4	2	50 to 100	25	*10 to 250
Power dissipation per gate (mW)	15	15	50	10	22.5	1	20	20	2	1	25	0.25	10 nW	*6 nW to 70 µW
Noise margin (V)	1.4	0.7	7.5	0.4	0.4	0.4	0.4	0.4	0.4	0.4	0.25	2	$0.3\ V_{DD}$	0.25
Noise immunity rating	good	poor	excellent	good	good	good	good	good	good	good	fair	excellent	excellent	fair
Fan-out	8	5	10	10	10	10	10	10	10	10	25	>50	>50	*depends upon injection current

of 50, while the fan-out for low-power TTL is 10. In situations where low-power dissipation and/or large fan-out are required, CMOS is the only choice.

I^2L is suitable for applications where low power and high gate density are important. For medium- or large-scale integrated systems, I^2L could be the least expensive of all available options.

12-12 INTERFACING DIFFERENT LOGIC TYPES

Interfacing Considerations

Connecting one logic gate to drive another, or connecting the output of a logic gate to an electronic circuit or device, is known as *interfacing*. It has already been demonstrated that a logic gate can easily drive several similar-type gates. Output and input current levels can be used to determine fan-out, while output and input voltage levels can be employed to calculate noise margins.

When a logic gate has to drive one or more different-type gates, the input and output voltages and current levels have to be carefully considered for each gate. Normally, some sort of interface circuit must be employed. This may consist of only a single resistor, or in some circumstances *a buffer circuit* or a *level translater* might be required. In all cases, the logic *0* and logic *1* output levels from the driving gate must be converted to satisfactory *0* and *1* input levels for the driven gate. Also, the *high*-level and *low*-level input currents for the driven gate have to be supplied from the driving gate.

Table 12-2 lists the input and output voltages and currents for standard TTL, CMOS with a 5 V supply, and CMOS with a 15 V supply. These quantities must be employed when considering the problems of interfacing TTL and CMOS gates.

TTL Driving 5 V CMOS

Consider the open-collector TTL-to-CMOS interface illustrated in Figure 12-17(a). The interface circuit consists solely of the pull-up resistor R_C, which is always required with open-collector TTL. The maximum logic *0* output voltage from the TTL gate is $V_{OL(max)} = 0.4$ V, while the maximum input that the CMOS gate can accept as a logic *0* is $V_{IL(max)} = 1.5$ V. Therefore, no problem exists with the logic *0* levels.

TABLE 12-2 INPUT AND OUTPUT VOLTAGE AND CURRENT LEVELS FOR STANDARD TTL AND CMOS LOGIC GATES

	$V_{IH(min)}$	$V_{IL(max)}$	$V_{OH(min)}$	$V_{OL(max)}$	$I_{IH(max)}$	$I_{IL(max)}$	$I_{OH(max)}$	$I_{OL(max)}$
Standard TTL	2 V	0.8 V	2.4 V	0.4 V	40 μA	1.6 mA	400 μA	16 mA
CMOS (5 V)	3.5 V	1.5 V	5 V	0 V	0	0	—	—
CMOS (15 V)	11 V	4.5 V	15 V	0 V	0	0	—	—

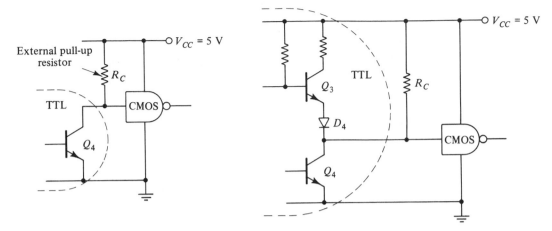

(a) Open-collector TTL driving 5 V CMOS   (b) Totem-pole output TTL driving 5 V CMOS

Figure 12-17   Open-collector TTL and TTL with totem-pole output can be interfaced directly to 5 V CMOS by the use of a pull-up resistor.

With the TTL output transistor *off* and zero input current to the CMOS gate, the TTL output voltage is $V_{OH(min)} \approx V_{CC} = 5$ V. This is satisfactory as a logic *1* input to the CMOS gate, which requires $V_{IH(min)} = 3.5$ V.

Now look at the circuit in Figure 12-17(b), which shows a TTL totem-pole output stage interfaced to a CMOS gate in exactly the same way as the open-collector TTL. In this case, when Q_4 is off, R_C pulls the cathode of D_4 up to V_{CC}, thereby biasing Q_3 *off*. The result is that the same *high* and *low* output levels are obtained from the totem-pole output stage as from the open-collector output. As already seen, these levels are compatible with the CMOS gate input.

The resistance of R_C in Figure 12-17 is determined exactly as in Example 12-2. Since there is virtually zero input current to the CMOS, the maximum current through R_C is the TTL $I_{O(max)}$ of 16 mA. As mentioned in Ex. 12-2, a smaller current level can be used to give a higher resistance for R_C, if the fastest possible switching time is not required. Note that, because CMOS requires zero input current, there is no loading problem when TTL is driving CMOS.

TTL Driving 15 V CMOS

The output of an open-collector TTL gate may be interfaced to CMOS which has a supply of $V_{DD} = 10$ V or 15 V, using a pull-up resistor. The method, as illustrated in Figure 12-18(a), is similar to that of Figure 12-17(a), except that R_C is connected to the CMOS supply. This requires that the TTL output transistor be able to survive the CMOS supply voltage. In fact, some open-collector TTL *buffers*, manufactured for this purpose, can operate with a collector supply of up to 30 V. Once again, the resistance of R_C is determined as in Example 12-2, except that now $V_{RC} = V_{DD} - V_{OL(max)}$.

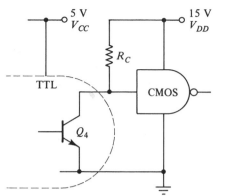

(a) Open-collector TTL driving 15 V CMOS

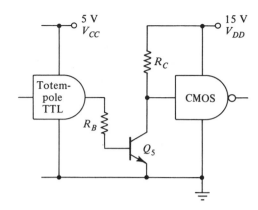

(b) Totem-pole output TTL driving 15 V CMOS

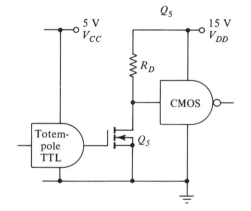

(c) Use of a MOSFET inverter to interface
TTL and CMOS

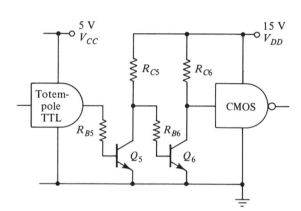

(d) Use of two cascaded inverters to
interface TTL and CMOS

Figure 12-18   Open-collector TTL with $V_{CC} = 5$ V can be interfaced to CMOS with a 15 V supply by the use of a pull-up resistor. For totem-pole output TTL, an open-collector *buffer* stage can be employed. Alternatively, a transistor or FET inverter, or two cascaded inverters, can be used as an interface circuit.

In the case of TTL with a totem-pole output stage, the best solution is to use an open-collector *buffer* stage and interface as in Figure 12-18(a). Alternatively, if an inversion of the TTL output is acceptable, a bipolar or MOSFET inverter stage might be employed, as illustrated in Figures 12-18(b) and (c), respectively. The direct-coupled bipolar inverter can be designed as discussed in Section 4-5. For the MOSFET inverter, the device must be selected to be *on* with a gate voltage of $V_{OH(min)}$ from the TTL gate, and to be off with $V_{OL(max)}$. If the inversion is not acceptable, two cascaded inverters can be used, as shown in Figure 12-18(d).

CMOS Driving TTL

For CMOS with $V_{DD} = 5$ V driving TTL with $V_{CC} = 5$ V, there should be no problem with output voltage levels, but the TTL low-level input current cannot be supplied by the CMOS gate. When biased *on*, the MOSFET channel resistance in CMOS logic is typically $R_{D(on)} = 1$ kΩ, and this is the output resistance of the CMOS gate. This is satisfactory for driving another CMOS gate with zero input current. But for driving just one TTL gate, the low-level output voltage from the CMOS is

$$V_{OL} = R_{D(on)} \times (I_{IL} \text{ to the TTL})$$

$$= 1 \text{ k}\Omega \times 1.6 \text{ mA}$$

$$= 1.6 \text{ V}$$

This is too high to be a satisfactory *low*-level input for the TTL, which requires $V_{OL(max)} = 0.8$ V. The 40 μA *high*-level input current to the TTL has no significant effect upon V_{OH} from the CMOS gate.

Figure 12-19(a) shows a simple method of interfacing CMOS to TTL where inversion of the CMOS output is acceptable. Q_1, R_B, and R_C constitute a direct-coupled inverter. When Q_1 is *off*, the $I_{IH(max)}$ of 40 μA to the TTL input flows through resistor R_C. The resistor is calculated so that $I_{IH(max)}$ produces a very small voltage drop across R_C, leaving $(V_{CC} - V_{RC})$ greater than $V_{IH(min)}$.

When Q_1 is *on*, it must sink the $I_{IL(max)}$ of 1.6 mA for the TTL gate. If additional TTL gates are to be driven by the CMOS gate, Q_1 must sink the total current. Resistor R_B must pass sufficient base current to saturate Q_1 when Q_1 is sinking the necessary TTL input current. Also, $R_{D(on)}$ must be taken into account when calculating R_B.

The circuit in Figure 12-19(a) can be a satisfactory interface for CMOS driving

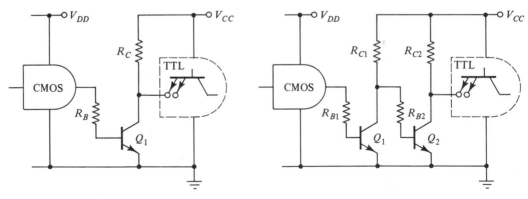

(a) Use of an inverter to drive
    TTL from CMOS

(b) Use of two cascaded inverters
    to drive TTL from CMOS

Figure 12-19 For CMOS driving TTL, a transistor inverter interface circuit must be used to sink the low-level TTL input current. Two cascaded inverters should be used where inversion of the CMOS output is not acceptable.

TTL, regardless of the levels of V_{DD} and V_{CC}. Where inversion of the CMOS output is not acceptable, two cascaded inverters should be employed, as in Figure 12-19(b).

REVIEW QUESTIONS AND PROBLEMS

12-1. Discuss the logic *0* and logic *1* input and output voltage levels for a logic gate. Also, define *noise margin*, and explain the factors that affect the noise immunity of a logic gate.

12-2. Explain *propagation delay time* for a logic gate, and discuss the factors that affect the switching speed of logic circuits. Sketch waveforms to show how propagation delay time is measured.

12-3. Estimate suitable maximum logic *0* and minimum logic *1* input levels for the RTL gate in Figure 12-1(a). Assume that $V_{CC} = 3$ V, $R_C = 1$ kΩ, $R_B = 1.5$ kΩ, and the transistors have $h_{FE} = 20$.

12-4. Estimate suitable maximum logic *0* and minimum logic *1* input levels for the DTL gate in Figure 12-3. Assume the transistors have $h_{FE} = 20$.

12-5. Discuss the input and output current levels for a logic gate, and define *fan-out*, *loading factor*, and *unit load*. Explain the relationship between propagation delay time and the fan-out of a gate.

12-6. Determine the loading factor for the RTL gate in Figure 12-1(a). Take $V_{CC} = 3$ V, $R_C = 1$ kΩ, $R_B = 1.5$ kΩ, and $h_{FE} = 20$.

12-7. Determine the loading factor for the DTL *NOR* gate in Figure 10-6. Take $V_{CC} = 5$ V, $V_{BB} = -2$ V, $R_C = 2$ kΩ, $R_1 = 20$ kΩ, and $h_{FE} = 20$.

12-8. Briefly discuss DTL, HTL, and RTL. Sketch one gate circuit for each.

12-9. Using illustrations, explain the principle of TTL. Discuss the reasons for the fast switching speed and good noise immunity of TTL.

12-10. Sketch the circuit of a typical integrated circuit TTL gate. Explain the function of each component.

12-11. For a standard TTL logic gate, define the quantities I_{OL}, I_{OH}, I_{IL}, and I_{IH}. State typical values for each quantity, and explain how the fan-out should be calculated.

12-12. Sketch an *open-collector TTL* output stage, and show how several gates of this type may be connected in *wired-AND* configuration. Explain.

12-13. Several open-collector TTL gates connected in *wired-AND* are to drive the inputs of three standard TTL gates with $V_{CC} = 5$ V. Determine a suitable resistance for the pull-up resistor.

12-14. Sketch the circuit diagram and logic symbol for a *tri-state TTL gate*. Explain how the circuit functions, and discuss its applications.

12-15. Explain the differences between standard TTL and: *high-speed TTL*, *low-power TTL*, *Schottky TTL*, and *low-power Schottky TTL*.

12-16. Sketch the complete circuit and logic symbol for an *ECL OR/NOR* gate. Carefully explain the operation of the circuit, and discuss the major advantages and disadvantages of ECL.

12-17. Sketch the circuits of *N-MOS NAND* and *NOR* gates. Explain how each circuit operates.

12-18. Sketch the circuits of *P-MOS NAND* and *NOR* gates. Discuss the characteristics of P-MOS and N-MOS logic.

12-19. Sketch the circuits of *CMOS NAND* and *NOR* gates. Carefully explain the operation of each circuit, and discuss the advantages and disadvantages of CMOS logic.

12-20. Sketch the circuit of an I^2L inverter. Explain the operation of the circuit, and compare it to a resistor-transistor inverter.

12-21. Sketch I^2L *NAND* and *NOR* gates, and explain how each operates.

12-22. Using illustrations, explain the construction of an I^2L gate, and discuss its advantages compared to other logic types.

12-23. Compare the various types of IC logic gates in terms of propagation delay time, power dissipation, noise immunity, and fan-out.

12-24. Discuss the problems involved in interfacing different types of logic gates.

12-25. Draw sketches to show how CMOS with a 5 V supply may be driven from (a) open-collector TTL, and (b) standard TTL. Explain.

12-26. Ten CMOS gates with $V_{DD} = 5$ V are to be driven from the output of one open-collector TTL gate. Determine a suitable-value pull-up resistor.

12-27. Show how CMOS with $V_{DD} = 10$ V can be driven by open-collector TTL with $V_{CC} = 5$ V. Design a suitable interface circuit.

12-28. Show how CMOS with $V_{DD} = 10$ V can be driven by standard TTL with $V_{CC} = 5$ V (a) if inversion of the TTL output is acceptable, and (b) if inversion is not acceptable.

12-29. Design suitable interface circuits for each case described in Question 12-28.

12-30. Show how standard TTL with $V_{CC} = 5$ V can be driven by CMOS with $V_{DD} = 10$ V (a) if inversion of the CMOS output is acceptable, and (b) if inversion is not acceptable.

12-31. Design suitable interface circuits for case (a) described in Question 12-30 if four TTL gates are to be driven.

Chapter 13

Bistable Multivibrators (Flip-Flops)

INTRODUCTION

The **bistable multivibrator**, or **flip-flop**, is a switching circuit with two stable states. Each of the two output terminals, identified as Q and $\overline{Q}$, is either *high* (at logic *1*) or *low* (at logic *0*). When Q is at logic *1*, $\overline{Q}$ is at logic *0*, and *vice versa*. A flip-flop can be switched from one state to the other by applying a suitable triggering input. Flip-flops are readily available in integrated circuit form; however, a knowledge of how to design simple transistor flip-flops and triggering circuits gives a thorough understanding of their operation. Flip-flops are essentially memory circuits: they remain in the last *set* or *reset* condition that they were triggered into. Digital counting circuits are cascades of several flip-flops. Integrated circuit flip-flops are categorized as **toggle, set-clear, clocked set-clear, JK**, and **master-slave**.

13-1 TRANSISTOR COLLECTOR-COUPLED BISTABLE

Bistable Operation

The collector-coupled bistable multivibrator circuit shown in Figure 13-1(a) has two stable states: Either Q_1 is *on* and Q_2 is *off*; or Q_2 is *on* and Q_1 is *off*. The circuit is completely symmetrical. Load resistors R_{C1} and R_{C2} are equal, and potential dividers (R_1, R_2) and (R'_1, R'_2) form identical bias networks at the transistor bases. Each transistor is biased from the collector of the other device. When either transistor is *on*, the other transistor is biased *off*.

Consider the condition of the circuit when Q_1 is *on* and Q_2 is *off*. With Q_2 *off*, there is no collector current flowing through R_{C2}. Therefore, as shown in Figure

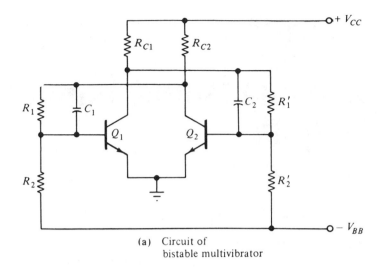

(a) Circuit of
     bistable multivibrator

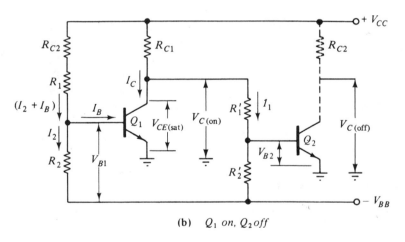

(b) Q_1 on, Q_2 off

Figure 13-1   Collector-coupled bistable multivibrator circuit (a), and circuit conditions
when Q_1 is *on* and Q_2 is *off* (b). Each transistor base is potential-divider biased from
the collector of the other transistor. When one transistor is *on*, the other is biased *off*.

13-1(b), R_{C2}, R_1, and R_2 can be treated as a potential divider biasing Q_1 base from V_{CC}
and $-V_{BB}$. With Q_1 *on* (in saturation), its collector voltage is $+V_{CE(sat)}$, and R'_1 and R'_2
bias V_{B2} below ground level. Since the emitters of the transistors are grounded, Q_2 is
off. The circuit can remain in this condition (Q_1 *on*, Q_2 *off*) indefinitely, as long as the
supply voltages are present. When Q_1 is triggered *off*, Q_2 switches *on*, and remains *on*
with its base biased via R_{C1}, R'_1, and R'_2. At this time, the base of Q_1 is biased negatively
from Q_2 collector and $-V_{BB}$. Thus, Q_1 remains *off* and Q_2 remains *on* indefinitely. The
output voltage at the collector of the *off* transistor is approximately V_{CC}.

Capacitors C_1 and C_2 operate as speed-up capacitors (see Section 4-4) to improve the switching speed of the transistors. However, in the bistable circuit, C_1 and C_2 are also termed *commutating* or *memory capacitors*.

Consider the conditions when Q_1 is *on* and Q_2 is *off*. Capacitor C_1 is charged to the voltage across R_1, and C_2 is charged to the voltage across R_1'. As will be seen when the design of a bistable circuit is considered, the voltage across R_1 (at the base of the *on* transistor) is greater than that across R_1' (at the base of the *off* transistor). Therefore, when Q_1 is *on*, C_1 is charged to a voltage greater than the voltage on C_2. Now, with the capacitor voltages in mind, consider what occurs when both transistors are triggered into an *off* state for a brief instant. With both transistors *off*, both collector voltages are approximately at the level of V_{CC}. Also, the base voltages are; $V_{B1} \approx V_{CC} - V_{cap1}$ and $V_{B2} \approx V_{CC} - V_{cap2}$. Since C_2 has a smaller charge than C_1, V_{B2} is greater than V_{B1}. When the triggering voltage is removed, one transistor must begin to switch *on*, and the one with the higher base voltage switches *on* first. Thus, Q_2 (the formerly *off* transistor) switches *on* before Q_1, and, in so doing, it biases Q_1 *off*. Once switchover occurs, C_2 becomes charged to a greater voltage than C_1, and the next trigger input will cause the circuit to change state once again.

It is seen that the charge on the capacitors enables them to "remember" which transistor was *on* and which was *off*, and facilitates the circuit changeover from one state to another.

Bistable Design

Design of a bistable multivibrator commences with a specification of the supply voltage and collector resistor. Alternatively, the collector current may be specified, or I_C may simply be taken as a level much larger than an output load current. As in the case of the monostable and Schmitt circuits, the bias resistances R_1 and R_2 must be chosen small enough to provide a stable bias level, yet large enough that they do not overload R_C. The rule of thumb that (bias current) $I_2 \approx \frac{1}{10} I_C$ can again be applied (see the discussion in Section 6-2), and the circuit design procedure is then fairly simple. When the value of R_C is calculated, the next *larger* standard resistance should be selected. This will ensure sufficient voltage drop across R_C to have the transistor in saturation. The bias resistances should normally be selected as the next standard resistance size *smaller* than that calculated. This will provide slightly more base current than required for saturation.

The voltage on the commutating capacitors must not change significantly during the turn-*off* time of the transistors. If the capacitors are allowed to discharge by 10% of the difference between maximum and minimum capacitor voltages, Eq. 2-11 may be applied:

$$t = 0.1\,CR$$

Therefore,

$$C = \frac{t_{(off)}}{0.1\,R}$$

In this case, $C = C_1 = C_2$, and $t_{\text{(off)}}$ is the turn-*off* time for the transistors; R is the resistance "seen" looking into the terminals of R_1 or R'_1. With one transistor *on*, the minimum value of R approximates to $R_1 \| R_2$. This gives

$$C_1 = C_2 = \frac{t_{\text{(off)}}}{0.1(R_1 \| R_2)} \tag{13-1}$$

As with other switching circuits, the presence of capacitors limits the maximum frequency at which the bistable circuit may be triggered. To determine the maximum triggering frequency, the *recovery time* for the capacitors must be calculated. This is the time for the capacitors to discharge from maximum voltage to minimum voltage, or *vice versa*. The maximum triggering frequency is then calculated as 1/(recovery time). Using Eq. 2-10, the recovery time is

$$t_{re} = 2.3\, CR$$

where again $R = (R_1 \| R_2)$ and maximum triggering frequency is

$$f_{\max} = \frac{1}{t_{re}} = \frac{1}{2.3\, C(R_1 \| R_2)} \tag{13-2}$$

EXAMPLE 13-1

Design a collector-coupled bistable multivibrator as in Fig. 13-2 to operate from a ± 5 V supply. Use 2N3904 transistors, with $I_{C(\text{sat})} = 2$ mA. Analyze the design to determine all voltage levels.

Solution   Refer to Figure 13-1(b),

$$V_{CE(\text{sat})} = 0.2 \text{ V (typically)}$$

$$R_{C1} = R_{C2} \approx \frac{V_{CC} - V_{CE(\text{sat})}}{I_{C(\text{sat})}} \text{ (i.e., neglecting } I_1)$$

$$= \frac{5 \text{ V} - 0.2 \text{ V}}{2 \text{ mA}}$$

$$= 2.4 \text{ k}\Omega \text{ (use 2.7 k}\Omega \text{ standard value)}$$

From the 2N3904 data sheet in Appendix 1-4, $h_{FE(\min)} = 70$; so

$$I_{B(\min)} = \frac{I_{C(\text{sat})}}{h_{FE(\min)}}$$

$$= \frac{2 \text{ mA}}{70} = 28.6 \text{ }\mu\text{A}$$

With Q_1 *on*,

$$V_{B1} = V_{R2} = V_{BE1} - V_{BB}$$

$$= 0.7 \text{ V} - (-5 \text{ V}) = 5.7 \text{ V}$$

$$I_2 \approx \frac{1}{10} I_C = \frac{2 \text{ mA}}{10} = 200 \text{ μA}$$

$$R_2 = \frac{V_{R_2}}{I_2} = \frac{5.7 \text{ V}}{200 \text{ μA}}$$

$$= 28.5 \text{ kΩ (use 27 kΩ standard value)}$$

Now I_2 becomes

$$I_2 = \frac{5.7 \text{ V}}{27 \text{ kΩ}} = 211 \text{ μA}$$

$$R_{C2} + R_1 = \frac{V_{CC} - V_{BE}}{I_2 + I_B}$$

$$= \frac{5 \text{ V} - 0.7 \text{ V}}{211 \text{ μA} + 28.6 \text{ μA}}$$

$$= 17.9 \text{ kΩ}$$

$$R_1 = (R_{C2} + R_1) - R_{C2}$$

$$= 17.9 \text{ kΩ} - 2.7 \text{ kΩ}$$

$$= 15.2 \text{ kΩ (use 15 kΩ standard value)}$$

Analysis

$$V_{C(\text{on})} = V_{CE(\text{sat})} = 0.2 \text{ V}$$

$$V_{C(\text{off})} = V_{CC} - V_{RC2} \text{ (for } Q_2 \text{ off)}$$

$$\text{Voltage across } (R_{C2} + R_1) = V_{CC} - V_{BE1}$$

$$= 5 \text{ V} - 0.7 \text{ V} = 4.3 \text{ V}$$

$$V_{RC2} = (V_{CC} - V_{BE1}) \times \frac{R_{C2}}{R_1 + R_{C2}}$$

$$= 4.3 \text{ V} \times \frac{2.7 \text{ kΩ}}{15 \text{ kΩ} + 2.7 \text{ kΩ}} = 0.66 \text{ V}$$

$$V_{C(\text{off})} = 5 \text{ V} - 0.66 \text{ V} = 4.34 \text{ V}$$

$$V_{B(\text{off})} = V_{CE(\text{sat})} - V_{R'1} \text{ (for } Q_2 \text{ off)}$$

$$\text{Voltage across } (R'_1 + R'_2) = V_{CE(\text{sat})} - V_{BB}$$

$$= 0.2 \text{ V} - (-5 \text{ V})$$

$$= 5.2 \text{ V}$$

$$V_{R'1} = (V_{CE(\text{sat})} - V_{BB}) \times \frac{R'_1}{R'_1 + R'_2}$$

$$= 5.2 \text{ V} \times \frac{15 \text{ k}\Omega}{15 \text{ k}\Omega + 27 \text{ k}\Omega} = 1.86 \text{ V}$$

$$V_{B(\text{off})} = 0.2 \text{ V} - 1.86 \text{ V} = -1.66 \text{ V}$$

Two-Inverter Bistable

Again referring to the circuit in Figure 13-1, it is seen that each transistor operates as a direct-coupled inverter. In fact, a bistable multivibrator can be constructed by cross-coupling two IC inverters, as illustrated in Figure 13-3. With Q_1 output *low* (at ground level), resistors R'_1 and R'_2 bias the input of Q_2 *low*. Consequently, Q_2 output is *high* (near V_{CC}), and resistors R_1 and R_2 bias the input of Q_1 *high*, to give a *low* output. Similarly, when Q_1 output is *high*, Q_2 output is *low*, and the potential dividers provide the appropriate input bias levels.

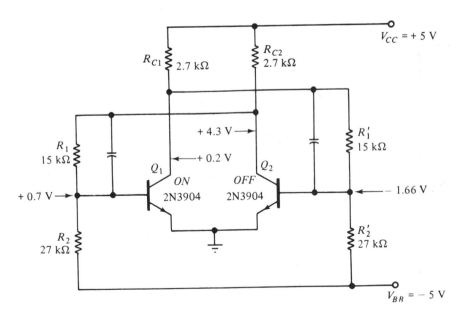

Figure 13-2 Component values and voltage levels for the bistable multivibrator circuit designed in Example 13-1.

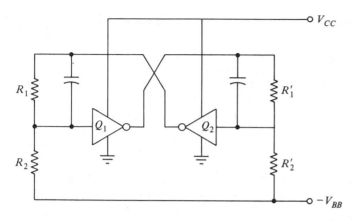

Figure 13-3 Bistable multivibrator constructed of two cross-coupled IC inverters. When Q_1 output is *high*, a *high* input is provided to Q_2. Thus, Q_2 output is *low*, and this biases Q_1 input *low*.

13-2 EMITTER-COUPLED BISTABLE

The emitter-coupled bistable multivibrator circuit (Figure 13-4) is the same as the collector-coupled circuit, except that an emitter resistor R_E has been added. Load resistors R_{C1} and R_{C2} are equal, as are capacitors C_1 and C_2, and potential dividers (R_1, R_2) and (R'_1, R'_2).

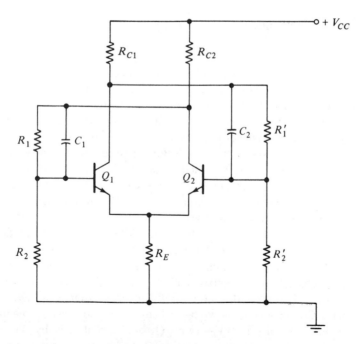

Figure 13-4 An emitter-coupled bistable circuit uses a common emitter resistor R_E. This controls the level of transistor current and permits the circuit to be designed for saturated or unsaturated operation.

The presence of R_E allows the circuit to be operated from a single-polarity supply. The emitter resistor also limits the collector current of the *on* transistor to any desired level, so that the transistor may be saturated or unsaturated. An emitter-coupled bistable has a lower output voltage swing than a collector-coupled circuit with the same V_{CC} level. Also, if the transistors are operated in an unsaturated condition, there is more power dissipation than with saturated transistors.

In designing a nonsaturated emitter-coupled bistable circuit, the voltage drop across R_E must be made several times the base-emitter voltage of the device. This is necessary to maintain reasonably stable bias conditions. A minimum V_{CE} of about 3 V should be designed into the circuit in order to avoid device saturation. A good rule of thumb is to divide $(V_{CC} - V_{CE})$ equally between V_{RC} and V_{RE}. Also, to avoid saturation, the *maximum* transistor h_{FE} must be employed in the design calculation. When R_C is calculated, the next *smaller* standard value should be selected, again to avoid saturation. R_1 and R_2 must also be carefully chosen with nonsaturation in mind. As with other multivibrator circuits, bias current I_2 should be approximately $\frac{1}{10} I_C$. This is to ensure that R_1 and R_2 are small enough to provide a stable bias voltage, but not so small that they overload R_C.

13-3 BISTABLE TRIGGERING

Asymmetrical Collector Triggering

Bistable multivibrator triggering circuits are normally designed to turn *off* the *on* transistor. The triggering may be *asymmetrical* or *symmetrical*. In *asymmetrical triggering*, two trigger inputs are employed, one to *set* the circuit in one particular state, and the other to *reset* to the opposite state. This process is sometimes referred to as *set-reset triggering*. Symmetrical triggering uses only one trigger input, and the state of the circuit is changed each time a suitable trigger pulse is applied.

Refer to the asymmetrical collector trigger circuit and waveforms shown in Figure 13-5, and assume that Q_1 is *on* and Q_2 is *off*. Two trigger circuits are provided at input terminals A and B, each consisting of a capacitor and two diodes. With Q_2 *off*, the voltage at its collector is approximately V_{CC}. The negative-going step input to terminal A, coupled via C_3, forward-biases D_1 and pulls its cathode down by ΔV. At this time, D_2 is reverse biased by the negative-going input and has no function to perform. (The function of D_2 will become apparent shortly.) The anode of D_1 is pulled down by $\Delta V - V_{F1}$; that is, by ΔV minus the diode forward voltage drop. Thus, Q_2 collector voltage is changed from approximately V_{CC} to $[V_{CC} - (\Delta V - V_{F1})]$. (See the waveforms in Figure 13-5.) Capacitor C_1 does not discharge instantaneously, but acts initially like a battery. Consequently, the voltage change at Q_2 collector also appears at Q_1 base. Q_1 base voltage initially is $+V_{BE}$ (with Q_1 *on*), and it falls by $\Delta V - V_{F_1}$. Clearly, the input voltage at C_3 causes the base of Q_1 to be pushed below the level of its emitter voltage.

When the step input is applied, C_3 immediately starts to charge via R_{C2}. (The

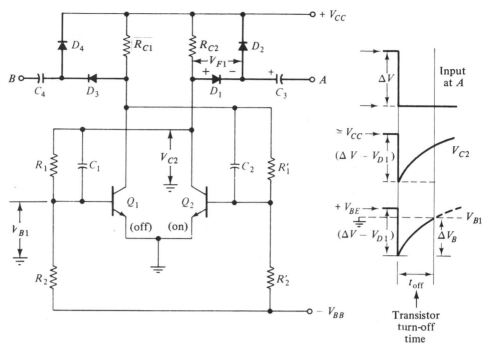

Figure 13-5 Asymmetrical collector triggering. A negative-going trigger input to terminal A switches Q_1 *off* and Q_2 *on*. A negative-going input to terminal B triggers Q_1 *on* and Q_2 *off*.

polarity is shown in Figure 13-5.) Both collector voltage V_{C2} and base voltage V_{B1} rise from their minimum levels, as illustrated. To ensure that Q_1 switches *off*, its base voltage must remain below the emitter voltage level for the transistor turn-*off* time $t_{(off)}$. This may be achieved by use of a large-valued coupling capacitor for C_3. However, it is best to choose C_3 as small as possible. The smallest suitable capacitor is one that will allow V_B to rise to the level of the emitter voltage during the transistor turn-*off* time. The waveform in Figure 13-5 shows V_{B1} rising to ground level during $t_{(off)}$.

When the triggering input becomes positive, returning to its normal dc level, D_1 is reverse biased and the state of the bistable circuit is unaffected. The charge on C_3, which resulted from the negative-going input, remains until the trigger input becomes positive. Diode D_2 is then forward biased by the capacitor charge, and C_3 is rapidly discharged via D_2. The triggering circuit now is ready to receive another negative-going input. However, with Q_1 already *off*, the state of the circuit will not be altered by a triggering input to terminal A. Instead, Q_2 must be triggered *off* by an input applied to terminal B.

Asymmetrical collector triggering, using transistors instead of diodes, is illustrated in Figure 13-6. A positive input applied to terminal A switches transistor Q_3 *on* into saturation. Collector current for Q_3 flows through resistor R_{C2}, pulling the collector of Q_2 to its saturated level. The base of Q_1 is now biased (via R_1 and R_2) below its emitter

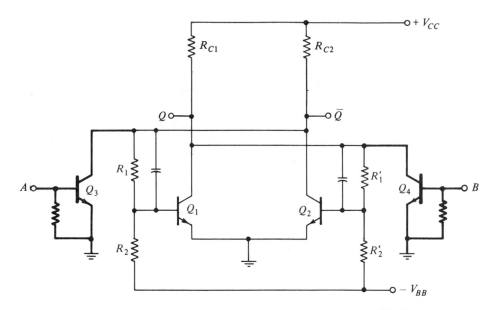

Figure 13-6 Asymmetrical collector triggering using transistors. A positive input to terminal A switches Q_1 *off*, and Q_2 *on*. A positive input to terminal B switches Q_1 *on*, and Q_2 *off*.

voltage, and Q_1 is *off*. Because Q_1 is *off*, its collector voltage is *high*, and consequently, Q_2 base is biased (via R'_1 and R'_2) above its emitter voltage level, so that Q_2 is *on*. When the positive input is removed from terminal A, Q_3 switches *off*, but Q_2 remains *on*, and Q_1 remains *off*. In a similar manner, a positive input to terminal B switches Q_4 *on*, to change the state of the flip-flop to Q_1 *on* and Q_2 *off*.

Symmetrical Collector Triggering

Symmetrical collector triggering is illustrated in Figure 13-7. With Q_1 *on* and Q_2 *off*, I_C flows through R_{C1}, causing V_{C1} to be approximately zero volts. Also, V_{C2} is approximately V_{CC}. The amplitude ΔV of the negative-going trigger input does not exceed the voltage drop across R_C (i.e., ΔV is less than V_{CC}), so that diode D_3 does not become forward biased. However, with $V_{C2} \approx V_{CC}$, diode D_1 is forward biased by the negative-going input. Thus, Q_1 base is pushed down, exactly as discussed for asymmetrical triggering, and Q_1 is turned *off*. With Q_1 *off*, V_{C1} rises to approximately V_{CC}, and V_{C2} drops to near zero. The next negative-going input forward-biases D_3, causing Q_2 base to be pushed below its emitter level. Hence, Q_2 switches *off* and the circuit returns to its original state.

It is clear that the circuit changes state each time a negative-going trigger voltage is applied. D_2 functions as before, becoming forward biased and discharging C_3 each time the input returns to its upper level. A resistor could function in place of D_2, but it would load the trigger signal and would take a relatively long time to discharge C_3.

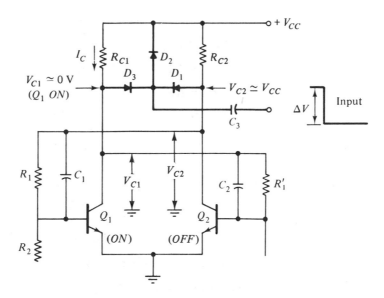

Figure 13-7 Symmetrical collector triggering. A negative-going trigger input changes the state of the flip-flop. The previously-*on* transistor is switched *off*, and the previously *off* device is turned *on*.

The design of collector triggering circuits mainly involves determination of the smallest suitable coupling capacitor. The allowable change in voltage at the base of the transistor to be switched *off* dictates the voltage through which the coupling capacitor may be charged.

Consider Figures 13-5 and 13-7 again. Note that, when the trigger voltage pulls the collector of Q_2 down to near ground level, capacitor C_1 begins to discharge via $R_1 \| R_2$. As already explained in Section 13-1, the commutating capacitor voltage should not be allowed to discharge by more than 10% of the difference between maximum and minimum capacitor voltages. Equation 13-1 can be applied to calculate C_1 and C_2.

EXAMPLE 13-2

The saturated collector-coupled flip-flop designed in Example 13-1 is to be triggered by the collector output of a previous similar stage. Design a suitable symmetrical collector triggering circuit.

Solution   The waveforms of the triggering voltage as it appears at various points in the circuit are shown in Figure 13-8. From Example 13-1, the collector voltage of the flip-flop changes from 4.3 V to 0.2 V. This change is used as an input triggering voltage. (See Figure 13-8.)

$$\Delta V_i = 4.3 \text{ V} - 0.2 \text{ V} = 4.1 \text{ V}$$

At the diode cathodes,

$$\Delta V_K \approx 4.1 \text{ V}$$

$$\Delta V_{C2} = \Delta V_K - V_{F1}$$

$$= 4.1 \text{ V} - 0.7 \text{ V} = 3.4 \text{ V}$$

$$\Delta V_{B1} = \Delta V_{C2} = 3.4 \text{ V}$$

To keep $V_{B1} < V_E$ during $t_{(off)}$,

let $\qquad\qquad\qquad \Delta V = \Delta V_{B1} - V_{BE}$

$$= 3.4 \text{ V} - 0.7 \text{ V} = 2.7 \text{ V}$$

Therefore, C_3 can charge by 2.7 V during $t_{(off)}$. Hence, for C_3,

$$\text{Initial voltage} = E_0 \approx 0 \text{ V}$$

$$\text{Final voltage} = e_c \approx \Delta V = 2.7 \text{ V}$$

$$\text{Charging voltage} = E = \Delta V_i - V_{F1} = 3.4 \text{ V}$$

$$\text{Charging resistance} \approx R_C = 2.7 \text{ k}\Omega$$

$$\text{Turn-}off\text{ time (for 2N3904)} = t_{(off)} = 250 \text{ ns}$$

From Eq. 2-9,

$$C_3 = \frac{t}{R_C \ln\left(\dfrac{E - E_o}{E - e_c}\right)}$$

$$= \frac{250 \text{ ns}}{2.7 \text{ k}\Omega \times \ln\left(\dfrac{3.4 \text{ V} - 0}{3.4 \text{ V} - 2.7 \text{ V}}\right)}$$

$$\approx 59 \text{ pF (use 62 pF standard value)}$$

The diodes required for the triggering are low-current devices capable of surviving a peak inverse voltage greater than V_{CC}. 1N914 diodes are more than adequate. (See the data sheet in Appendix 1-1.)

EXAMPLE 13-3

Determine the capacitance of suitable commutating capacitors for the flip-flop designed in Example 13-1 when collector triggering is employed. Also, calculate the maximum triggering frequency for the circuit.

Solution   By Eq. 13-1,

$$C_1 = C_2 = \frac{t_{(off)}}{0.1(R_1 \| R_2)}$$

$$= \frac{250 \text{ ns}}{0.1(15 \text{ k}\Omega \| 27 \text{ k}\Omega)} = 259 \text{ pF (use 270 pF standard value)}$$

By Eq. 13-2,

$$f_{(max)} = \frac{1}{2.3\, C(R_1 \,\|\, R_2)}$$

$$= \frac{1}{2.3 \times 270\ \text{pF}(15\ \text{k}\Omega \,\|\, 27\ \text{k}\Omega)}$$

$$= 167\ \text{kHz}$$

Base Triggering

Base triggering circuits are subdivided into *asymmetrical base triggering*, *symmetrical base triggering*, and *collector-steered base triggering* circuits. The first two of these are shown schematically in Figure 13-9. In Figure 13-9(a), a negative-going input coupled

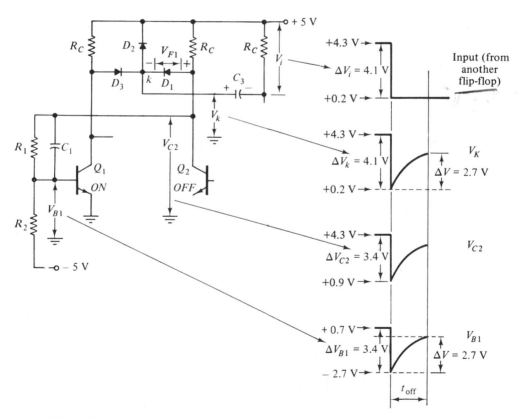

Figure 13-8 Triggering voltage waveforms for the symmetrical collector trigger circuit designed in Example 13-2.

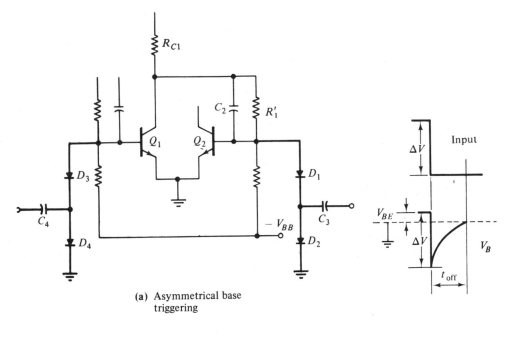

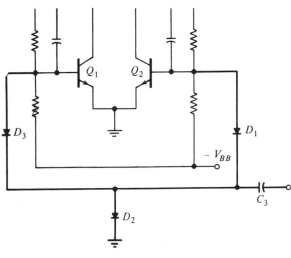

(a) Asymmetrical base
triggering

(b) Symmetrical base
triggering

Figure 13-9 Asymmetrical and symmetrical base triggering. In circuit (a), separate negative-going inputs switch each transistor *off*. In (b), a single negative-going input triggers the flip-flop from one state to another.

via C_3 forward-biases D_1 and pulls Q_2 base below its emitter voltage level. C_3 charges via R_{C1} (C_2 bypasses R_1') and allows the base voltage to rise to ground over a time period equal to the transistor turn-*off* time. C_3 is discharged via D_2 when the input becomes positive to return to its normal dc level.

In Figure 13-9(a), the negative-going input to C_3 can only switch Q_2 *off*. To turn Q_1 *off*, a negative-going input must be coupled via C_4 and D_3. Positive-going inputs at either terminal reverse-bias D_1 and D_3 and have no effect on the bistable circuit.

For the *symmetrical base triggering circuit* in Figure 13-9(b), D_2 is common and the input is applied to the common cathode terminal of D_3 and D_1. If Q_1 is *off*, its base voltage is biased negatively with respect to ground. Thus, if the input signal is kept small enough, D_3 does not become forward biased. Q_2 base is at $+V_{BE}$ with respect to ground (with Q_2 *on*), so that the negative-going input forward-biases D_1 and pulls the transistor base below its emitter level. When Q_2 is *off* and Q_1 is *on*, D_3 is forward biased by the trigger input, and D_1 remains reverse biased. D_2 becomes forward biased only when the input becomes positive. At this time, C_3 is discharged.

The amplitude of the trigger input to the symmetrical base triggering circuit is very critical. The input must be greater than 0.7 V in order to switch *off* the *on* transistor. If the base voltage of the *off* transistor is, say, -1.5 V, then the input voltage amplitude should be less than 1.5 V to avoid affecting the *off* transistor. The collector-steered base triggering circuit overcomes the problem of critical triggering amplitude.

In the *collector-steered base triggering circuit*, shown in Figure 13-10, diode D_2 has its anode connected to Q_2 base and its cathode connected via *steering resistance R_4* to Q_2 collector. Similarly, the anode of D_1 is connected to Q_1 base, and its cathode is connected via R_3 to Q_1 collector. The common triggering signal is applied to D_1 and D_2 cathodes via separate capacitors C_3 and C_4, respectively.

Consider the conditions for the circuit in Fig. 13-10 when Q_1 is *off* and Q_2 is *on*. With Q_1 *off*, the voltage at its collector is a little lower than V_{CC}. Therefore, the cathode of D_1 is approximately at V_{CC}. In this case, triggering inputs with amplitude less than V_{CC} will *not* forward-bias D_1. Since Q_2 is *on*, its collector voltage is $V_{CE(sat)}$ above ground level; consequently, the cathode of D_2 is at $+V_{CE(sat)}$. A negative-going trigger input with an amplitude of a few volts will forward-bias D_2 and pull the base of Q_2 below ground. Thus, to cause the circuit to correctly change state, the trigger voltage can have an amplitude anywhere between about 2 V and V_{CC}. The triggering voltage at each of the bases and at the diode cathodes are illustrated by the waveforms in Figure 13-10.

For collector-steered base triggering (Figure 13-10), R_3 and R_4 should be selected much larger than R_C. This is to ensure that the steering resistors do not constitute a significant load on R_{C1} and R_{C2}. When Q_2 base voltage is pulled down by the input trigger voltage, commutating capacitor C_2 commences to charge via R_C. Since Q_2 is to be switched *off*, C_2 voltage can be allowed to increase slightly. However, the charge on C_1 (at the base of the transistor that is to switch *on*) should not change significantly. Once again, Eq. 2-11 applies.

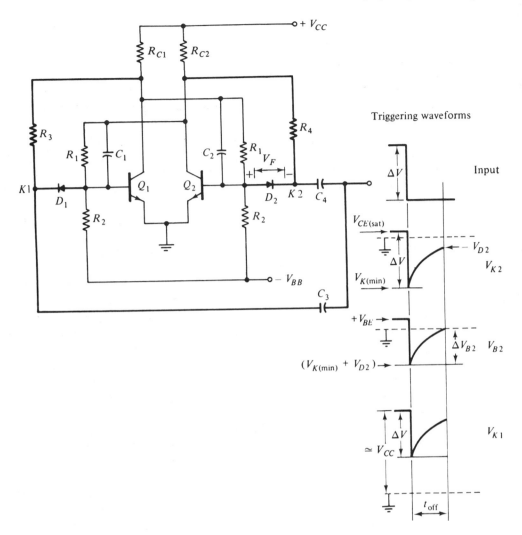

Figure 13-10 Collector-steered base triggering circuit and waveforms. This facilitates reliable triggering, with relatively large triggering voltages.

Level Triggering and Edge Triggering

Some flip-flops are triggered by dc voltage levels applied at the triggering input terminals. Other flip-flops require the fast rise or fall time at a pulse edge to trigger from one state to another. The circuit in Figure 13-6 is *level triggered*. The flip-flop will change state just as soon as a trigger input is applied at one of the input terminals. Thus, it might seem to be triggered by the leading edge of the input pulse. But it is the dc input level, turning Q_3 or Q_4 *on*, that causes the change of state; the rise time of the input pulse is of no consequence at all in this case.

The flip-flops in Figures 13-5 and 13-7 are *edge triggered*, as are the circuits in Figures 13-9 and 13-10. For these circuits, the dc voltage level at the triggering input terminals is not important. When the voltage level at the triggering inputs changes rapidly, the change is coupled via the capacitors to the transistor collectors or bases. This rapid change in voltage level causes the flip-flop to change state.

13-4 THE *T* FLIP-FLOP

When used as a logic element, the bistable multivibrator is usually termed a *flip-flop*. Flip-flops are available as integrated circuits in a wide variety of forms and combinations. The simplest of all is the *T flip-flop*, or *toggle* flip-flop.

The *T* flip-flop is a bistable multivibrator symmetrically triggered by an input to one terminal. Any one of the symmetrically triggered bistable circuits already studied (e.g., the circuit in Fig. 13-7) can be employed as a *T* flip-flop. The logic symbol for a *T* flip-flop is shown in Figure 13-11(a). The Q and $\bar{Q}$ terminals are the circuit output points. These would be the collector terminals of transistors Q_1 and Q_2 for the circuit in Figure 13-7. The Q output terminal is recognized as the *normal* output of the flip-flop, and the $\bar{Q}$ output is identified as the *inverted* output.

The *T* terminal is the input trigger, or *toggling*, terminal for the flip-flop. Note

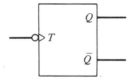

(a)   Logic symbol for T flip-flop
triggered by a negative-going edge.

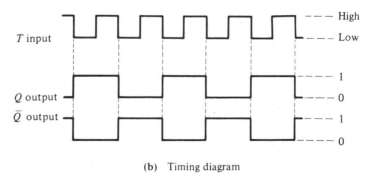

(b)   Timing diagram

Figure 13-11   Logic symbol and timing diagram for a *T* flip-flop, or *toggle* flip-flop, triggered by a negative-going input. The flip-flop changes state each time the input changes from *high* to *low*.

that the small circle, or *bubble*, at the toggle input on the logic symbol indicates that a negative-going signal is required to trigger this flip-flop. The arrowhead at the *T* input identifies this as an edge-triggered flip-flop, rather than a level-triggered circuit.

Figure 13-11(b) shows the timing diagram for the flip-flop. Each time that the *T* input goes from a *high* to a *low* level, the *Q* and $\overline{Q}$ outputs are seen to change state.

In Figure 13-12, the logic symbol and timing diagram are shown for a flip-flop which is triggered by a positive-going edge. In this case, the *Q* and $\overline{Q}$ outputs change state at the instant that the *T* input changes from *low* to *high*. The bistable circuit in Figure 13-7 could be converted into a positive-going edge-triggered flip-flop by connecting a normally *off* inverter (see Section 4-6) at the triggering input.

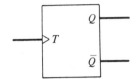

(a)   Logic symbol for T flip-flop triggered by a positive-going edge.

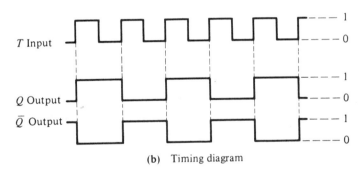

(b)   Timing diagram

Figure 13-12   Logic symbol and timing diagram for a *T* flip-flop, or *toggle* flip-flop, triggered by a positive-going input. The flip-flop changes state each time the input changes from *low* to *high*.

13-5 THE *SC* FLIP-FLOP

The *set-clear* or *SC* flip-flop, which has the logic symbol illustrated in Figure 13-13(a), could have a circuit similar to the asymmetrically triggered flip-flop in Figure 13-6, except that the circuit in Figure 13-6 is triggered by positive-going levels, while that in Figure 13-13 requires negative-going inputs. Previous consideration of the circuit in Figure 13-6 showed that it can be triggered into one particular state (i.e., it can be *set*) when a trigger voltage is applied to one input terminal. Also, it can be triggered (or

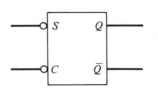

(a) Logic symbol for *SC* flip-flop triggered
by negative-going signals

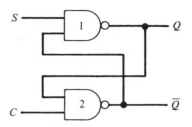

(b) Cross-coupled *NAND* gates as
SC flip-flop

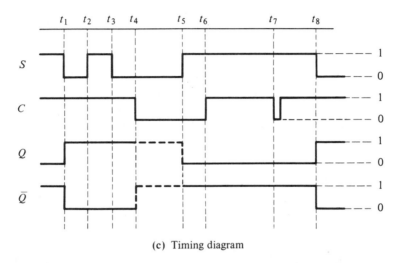

(c) Timing diagram

Figure 13-13 *SC*, or *set-clear*, flip-flop. A negative-going input to terminal *S* sets
the flip-flop into a state of $Q = 1$ and $\overline{Q} = 0$. A negative-going input to terminal *C*
clears the circuit to $Q = 0$ and $\overline{Q} = 1$.

cleared) into the opposite state by a voltage applied to the other input terminal. Another
name for the *SC* flip-flop is the *RS*, or *reset-set*, flip-flop.

The cross-coupled *NAND* gates in Figure 13-13(b) function as an *SC* flip-flop
triggered by negative-going inputs. Assume that both *S* and *C* inputs are *high*, and that
$Q = 1$ and $\overline{Q} = 0$. *Q* and *C* are the inputs to gate 2, so, with both *Q* and *C* at logic
1, the output of gate 2 is logic *0*, i.e., $\overline{Q} = 0$, as assumed. With gate 1 inputs at $\overline{Q} = 0$ and $S = 1$, the output of gate 1 is $Q = 1$, as assumed. The flip-flop is now in its *set*
condition.

Now suppose that a *0* input is applied to terminal *S*. The output of gate 1 remains
at $Q = 1$, because a (negative-going) input to the *set* terminal does not change the
flip-flop from its *set* condition. With *S* remaining at *1*, and a *0* input applied to *C*, the
output of gate 2 becomes $\overline{Q} = 1$. Both *S* and $\overline{Q}$ are at logic *1*, and consequently, *Q*
switches to *0*.

If a *0* input level is applied to both *S* and *C* in the cross-coupled *NAND* gate flip-flop, both outputs go to *1*. For other negative-going level-triggered *SC* flip-flop circuits, the output may be indeterminate with two *0* input levels. These conditions are unacceptable for any flip-flop, and must be avoided.

Following the timing diagram waveforms in Figure 13-13(c) from left to right,

At t_1, Q and $\overline{Q}$ change to *set* state ($Q = 1$, $\overline{Q} = 0$) when *S* goes from *1* to *0*.

At t_2, Q and $\overline{Q}$ do not change state when *S* goes from *0* to *1*.

At t_3, Q and $\overline{Q}$ do not change state when *S* goes from *1* to *0* because the flip-flop is already in the *set* condition.

At t_4, Q and $\overline{Q}$ are both *1* or indeterminate when *S* is *0* and *C* goes from *1* to *0*. (Unacceptable condition)

At t_5, Q and $\overline{Q}$ change to the cleared state when *S* is *1* and *C* is *0*.

At t_6, Q and $\overline{Q}$ do not change state when *C* goes from *0* to *1*.

At t_7, Q and $\overline{Q}$ do not change state when *C* goes from *1* to *0* and back to *1*, because the flip-flop is already in the *cleared* condition.

At t_8, Q and $\overline{Q}$ go into the *set* condition when *S* goes from *1* to *0*.

The *SC* flip-flop logic symbol in Figure 13-14(a) has no bubbles at the *S* and *C* inputs, showing that it requires positive-going inputs for triggering as in Figure 13-6. The timing diagram for this circuit is similar to that in Figure 13-13(c), except that the *S* and *C* waveforms must be inverted, to show that the outputs change state only when the input levels change from *0* to *1*.

Figure 13-14(b) shows two cross-coupled *NOR* gates, which function as an *SC* flip-flop triggered by a positive-going input. Note the reversed position of terminals Q and $\overline{Q}$ relative to the logic symbol in Figure 13-14(a). Assume that both inputs are at *logic 0*, and that the circuit is in its *set* condition of $Q = 1$ and $\overline{Q} = 0$. Then *C* and $\overline{Q}$ inputs to gate 2 are at *0*, giving the output of $Q = 1$. With one input to gate 1 equal to *logic 1*, its output is $\overline{Q} = 0$. Further consideration demonstrates that a positive-going input to *C* changes the state of the circuit to *clear*, ($Q = 0$, $\overline{Q} = 1$). Also, a subsequent positive-going input to *S* changes the state of the circuit back to the *set* condition.

Figure 13-14(c) shows a *truth table* for the flip-flop of Figure 13-14(a). This is an alternative to the timing diagram as a method of describing the effect of various input combinations. When *S* and *C* are both *0*, there is no change in the flip-flop outputs; the flip-flop could be in either the *set* or the *clear* state. When a *1* is applied to *S*, and *C* remains *0*, the outputs go to the *set* state ($Q = 1$, $\overline{Q} = 0$). When a *1* is applied to *C*, and *S* remains at *0*, the circuit is *cleared* to ($Q = 0$, $\overline{Q} = 1$). When a *1* is applied to both *S* and *C* at once, both outputs in the cross-coupled *NAND* gate flip-flop go to *0*. For other positive-going level triggered *SC* flip-flop circuits the output is indeterminate when both inputs are *1*. Once again, these unacceptable conditions for a flip-flop must be avoided.

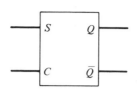

(a) Logic symbol for *SC* flip-flop triggered by positive-going inputs.

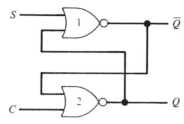

(b) Cross-coupled *NOR* gates as *SC* flip-flop

S	C	Q	$\overline{Q}$
0	0	No change	No change
1	0	1	0
0	1	0	1
1	1	Indeterminate	

(c) Truth table

Figure 13-14 *SC* flip-flop triggered by positive-going inputs. A positive-going input to terminal *S sets* the flip-flop into a state of $Q = 1$ and $\overline{Q} = 0$, while a positive-going input to terminal *C clears* the circuit to $Q = 0$ and $\overline{Q} = 1$.

13-6 THE CLOCKED *SC* FLIP-FLOP

A *clocked SC flip-flop* (also known as a *reset-set-toggle*, *RST* flip-flop, or *gated RS* flip-flop) is one that combines the triggering facilities of a toggle flip-flop with the set-clear arrangement of an *SC* flip-flop. *Clock* is the logic term for an accurate trigger frequency source. In Figure 13-15(a), the toggle input is identified as *CLK* for a clock.

From the timing diagram in Figure 13-15(b), it is seen that the outputs change state only when the clock input goes from *0* to *1*. With the outputs in the *clear* condition ($Q = 0$, $\overline{Q} = 1$), and with $S = 1$ and $C = 0$, the output changes to the *set* condition at the first positive-going edge of the clock. The circuit then remains in the *set* condition even when *S* goes to *0*. When *C* has a *1* input applied (and *S* is at *0*), the outputs return to the *clear* condition as soon as the clock goes from *0* to *1* again. With both *S* and *C* inputs at *0*, the clock has no further effect on the outputs. If both *S* and *C* inputs have a *1* applied simultaneously, the state of the flip-flop would be indeterminante. Thus, this condition is to be avoided.

It is evident that the *S* and *C* inputs are control inputs that determine which state the flip-flop will assume when a clock signal is applied. In the absence of the clock signal, the *S* and *C* inputs have no effect on the flip-flop. The clock input triggers the

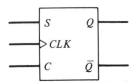

(a)   Logic symbol for clocked SC flip-flop

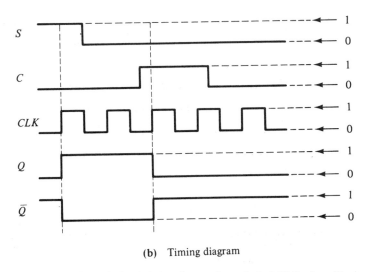

(b)   Timing diagram

Figure 13-15   Logic symbol and timing diagram for a *clocked SC* flip-flop. The input levels at the *S* and *C* terminals determine the state that the flip-flop is triggered into by the clock pulse.

flip-flop into the state dictated by the *S* and *C* input levels. The control inputs must be present for a minimum time, known as the *set-up time*, prior to the arrival of a clock pulse, if the flip-flop is to be triggered into the desired state. The *S* and *C* inputs must also remain constant for a minimum time, termed the *holding time*, after arrival of the clock pulse.

13-7 THE *D* FLIP-FLOP

The *D flip-flop* (or *data flip-flop*) is similar to the clocked *SC* flip-flop, except that it has a single control input, terminal *D* in Figure 13-16(a). Figure 13-16(b) shows that the equivalent circuit of the *D* flip-flop is a clocked *SC* flip-flop with an inverter connected between the *S* and *C* terminals. The input to the *clear* terminal is now always the inverse of the input to the *set* terminal. Consequently, when the flip-flop is triggered,

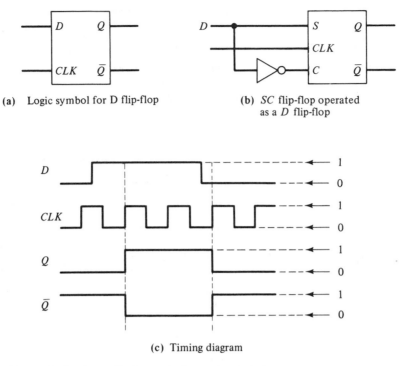

(a) Logic symbol for D flip-flop

(b) *SC* flip-flop operated as a *D* flip-flop

(c) Timing diagram

Figure 13-16 The *D* flip-flop is similar to the *clocked SC* flip-flop, except that a single *D* terminal functions as a *set* input. When triggered by the clock pulse, the *Q* output assumes the level at the *D* input.

the *Q* output always assumes the logic level at the *D* (*set*) input, and the $\overline{Q}$ output assumes the inverse.

The timing diagram in Figure 13-16(c) illustrates the action of the *D* flip-flop. When *D* = *1*, the *Q* output switches to *1* when the clock input goes in a positive direction. When *D* = *0*, *Q* changes to *0* on receipt of a clock input. The $\overline{Q}$ output always assumes the opposite state to that of the *Q* output.

13-8 THE *JK* FLIP-FLOP

The *JK flip-flop* performs identically to the clocked *SC* flip-flop with one exception: when both the *J* and *K* inputs are at logic *1*. In this case, the flip-flop toggles each time a clock input is applied. Figure 13-17 shows the logic symbol and timing diagram for the *JK* flip-flop. At time t_1, the clock input triggers the flip-flop as it goes from *0* to *1*. At this time, the control inputs are *J* = *1* and *K* = *0*. The flip-flop outputs switch to *Q* = *J* = *1* and $\overline{Q}$ = *K* = *0*. At t_2, *J* = *0* and *K* = *1*, and the outputs are changed to *Q* = *J* = *0* and $\overline{Q}$ = *K* = *1*. At t_3, t_4, t_5, and t_6, both control inputs are at

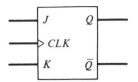

(a) Logic symbol for *JK* flip-flop

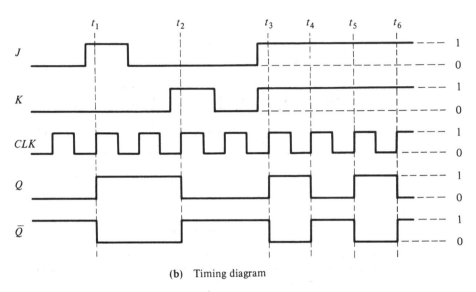

(b) Timing diagram

Figure 13-17 The *JK* flip-flop is similar to the *clocked SC* flip-flop, except that when both the *J* and *K* inputs are at *1*, the flip-flop can be toggled by the clock input.

1, and consequently, the flip-flop toggles from one state to the other each time the clock input goes in a positive direction.

Appendix 1-17 contains the data sheet for a TTL *7476 JK* flip-flop (and others). Note that, in addition to the input and output terminals already discussed, the logic symbol on the data sheet shows *preset* and *clear* terminals. These are inputs which operate independently of the *J*, *K*, and *clock* terminals, and which allow the flip-flop to be set in any desired state prior to a clock signal being applied. Note also that only the *preset* and *clear* terminals have small circles. Thus, these two are affected by negative-going signals, while all other inputs are triggered by positive-going inputs.

13-9 MASTER-SLAVE FLIP-FLOPS

With a clocked *SC* flip-flop or a *JK* flip-flop, the outputs assume the control input levels during the time that the flip-flop is enabled by the clock. In some circumstances, changes of the control input levels can occur during the enable time. Such changes can make the flip-flop output indeterminte, so the control inputs should preferably remain constant during the enable time. The *master-slave flip-flop* maintains constant inputs (during the enable time) by using one flip-flop (the *master*) to control another flip-flop (the *slave*).

The master-slave arrangement is illustrated in Figure 13-18(a), where it is seen that the Q and $\overline{Q}$ outputs of the master are fed as S and C inputs to the slave. When the slave is triggered, it assumes the output levels of the master. Note that an inverter is employed to make the clock input to the slave the inverse of that applied to the master.

The timing diagram in Figure 13-18(b) illustrates the operation of the master-slave flip-flop. During time t_1, the clock input to the master is a positive pulse. This enables the master, so that the Q_1 and $\overline{Q}_1$ outputs assume the levels of the S_1 and C_1 inputs. During this time, the inverted clock input applies a zero level to the slave, which disables the slave, so that the output levels (Q_2 and $\overline{Q}_2$) remain constant.

During time t_2, the master is disabled by the zero level clock signal, and its

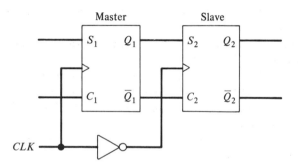

(a) Construction of a master-slave flip-flop

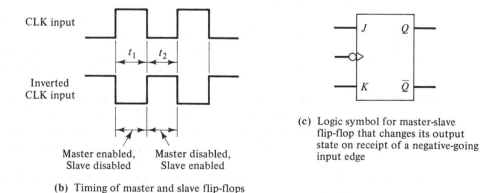

(b) Timing of master and slave flip-flops

(c) Logic symbol for master-slave flip-flop that changes its output state on receipt of a negative-going input edge

Figure 13-18   A *master-slave* flip-flop is a combination of two flip-flop circuits. The *master* flip-flop determines the condition that the *slave* will be switched into when the clock signal is received.

outputs (Q_1 and $\overline{Q_1}$) remain constant. The slave is enabled during t_2, by the positive (inverted) clock input. The slave outputs now assume the input levels applied to S_2 and C_2 from Q_1 and $\overline{Q_1}$. Because the master is disabled, the slave inputs remain constant while it is enabled (during t_2). Thus, although the data levels at S_1 and C_1 have been transferred to the slave outputs, the slave flip-flop is always isolated from the S_1 and C_1 inputs.

For the master-slave flip-flop illustrated in Figure 13-18(a), the outputs change state when the clock signal goes from *1* to *0*, i.e., when the input edge is negative going. Thus, the (complete) flip-flop is identified as a negative-edge-triggered circuit. As illustrated in Figure 13-18(c), the logic symbol for the flip-flop has a *bubble* and an arrowhead at the clock input, to show that it is negative-edge triggered.

REVIEW QUESTIONS AND PROBLEMS

13-1. Sketch the circuit of a collector-coupled bistable multivibrator employing *npn* transistors. Explain how the circuit operates.

13-2. Sketch the circuit of a collector-coupled bistable multivibrator employing *pnp* transistors. Explain the circuit operation.

13-3. Sketch the circuit of an emitter-coupled bistable multivibrator employing *npn* transistors. Explain how the emitter resistor affects the circuit operation.

13-4. Sketch the circuit of a bistable multivibrator using two IC inverters. Explain how the circuit operates.

13-5. Design a collector-coupled bistable multivibrator to operate from a ±6 V supply. Use 2N3904 transistors, and make $I_{C(\text{sat})} \approx 1$ mA.

13-6. Design a collector-coupled bistable multivibrator using 2N3906 transistors. The supply voltage is ±9 V, and the collector current is to be approximately 2 mA.

13-7. Sketch a circuit for asymmetrical collector triggering of a bistable multivibrator. Show the triggering waveforms, and explain how the circuit functions.

13-8. Sketch a circuit for symmetrical collector triggering of a bistable multivibrator. Show the waveforms, and explain how the triggering circuit functions.

13-9. The bistable multivibrator designed for Problem 13-5 is to use symmetrical collector triggering. The trigger input is to be the collector output of a previous similar stage. Design a suitable triggering circuit.

13-10. Calculate the capacitance of the commutating capacitors for the circuits of Problems 13-5 and 13-9. Also, determine the maximum triggering frequency that should be employed.

13-11. Sketch the circuits for asymmetrical and symmetrical base triggering of a bistable multivibrator. Explain the operation of each circuit, and discuss the major disadvantage of symmetrical base triggering.

13-12. Sketch a collector-steered base-triggering circuit. Show the triggering waveforms, and explain how the circuit functions.

13-13. Design a collector-steered base-triggering circuit for the flip-flop designed for Problem 13-5. The triggering input is to be the collector output of a previous similar stage.

13-14. Determine the capacitance of suitable commutating capacitors for the circuits of Problems 13-5 and 13-13. Also, calculate the maximum triggering frequency.

13-15. Sketch the logic symbol for a *T* flip-flop triggered by a positive-going input. Also, sketch the timing diagram and briefly explain the operation of the flip-flop.

13-16. Sketch the logic symbol for an *SC* flip-flop triggered by a positive-going input. Also, sketch the timing diagram and explain the flip-flop operation.

13-17. Sketch the circuit of an *SC* flip-flop using two *NAND* gates. Explain how the flip-flop operates.

13-18. Sketch the circuit of an *SC* flip-flop using two *NOR* gates. Explain the circuit operation.

13-19. Sketch the logic symbol for *T* and *SC* flip-flops which are triggered by negative-going inputs. Also, sketch the timing diagrams and write a truth table for each flip-flop.

13-20. Sketch the logic symbol for a *clocked SC* flip-flop triggered by a positive-going input. Also, sketch the timing diagram and explain the flip-flop operation. Explain *set-up time* and *holding time*.

13-21. Sketch the logic symbol and timing diagram for a *D* flip-flop, and explain its operation. Show how a clocked *SC* flip-flop may be converted into a *D* flip-flop.

13-22. Sketch the logic symbol and timing diagram for a *JK* flip-flop. Explain how the *JK* flip-flop differs from the *SC* and *D* flip-flops.

13-23. Discuss the *master-slave* flip-flop. Show how it is constructed, and explain its operation.

Chapter 14

Digital Counting and Measurement

INTRODUCTION

Because the bistable multivibrator, or flip-flop, has two stable states, it can be used to count up to two. A cascade of four flip-flops can count up to 16. The **scale-of-16 counter** can be modified to produce a **decade counter**, which has an output in the form of a **binary number**. For counting in decimal form, the binary number must be converted to drive a numerical display. Decade counters and their numerical displays can be cascaded to construct systems for counting to hundreds, thousands, tens of thousands, etc.

If a pulse waveform is fed to the input of a digital counter for a time period of exactly one second, the counter indicates the frequency of the waveform. Suppose the counter registers 1000 at the end of a second; then the frequency of the input is 1000 **pulses per second**. Essentially, a digital frequency meter is a digital counter combined with an accurate timing system. The timing system usually is such that the input frequency is sampled repeatedly. This necessitates the use of a **latch**, which keeps the display constant while the frequency remains unchanged.

A dc voltage can be converted to a frequency which is directly proportional to the voltage. This frequency can be measured by a digital frequency meter, and the output read as a voltage. Several methods of converting from voltage to frequency are available.

14-1 FLIP-FLOPS IN CASCADE

The schematic diagram of four flip-flops (FF) connected in cascade is shown in Figure 14-1. Each flip-flop is a collector-coupled circuit, and each has symmetrical collector triggering. Negative-going input pulses are applied to FF_A via coupling capacitor C_1.

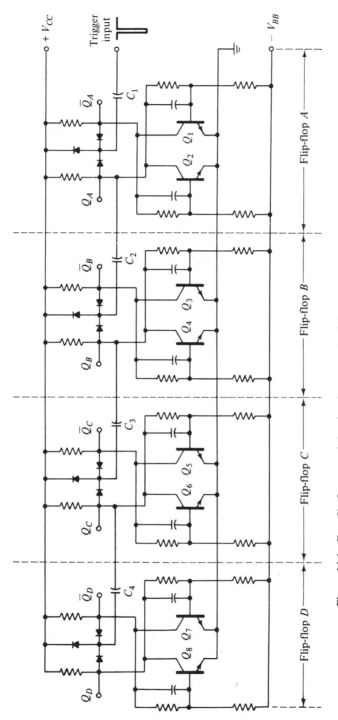

Figure 14-1 Four flip-flops cascaded to function as a scale-of-sixteen counter. Flip-flop A is toggled by the negative-going trigger input pulse. Flip-flop B is toggled by the Q_A output of flip-flop A as it goes negative. Flip-flops C and D are toggled by Q_B and Q_C outputs, respectively.

Each time an input pulse is applied, FF_A will change state. The triggering circuit for FF_B is coupled via capacitor C_2 to transistor Q_2 in FF_A. When Q_2 switches *off*, its collector voltage rises, applying a positive voltage step to C_2. Since a negative-going voltage is required to trigger these flip-flops (see Chapter 13), FF_B is not affected by the positive-going voltage. When Q_2 switches *on*, its collector voltage drops, thus applying a negative voltage step to FF_B via C_2. This negative voltage change triggers FF_B. In a similar way, FF_C is triggered from FF_B, and FF_D is triggered from FF_C. Thus, each flip-flop is triggered from each preceding stage.

The four-stage cascade in Figure 14-1 can have a number of combinations of flip-flop states. In Figure 14-2, the flip-flops are shown in block form with the arrowheads indicating that each is triggered from the previous stage. The state of each of the four flip-flops is best indicated by using the *binary* number system, where *0* represents a voltage at or near ground level and *1* represents a positive voltage level. (See Figure 14-2.) When a transistor is *on*, its collector voltage is *low* and is represented by *0*. An *off* transistor, on the other hand, has a high collector voltage and is designated *1*. In

(a)   Block diagram

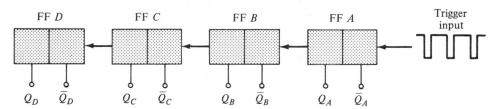

(b)   Truth table

				Binary count	Decimal count
0	0	0	0	0	0
0	0	0	1	1	1
0	0	1	0	10	2
0	0	1	1	11	3
0	1	0	0	100	4
0	1	0	1	101	5
0	1	1	0	110	6
0	1	1	1	111	7
1	0	0	0	1000	8
1	0	0	1	1001	9
1	0	1	0	1010	10
1	0	1	1	1011	11
1	1	0	0	1100	12
1	1	0	1	1101	13
1	1	1	0	1110	14
1	1	1	1	1111	15
0	0	0	0	0000	16

Figure 14-2   Block diagram and truth table for a four-flip-flop scale-of-16 counter. Reading the logic level of the Q_A, Q_B, Q_C, and Q_D outputs, it is seen that the counter can be toggled through 16 different states.

the decimal system, counting goes from *0* to *9*, and then the next count is indicated by *0* in the first column and *1* in the next leftward column. In the binary system, the count in all columns can go only from *0* to *1*. Thus, the count for *1* in both binary and decimal systems is *1*; in the binary system, the count for decimal 2 is indicated by *0* in the first column and *1* in the next leftward column. Thus, *binary 10* is equivalent to *decimal 2*. The next count in a binary system is *11* and is followed by *100*. The table of *0*'s and *1*'s showing the state of the flip-flops at each count is the *truth table* for the circuit.

Refer to Figures 14-1 and 14-2, and suppose that, before any pulses are applied, the state of the flip-flops is such that all even-numbered (i.e., left-hand) transistors are *on*. Reading only the Q outputs in Figure 14-2 from left to right, the binary count is *0000*. At this time the decimal count is *0* and the binary count is *0*.

The first trigger pulse causes Q_1 (Fig. 14–1) to switch *on* and Q_2 to switch *off*. Thus, output Q_A (Fig. 14–2) reads as *1* (positive), and the binary count and decimal count are both *1*. The second input trigger pulse causes FF$_A$ to change state again, so that Q_1 goes *off* and Q_2 switches *on*. When Q_2 switches *on*, a negative step is applied from Q_A to FF$_B$, triggering Q_3 *on* and Q_4 *off* to give $Q_B = 1$ and $Q_A = 0$. Now the binary count is *10*, and the decimal count is *2* (again, reading only the Q outputs). The third input pulse triggers Q_1 *on* and Q_2 *off* once again. This produces a positive output from FF$_A$, which does not affect FF$_B$. At this time, the binary count is *11*, for a decimal count of *3*. The fourth trigger pulse applied to the input switches Q_1 *off* and Q_2 *on*. Q_2 coming *on* produces a negative step at Q_A, which causes Q_3 to go *off* and Q_4 to switch *on*. Q_4 switch-*on*, in turn, produces a negative voltage step at Q_B, which switches Q_5 *on* and Q_6 *off*, to give $Q_C = 1$. Now the binary count is read from the flip-flop outputs as *100*, and the decimal count is *4*.

The counting process is continued with each new pulse until the maximum binary count of *1111* is reached. This occurs when 15 input pulses have been applied. The sixteenth input switches Q_2 *on* once more, producing a negative pulse which triggers Q_4 *on*. Q_4 output is a negative pulse which triggers Q_6 *on*, and Q_6 output triggers Q_8 *on*. Thus, the four flip-flops have returned to their original states, and the binary count has returned to *0000*. Including the zero condition, it is seen that the four flip-flops in cascade have 16 different states. Therefore, the circuit is termed a *scale-of-16 counter*.

The collector voltage levels for the scale-of-16 counter are shown as waveforms in Figure 14-3. The waveform for $\overline{Q_A}$ shows that transistor Q_1 is initially *off*; its collector voltage is high and therefore is designated *1*. Q_2 is initially *on*, with its collector voltage at $Q_A = 0$. Each time a trigger pulse is applied, $\overline{Q_A}$ and Q_A change state. $\overline{Q_B}$ and Q_B are initially *1* and *0*, respectively, and they change state each time Q_A goes from *1* to *0*, that is, when FF$_A$ produces a negative-going output. This occurs on every second input pulse. $\overline{Q_C}$ starts as *1* and Q_C as *0*, and they change state only when Q_B goes from *1* to *0*, which is at every fourth input pulse. Finally, the waveforms for $\overline{Q_D}$ and Q_D show that initially $\overline{Q_D}$ is *1* and Q_D is *0*, and that they change state when Q_C goes negative, that is, at every eighth input pulse. On the sixteenth input pulse all flip-flops change state, and the transistor collector voltages return to their original levels.

The scale-of-16 counter can be used to divide the input pulse frequency by a factor of 16. Reference to the collector waveforms in Figure 14-3 shows that a negative-going voltage is produced at Q_D after 16 input pulses. Another negative-going step will occur at Q_D after another 16 input pulses. Hence, the name *divide-by-16 counter* is sometimes applied to this circuit. An output taken from FF_C will produce a pulse frequency which is the input PRF divided by eight. Similarly, the output of FF_B divides the input by four.

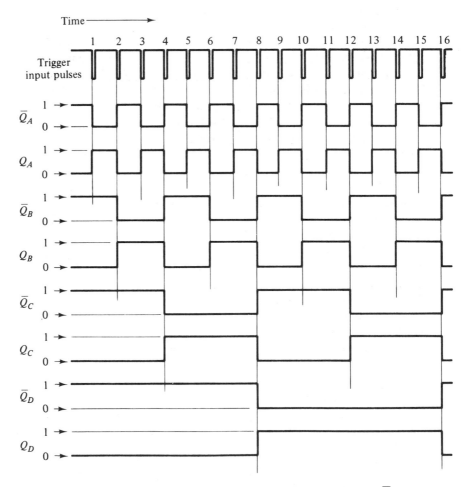

Figure 14-3 Output waveforms for a scale-of-16 counter. Outputs Q_A and $\overline{Q}_A$ change state each time an input pulse is received. Q_B and $\overline{Q}_B$ change at every second input pulse. Q_C and $\overline{Q}_C$ change at every fourth input pulse, and Q_D and $\overline{Q}_D$ change on every eighth input pulse.

14-2 DECADE COUNTER

The scale-of-16 counter has many applications. However there are also a great many instances in which a *scale-of-10*, or *decade counter* is required. A decade counter also requires the use of a cascade of four flip-flops. Three flip-flops would count only up to seven, and then on the eighth pulse the count would revert to the *000* starting condition. This can be seen in Figure 14-2. Therefore, to produce a decade counter, a scale-of-16 must be modified to eliminate six of the 16 states. This can be done by eliminating either the first six states or the last six states, or perhaps by eliminating some of the intermediate states.

When the first six states of a scale-of-16 counter are to be eliminated, the counter must always have an initial condition of *0110* (decimal 6 in Figure 14-2). To obtain this condition, transistors Q_4 and Q_6 (in Fig. 14-1) must be in the *off* state. Q_4 and Q_6 can be reset to *off* by the asymmetrical base-triggering circuit shown in Figure 14-4. (Asymmetrical base triggering is discussed in Sec. 13-3.) When Q_8 switches *on*, Q_D drops to *0*, providing a negative step which forward-biases D_1 and D_3, and triggers Q_6 and Q_4 *off*. Figures 14-2 and 14-3 show that Q_8 switches *on* when the sixteenth input pulse is applied. Therefore, at the end of the count of 16, the flip-flops are reset to *0110*. The block diagram, truth table, and collector waveforms for the decade counter are shown in Figure 14-5.

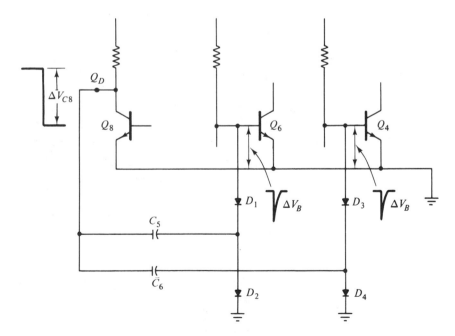

Figure 14-4   Method of resetting transistors Q_4 and Q_6 to *off* each time Q_D goes negative. The negative-going output from Q_D is coupled via C_5 and D_1 to the base of Q_6, and via C_6 and D_3 to the base of Q_4.

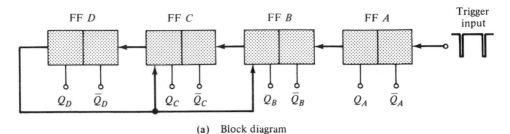

(a) Block diagram

Q_D	Q_C	Q_B	Q_A	Decimal count
0	1	1	0	0
0	1	1	1	1
1	0	0	0	2
1	0	0	1	3
1	0	1	0	4
1	0	1	1	5
1	1	0	0	6
1	1	0	1	7
1	1	1	0	8
1	1	1	1	9
Reset 0	0	0	0	10
0	1	1	0	0 =

(b) Truth table showing state of transistor collector

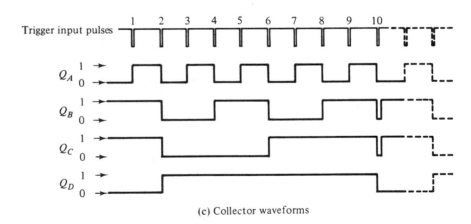

(c) Collector waveforms

Figure 14-5 Block diagram, truth table, and output waveforms for four flip-flops connected as a decade counter. The feedback from Q_D to reset flip-flops B and C eliminates the first six states of the scale-of-16 counter, converting it into a decade counter.

In Figure 14-5(a), the line from FF_D to FF_B and FF_C indicates that these flip-flops are reset by the output from FF_D. The initial state of the four flip-flops (i.e., decimal *0*) is read in Figure 14-5(b) as *0110*. The first input pulse now changes the state of FF_A, causing Q_A to switch to *1* (i.e., high positive), and the condition of the counter is *0111*. This is also illustrated by the collector waveforms in Figure 14-5(c). The second input pulse (decimal *2*) again changes the state of FF_A, this time causing Q_A to switch to *0*. The output from Q_A is a negative step which triggers FF_B, switching Q_B to *0*. This, in turn, produces a negative step which triggers FF_C from $Q_C = 1$ to $Q_C = 0$. The output from FF_C triggers FF_D. Counting continues in this way, exactly as explained for the scale-of-16 counter, until the tenth pulse. The ninth pulse sets the counter at *1111*, and the tenth pulse changes it to *0000*. However, as Q_D switches *low*, it provides the negative output step which resets FF_B and FF_C. The flip-flops have then returned to their initial conditions of *0110*, and, clearly, the circuit has only ten different states.

The waveforms in Figure 14-5(c) indicate that a negative output step is generated at Q_D each time the tenth input pulse is applied. Thus, the decade counter can be employed as a *divide-by-10 counter*. Before counting begins, a decade counter (and a scale-of-16 counter) must have its flip-flops set in the correct starting condition. This can be accomplished by the manual resetting arrangement shown in Figure 14-6(a). When switch S_1 is closed, the diodes are forward biased and the transistor bases are pulled below ground level. Thus, transistors Q_1, Q_4, Q_6, and Q_7 are switched *off*, giving the desired *0110* initial condition for the decade counter.

The flip-flops can also be reset automatically in their initial condition by the CR circuit addition in Figure 14-6(b). When the supply voltages are first switched *on*, the capacitor behaves as a short circuit. Therefore, the diode cathode voltages are at $-V$, and the transistors are biased *off*. After a brief time period, C_1 charges to $+V$ via resistor R. Now the diodes are all reverse biased, and the reset circuit has no further effect.

Figure 14-7 shows a further modification of the circuit for resetting the flip-flops. Diode D_5 serves to isolate R and C from the rest of the reset circuit. When the cathodes of D_1 to D_4 are pulled down, D_5 is reverse biased. D_6 and D_7, together with coupling capacitor C_2, form a triggering circuit. A negative-going voltage step applied to C_2 generates a negative pulse at the cathode of D_6. This forward-biases D_6, D_1, D_2, D_3, and D_4, causing the flip-flops to reset. Thus, in addition to being reset to its starting condition when the supply is switched *on*, the counter can be reset to *zero* at any time by the application of a negative voltage step.

14-3 INTEGRATED CIRCUIT COUNTERS

The block diagrams and voltage waveforms for a 7493A TTL integrated circuit *binary counter* (or *scale-of-16 counter*) are shown in Figure 14-8. The block diagram shows that the counter consists of four *JK* flip-flops connected in cascade. The Q output of each flip-flop is connected as a trigger input to the *clock* terminal (*CK*) of the next stage. The triggering signal is applied to *input A*, which is the clock terminal of the

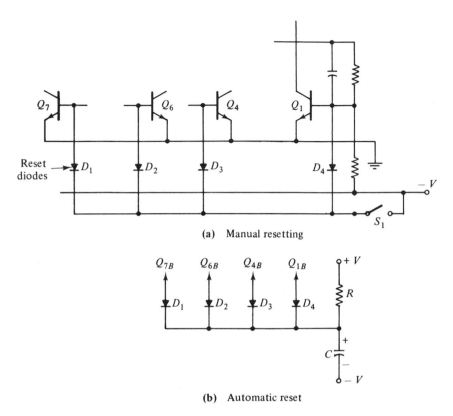

(a)   Manual resetting

(b)   Automatic reset

Figure 14-6   Manual and automatic methods of resetting the appropriate counter transistors to their initial *off* condition, prior to commencement of counting. The transistor bases are briefly pulled to $-V$.

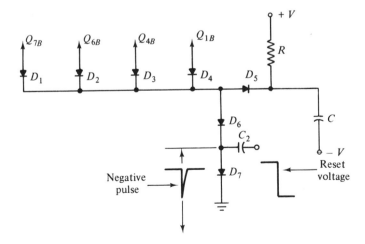

Figure 14-7   Circuit for resetting counter transistors to the initial *off* condition by means of a negative-going pulse input. Here again, the transistor bases are pulled below the levels of their emitters.

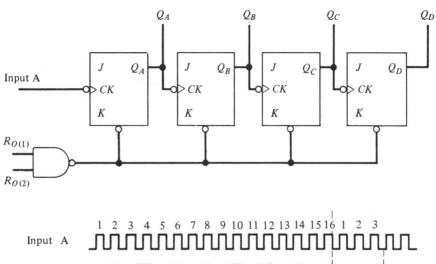

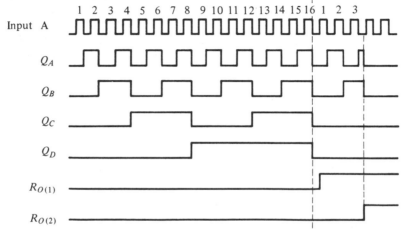

Figure 14-8   Block diagram and waveforms for the 7493 integrated circuit scale-of-16 counter. As in Figures 14-1 and 14-2, this is a cascade of four flip-flops, with the first being toggled by the input pulse, and each successive flip-flop toggled from the previous stage.

first stage. Four outputs are available: Q_A, Q_B, Q_C, and Q_D. Separate (unidentified) *reset* terminals are provided on each flip-flop (at the bottom of each block). These are activated by *NAND* gate inputs $R_{o(1)}$ and $R_{o(2)}$. The block diagram is also shown in a slightly different form on the data sheet for this counter in Appendix 1-24. For *JK* flip-flops to be toggled by the *clock* input, the J and K terminals must have *logic 1* input levels. (See Section 13-8.) Where the J and K terminals are shown unconnected, as in Figure 14-8, it is assumed that they are connected to the positive supply voltage.

The waveforms demonstrate that Q_A changes state each time the triggering signal to input A goes negative. Similarly, Q_B changes state when Q_A goes negative, Q_C changes state when Q_B goes negative, and Q_D changes state every time Q_C goes negative. The result of this, as already explained in Section 14-1, is that the counter outputs have 16 different states. Note also from the waveforms that the outputs are all reset to *low*, and counting ceases, when *high* inputs are applied to $R_{o(1)}$ and $R_{o(2)}$.

Figure 14-9 shows the block diagram for the 7492A IC *divide-by-12 counter*. Once again, the diagram is drawn in a slightly different form from that shown on the data sheet in Appendix 1-24. The scale-of-12 differs from the scale-of-16 block diagram as follows: the Q_B output of stage 2 is connected to the *J* terminal of stage 3 (instead of to the clock input); the clock input of stage 3 is triggered from Q_A, which is also connected to the clock input of stage 2; and the *J* terminal of stage 2 is connected to the $\bar{Q}$ output of stage 3.

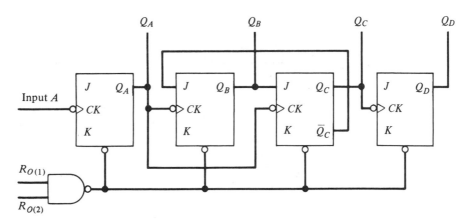

Figure 14-9   Block diagram and waveforms for the 7492A integrated circuit scale-of-12 counter. This is a 7493 scale-of-16 counter, modified to eliminate four of the states.

With the arrangement described above, stage 2 output (Q_B) triggers into whatever state is applied at the *J* terminal. When the *J* input is *high*, Q_B is triggered into a *high* state when its clock terminal receives a negative-going signal. The next negative-going clock signal has no further effect on Q_B unless the *J* input has changed to a *low* level. (The *JK* flip-flops are in fact behaving as *D* flip-flops.) Similarly, output Q_C can be triggered into a different state only when the *J* input of stage 3 has changed from *high* to *low*, or *vice versa*.

By careful consideration of the *count sequence* tables for the 92A and 93A counters in Appendix 1-24, it is found that four states of the divide-by-16 are eliminated in the divide-by-12 unit. Similarly, by the method already described in Section 14-2, and by other interconnection methods, a scale-of-16 IC counter can be employed as a decade counter.

14-4 DIGITAL DISPLAYS OR READOUTS

Light-emitting Diode Display

Charge carrier recombination occurs at a *pn*-junction as *electrons* cross from the *n*-side and recombine with *holes* on the *p*-side. When recombination takes place, the charge carriers give up energy in the form of heat and light. If the semiconductor material is translucent, the light is emitted and the junction is a light source, that is, a *light-emitting diode* (LED).

Figure 14-10(a) shows a cross-sectional view of a typical LED. Charge carrier recombinations take place in the *p*-type material; therefore, the *p*-region becomes the surface of the device. For maximum light emission, a metal film anode is deposited around the edge of the *p*-type material. The cathode connection for the device usually is a gold film at the bottom of the *n*-type region; this helps reflect the light to the surface. Semiconductor material used for LED manufacture is *gallium arsenide phosphide* (Ga AsP), which emits either red or yellow light, or *gallium arsenide* (Ga As) for green or red light emission.

The LED circuit symbol D_1 is shown in Figure 14-10(b). Figure 14-10(c) illustrates the arrangement of a seven-segment LED numerical display. Passing a current through the appropriate segments allows any numeral from 0 to 9 to be displayed. The actual LED device is very small, so, to enlarge the lighted surface, plastic *light pipes* are often employed, as illustrated in Figure 14-10(d). The typical voltage drop across a forward-biased LED is 1.2 V, and typical forward current for reasonable brightness is about 20 mA. This relatively large current requirement is a major disadvantage of LED displays. Some advantages of LEDs over other types of displays are (1) the ability to operate from a low-voltage dc supply, (2) ruggedness, (3) rapid switching ability, and (4) small physical size. The data sheet for a typical seven-segment LED display is shown in Appendix 1-22.

The simple transistor switch shown in Figure 14-10(b) is a suitable *on/off* control for LEDs. Q_1 is driven into saturation by input current I_B. Resistor R_C limits the current through the devices.

EXAMPLE 14-1

The LED shown in Figure 14-10(b) is to have a minimum forward current of 20 mA. The diode has a forward voltage drop of 1.2 V, and transistor Q_1 has $h_{FE(min)} = 100$. Using $V_{CC} = 5$ V and $V_i = 5$ V, determine suitable values for R_C and R_B.

Solution

$$V_{CC} = V_{D1} + I_C R_C + V_{CE(sat)}$$

$$R_C = \frac{V_{CC} - V_{D1} - V_{CE(sat)}}{I_C}$$

$$= \frac{5\text{ V} - 1.2\text{ V} - 0.2\text{ V}}{20\text{ mA}}$$

$$= 180\ \Omega \quad \text{(standard value)}$$

$$I_B = \frac{I_C}{h_{FE(min)}}$$

$$= \frac{20\text{ mA}}{100} = 200\ \mu\text{A}$$

$$V_i = I_B R_B + V_{BE}$$

$$R_B = \frac{V_i - V_{BE}}{I_B}$$

$$= \frac{5\text{ V} - 0.7\text{ V}}{200\ \mu\text{A}} = 21.5\text{ k}\Omega \quad \text{(use 18 k}\Omega\text{ standard value)}$$

Liquid Crystal Displays

A *liquid crystal cell display* (LCD) is usually arranged in the same seven-segment numerical format as the LED display. There are two types of liquid crystal display, the *dynamic scattering type* and the *field effect type*. The construction of a dynamic scattering type liquid crystal cell is illustrated in Figure 14-11(a). The liquid crystal material may be one of several organic compounds which exhibit the optical properties of a crystal, although they remain in liquid form. Liquid crystal is layered between glass sheets with transparent electrodes deposited on the inside faces. When a potential is applied across the cell, charge carriers flowing through the liquid disrupt the molecular alignment and produce turbulence. When not activated, the liquid crystal is transparent. When activated, the molecular turbulence causes light to be scattered in all directions, so that the cell appears quite bright. The phenomenon is termed *dynamic scattering*.

The construction of a *field effect* liquid crystal display is similar to that of the dynamic scattering type, except that two thin polarizing optical filters are placed at the inside surface of each glass sheet. The liquid crystal material in the field effect cell is also a different type from that employed in the dynamic scattering cell. Known as *twisted nematic*, this liquid crystal material actually twists the light passing through the cell when the cell is not energized. This allows light to pass through the optical filters, and the cell appears bright. (It can also be made to appear dark.) When the cell is energized, no twisting of the light occurs and the cell remains dull.

Liquid crystal cells may be *transmittive* or *reflective*. In the *transmittive-type* cell, both glass sheets are transparent, so that light from a rear source is scattered in the

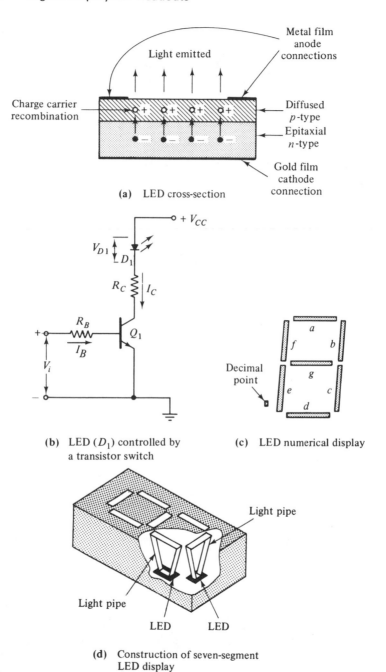

(a)   LED cross-section

(b)   LED (D_1) controlled by a transistor switch

(c)   LED numerical display

(d)   Construction of seven-segment LED display

Figure 14-10   Light-emitting diode (LED) cross-section, control circuit, and seven-segment numerical display. The LED typically has a forward voltage drop of 1.2 V and requires a forward current of 20 mA to produce a bright glow.

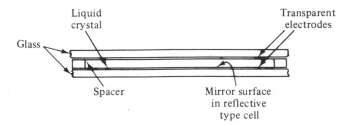

Glass

Liquid crystal

Transparent electrodes

Spacer

Mirror surface in reflective type cell

(a) Construction of liquid crystal cell

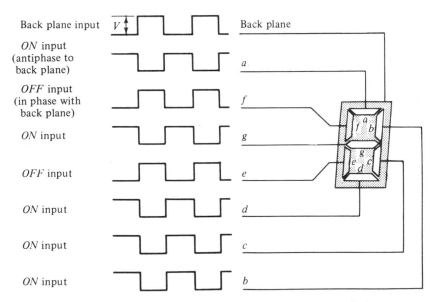

Back plane input V Back plane

ON input (antiphase to back plane)   a

OFF input (in phase with back plane)   f

ON input   g

OFF input   e

ON input   d

ON input   c

ON input   b

(b) Square wave drive method for liquid crystal cell 7-segment display

Figure 14-11   Construction and electrical drive arrangement for a liquid crystal display (LCD). A seven-segment LCD requires only microamps of current, but must have an ac voltage supply. A 15 V square wave is typically used to energize an LCD.

forward direction when the cell is activated. The *reflective-type* cell has a reflecting surface on one of the glass sheets. In this case, incident light on the front surface of the cell is dynamically scattered by an activated cell. When activated, both the transmittive- and reflective-type cells appear quite bright even under high ambient light conditions.

Since liquid crystal cells are light reflectors or transmitters rather than light generators, they consume very small amounts of energy. The only energy required by the cell is that needed to activate the liquid crystal. The total current flow through four small seven-segment displays is typically about 25 μA for dynamic scattering cells, and 300 μA for field effect cells. However, the LCD requires an ac voltage supply, in the form of either a sine wave or a square wave. This is because a direct current produces a plating of the cell electrodes, which could damage the device. A typical supply for a dynamic scattering LCD is a 30 V peak-to-peak square wave with a frequency of 60 Hz. A field effect cell typically uses 8 V peak-to-peak.

Figure 14-11(b) illustrates the square wave drive method for liquid crystal cells. The *back plane*, which is one terminal common to all cells, is supplied with a square wave. The other cell terminals each have square waves applied which are either in phase or in antiphase with the back plane square wave. Those cells with waveforms in phase with the back plane waveform [cells *e* and *f* in Figure 14-11(b)] have no voltage developed across them; therefore they are *off*. The cells with square waves in antiphase with the back plane input have an ac voltage developed across them (e.g., positive square waves with 15 V peak effectively produce 30 V peak to peak when in antiphase). Therefore, the cells which have square wave inputs in antiphase with the back plane are energized and appear bright.

The data sheet for the series 1603-02 liquid crystal display manufactured by Industrial Electronic Engineers, Inc., is shown in Appendix 1-23. The maximum power consumption is listed as 20 μW per segment, giving 140 μW per numeral when all seven segments are energized. Comparing this to the typical 400 mW per numeral for a LED display (see Appendix 1-22), the major advantage of liquid crystal displays is obvious.

The series 1603-02 LCD is described in the data sheet as a *3½ decade display*. This means that the three right-hand units are complete seven-segment units, while the fourth (left-hand) unit is only a single segment which indicates numeral 1 when energized. This unit is referred to as a half-unit, and the entire display is then described as a 3½ decade display. The maximum number that can be indicated by such a display is 1999.

Digital Indicator Tube

The basic construction of a *digital indicator tube* is shown in Figure 14-12(a), and its schematic symbol is illustrated in Figure 14-12(b). (Other names applied to this device are *cold cathode tube* and *glow tube*.) A flat metal plate with a positive voltage supply functions as an anode, and there are 10 separate wire cathodes, each in the shape of a numeral from 0 to 9. The electrodes are enclosed in a gas-filled glass envelope with connecting pins at the bottom. *Neon* gas usually is employed, and it gives an orange-red glow when the tube is activated; however, other colors are available with different gases.

When a voltage is applied across the anode and one cathode, electrons accelerated from the cathode cause secondary electrons to be emitted when they strike. The secondary electrons cause ionization and electron-atom recombination in the region close to the cathode. This results in energy being released in the form of light and produces a visible glow around the cathode. Since the cathodes are in the shape of numerals, a glowing numeral appears, depending upon which cathode is energized. A transistor gate is usually employed at each cathode, so that the desired numeral can be switched *on* by a small input voltage.

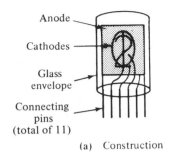

(a)   Construction

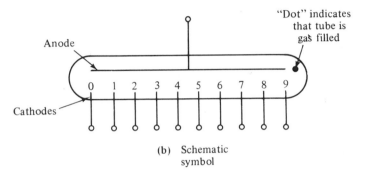

(b)   Schematic
symbol

Figure 14-12 Construction and schematic symbol for a digital indicator tube. Each cathode in the gas-filled tube is shaped in the form of a numeral. When a suitable anode-to-cathode voltage is applied (typically 140 V), the selected cathode glows.

The circuitry for driving digital indicator tubes is simpler than that for seven-segment devices. However, high voltages (140 V to 200 V) are required for these tubes, and in general they are much larger than comparable semiconductor seven-segment devices.

Seven-Segment Gas Discharge Displays

Gas discharge displays are also available in seven-segment format. Integrated circuits have been developed to drive these devices and to handle the high dc voltages involved. The mechanical construction of a seven-segment gas discharge display is illustrated in Figure 14-13(a). Note that separate cathodes are provided in seven-segment (and decimal point) form on a base. Each seven-segment group has a single anode deposited as a transparent metal film on the covering face plate. The gas is contained in the space between the anodes and cathodes, and rear connecting pins are provided for all electrodes. A *keep-alive cathode* is also enclosed with each group of segments. A 50 μA current maintained through the *keep-alive cathode* provides a source of ions which improves the switch-*on* speed of the display. The circuit symbol in general use for seven-segment gas discharge displays is shown in Figure 14-13(b).

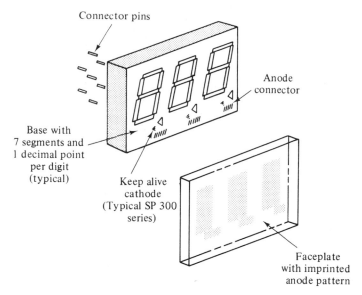

(a)   Construction of 7-segment
      gas discharge display
      (*Courtesy of Beckman Instruments, Inc.*)

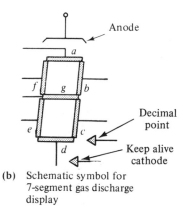

(b)   Schematic symbol for
      7-segment gas discharge
      display

Figure 14-13 Construction and schematic symbol for a seven-segment gas discharge display. The device is actually a flat, gas-filled tube, with the cathodes in seven-segment form. When energized, the selected cathodes glow.

The supply voltage required to operate gas discharge displays ranges from about 140 V to 200 V, and this is the most serious disadvantage of these devices. High-voltage transistors must be employed as switches for the cathodes, and usually a separate high-voltage supply must be provided. Offsetting the disadvantage of high voltages is the fact that bright displays can be achieved with tube currents as low as 200 μA. Thus, the drain on power supplies is minimal.

Fluorescent Display

The *fluorescent display* illustrated in Figure 14-14 is similar to the seven-segment gas discharge display in that both are electron tube devices. However, the fluorescent display is a *vacuum* tube device with a filament-type cathode, and a (wire mesh) grid to control the flow of the electrons from the cathode to the anodes (or plates). The seven electrically separate anodes are coated with fluorescent phosphorus, so that they glow brightly (blue-green) when struck by electrons.

The major advantage of the fluorescent display is that it can operate with normal solid-state supply voltage levels ($V_{CC} = 10$ V to 40 V), whereas gas discharge devices required 140 V to 200 V supplies. The required (ac or dc) filament current is on the order of 12 mA for small-size fluorescent displays. The grid, which is typically biased to -3 V, is pulsed to approximately $+2$ V in synchronism with $+V_{CC}$ on the selected anode segments. Some grid current flows, but the total power dissipation for each seven-segment numeral is around 9 mW for small-size displays.

14-5 BCD-TO-DECIMAL CONVERSION

The output of a decade counter can be read as a binary number if collector voltages are taken as *1* when high and *0* when low. In this case, the binary number is in a form which is termed *binary coded decimal* (*BCD*). For display purposes, it is necessary to convert this BCD number to decimal. Figure 14-15 shows the various binary states of a decade counter, and the circuitry required to convert each state to a decimal display.

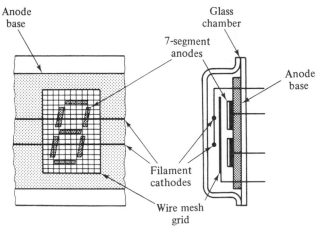

Anode base

Glass chamber

7-segment anodes

Anode base

Filament cathodes

Wire mesh grid

Front view   Cross-section

Figure 14-14 Construction of a seven-segment fluorescent display. This is a vacuum tube device, with phosphorus-coated anodes in seven-segment form. These glow when struck by electrons. Typical supply voltage is 10 V to 40 V.

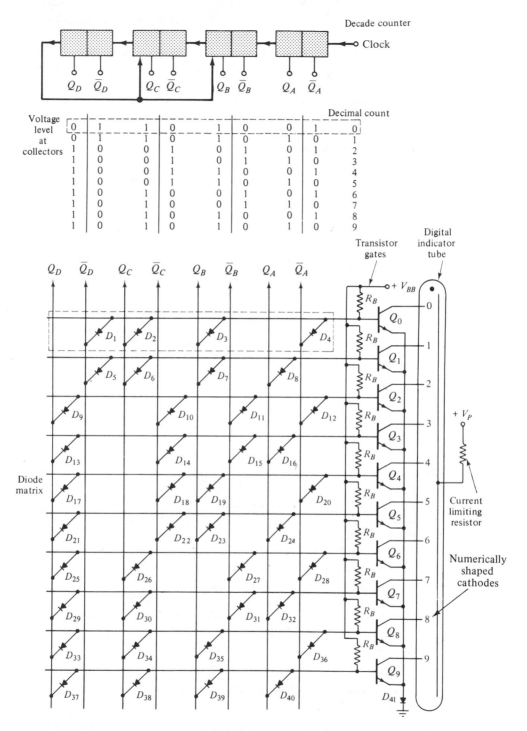

Figure 14-15 Decade counter block diagram and diode matrix for BCD-to-decimal conversion. At the count of zero, decade counter outputs Q_A, Q_B, Q_C, and Q_D are *high* (logic *1*), and diodes D_1, D_2, D_3, and D_4 are reverse biased. Thus, transistor Q_0 is biased *on* via resistor R_B, and the *0* cathode of the indicator tube glows.

389

The *diode matrix* consists of diodes D_1 to D_{40} which have their cathodes connected to the collectors of the transistors in the decade counter. The anodes of the diodes are connected to the bases of gate transistors Q_0 to Q_9. When one transistor switches *on*, it grounds the selected cathode in the digital indicator tube. Since the anode of the indicating tube has a positive supply, anode current flows when one of the cathodes is grounded, and the cathode then glows. The transistor emitters are connected in common, to ground via diode D_{41}. The presence of D_{41} ensures that the base voltage of each gate transistor (Q_0 through Q_9) has to be approximately 2 V_{BE} above ground level for it to switch *on*. Each gate transistor has four diodes connected to its base. When the cathode of one or more of these diodes is at *0* (i.e., near ground level), the transistor base is held below the switching voltage. In this condition, the transistor cannot switch *on*. When the cathodes of all four diodes connected to the base of any gate transistor are at *1*, the diodes are reverse biased, and the transistor is biased *on* via base resistance R_B.

Consider the output voltage levels and decimal count for the decade counter illustrated in Figure 14-15 for a decimal count of *0*. Outputs Q_D through $\overline{Q}_A$ read as *01 10 10 01*. Now, consider diodes D_1, D_2, D_3, and D_4, which are connected to the base of Q_0. The cathode of D_1 is connected to $\overline{Q}_D$. The voltage at $\overline{Q}_D$ is logic *1*; therefore, D_1 is reverse biased. The cathode of D_2 is connected to Q_C, which is also at *1*, so D_2 is also reverse biased. The cathode of D_3 is connected to Q_B, and since Q_B is at *1*, D_3 is reverse biased. Finally, D_4 has its cathode connected to Q_A, which is also at *1*. Thus, all four diodes at the base of Q_0 are reverse biased. Base current flows from V_{BB} via R_B into the base of Q_0. With Q_0 *on*, the *0* cathode in the digital tube glows, indicating that the decimal count is *0*.

For a correct *0* indication, all other gate transistors (Q_1 through Q_9) must be biased *off*. To check that this is the case, it is necessary to identify only one forward-biased diode at the base of each transistor. Consider diodes D_5 to D_8 at the base of Q_1. The cathodes of D_5, D_6, and D_7 are connected to decade counter outputs which are at *1* when the decimal count is *0*. Therefore, all three are reverse biased. D_8 cathode is connected to Q_A, which is at *0*. Consequently, D_8 is forward biased, and transistor Q_1 is held in the *off* condition. For Q_2, the cathodes of D_9, D_{10}, and D_{11} are connected to decade counter outputs which are at *0* while the decimal count remains *0*. Thus, Q_2 is biased *off*. Other diodes with *0* at their cathodes when the decade counter is in its decimal *0* condition are D_{13}, D_{14}, D_{15}, D_{16}, D_{17}, D_{18}, D_{21}, D_{22}, D_{24}, D_{25}, D_{27}, D_{29}, D_{31}, D_{32}, D_{33}, D_{37}, and D_{40}. It is seen that while the decimal count is *0*, all gate transistors except Q_0 have at least one forward-biased diode at their bases. Therefore, only Q_0 is biased *on*, and only the *0* cathode glows in the digital indicator tube.

A careful examination of the circuit conditions for any given decimal count shows that only the correct cathode is energized. All other cathodes have their transistor gates biased *off*.

EXAMPLE 14-2

In Figure 14-15, identify the forward-biased and reverse-biased diodes for a decimal count of 5.

Solution   For decimal 5, the outputs of the decade counter read *10 01 10 10*.
At the base of transistor Q_5, diodes D_{21}, D_{22}, D_{23}, and D_{24} have their cathodes connected to Q_D, $\overline{Q_C}$, Q_B, and Q_A, respectively. All these outputs are at logic *1*; therefore, diodes D_{21} to D_{24} are reverse biased, Q_5 is biased *on*, and cathode 5 in the digital tube glows.
Other reverse-biased diodes are D_3, D_7, D_8, D_9, D_{10}, D_{13}, D_{14}, D_{16}, D_{17}, D_{18}, D_{19}, D_{25}, D_{29}, D_{32}, D_{33}, D_{35}, D_{37}, D_{39}, and D_{40}.
The forward-biased diodes and their associated transistor gates are D_1, D_2, D_4—Q_0; D_5, D_6—Q_1; D_{11}, D_{12}—Q_2; D_{15}—Q_3; D_{20}—Q_4; D_{26}, D_{27}, D_{28}—Q_6; D_{30}, D_{31}—Q_7; D_{34}, D_{36}—Q_8; D_{38}—Q_9.

Figure 14-16 is a logic diagram for BCD-to-decimal conversion. The flip-flop blocks have terminals as follows: *clock input CLK*, *set S*, and *reset R*; and outputs identified by *set* conditions of *0* and *1* as illustrated. The triggering input pulses are applied to the *clock (CLK)* terminal of FF$_A$. Each succeeding flip-flop has its *CLK* terminal connected to the *0* output of the preceding stage. This arrangement can be compared to the cascaded flip-flops in Figure 14-1, where each stage is triggered from the second transistor in the previous stage. The *reset* terminals of FF$_B$ and FF$_C$ are connected to the output of FF$_D$. This corresponds with the *reset* circuitry in Figure 14-4. In Figure 14-16, the output terminals of the decade counter are identified as $\overline{Q_A}$, Q_A, etc., to show the correspondence with Figure 14-15.

Diodes D_1, D_2, D_3, and D_4 and transistor Q_0 in Figure 14-15 constitute a *NAND* gate. In Figure 14-16, these diodes are replaced by the *NAND* gate symbol. Thus, gate G_0 represents transistor Q_0 and diodes D_1 to D_4. Also, gate G_1 represents transistor Q_1 and diodes D_5 to D_8, G_2 represents Q_2 and D_9 to D_{12}, etc. In Figure 14-16, the input terminals of *NAND* gate G_0 are connected to the decade counter terminals in the same configuration as D_1 to D_4 in Figure 14-15. Thus, the inputs to G_0 are $\overline{Q_D}$, Q_C, Q_B, and $\overline{Q_A}$. Similarly, gate G_1 in Figure 14-16 is connected to the same decade counter terminals as D_5 to D_8 in Figure 14-15. The output of each *NAND* gate is connected to the appropriate cathode of the digital indicator tube. It is important to note that the *NAND* gates should have open-collector outputs (see Section 12-4) in order to control the digital tube. The output stages of the *NAND* gates must also be able to survive the high operating voltages of the tube.

EXAMPLE 14-3

From the collector voltage levels shown in Figure 14-15, determine the input terminal connections for *NAND* gate G_9 in Figure 14-16.

Solution   Gate G_9 should provide a *low* output only when the decimal count is 9. For G_9 to produce a *low* output, all its input terminals must be at logic *1*. From Figure 14-15, at the decimal count of 9, decade counter outputs Q_D, Q_C, Q_B, and Q_A are all at logic *1*. Therefore, the inputs of G_9 should be connected to those output terminals of the decade counter in Figure 14-16.

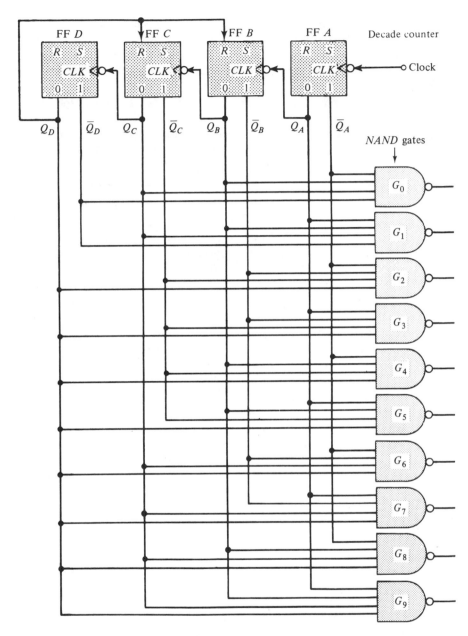

Figure 14-16 Logic diagram for BCD-to-decimal conversion. This corresponds to the decade counter and diode matrix in Figure 14-15. For example, diodes D_{37} through D_{40}, together with transistor Q_9, constitute a *NAND* gate. In Figure 14-16, they are replaced by *NAND* gate G_9.

14-6 SEVEN-SEGMENT DECODER DRIVER

The BCD-to-decimal conversion circuitry already discussed is suitable for driving a digital indicating tube. However, it is not suitable for driving a seven-segment display. BCD-to-decimal conversion is necessary for a seven-segment display, but in addition, another diode matrix is required to convert from decimal to seven-segment format. The required diode matrix configuration is shown in Figure 14-17.

For the seven-segment display shown, the anodes of the light-emitting diodes are connected in common, and have a positive supply voltage $+V$. The segments are lettered *a* through *g*, and terminals are provided for connecting to each segment. *NAND* gates G_0 through G_9 from Figure 14-16 are shown again in Figure 14-17. When the decimal count is *0*, the output of gate G_0 is *low*; when the count is *1*, the output of G_1 is *low*, etc. For a *0* indication, LED segments *a*, *b*, *c*, *d*, *e*, and *f* should be energized. Therefore, those segments are connected via diodes D_{42} through D_{47} to the output of gate G_0. Since the cathodes of the diodes are connected to the gate output, they are forward biased when G_0 output is *low*, giving a seven-segment display indication of 0.

When the decimal count is *1*, segments *b* and *c* should be energized. These segments are connected via diodes D_{48} and D_{49} to the output of *NAND* gate G_1. At the output of gate G_2, diodes D_{50} through D_{54} connect to LED segments *a*, *b*, *d*, *e*, and *g*. When the output of gate G_2 is *low*, these segments are energized and display the numeral *2*, as illustrated in the figure. At this time, only diodes D_{50} through D_{54} are conducting. No other diodes are conducting because only gate G_2 has a *low* output. Figure 14-17 shows the LED segments that are energized for each decimal count.

Consideration of the diode matrix in Figure 14-17 reveals that the diodes connected to each LED segment constitute an *AND* gate. For segment *e*, for example, diodes D_{44}, D_{52}, D_{72}, and D_{81} provide a *low* level at the segment cathode (to energize the segment) when any one of the diodes has a low input level. The segment cathode is *high* only when the inputs to all four diodes are *high*. The diode matrix can be replaced by a group of *AND* gates, as illustrated in the logic diagram of Figure 14-18. The *AND* gate inputs for each segment are the same as the cathode connections for the diodes associated with the segments in Figure 14-17.

EXAMPLE 14-4

Determine the input terminal connections for the *AND* gate connected to segment *b* in Figure 14-18.

Solution   Consideration of the 0 to 9 display arrangements shows that segment *b* must be energized for display of numerals 0, 1, 2, 3, 4, 7, 8, and 9. Therefore, the *AND* gate input for segment *b* must be connected to the outputs of *NAND* gates G_0, G_1, G_2, G_3, G_4, G_7, G_8, and G_9.

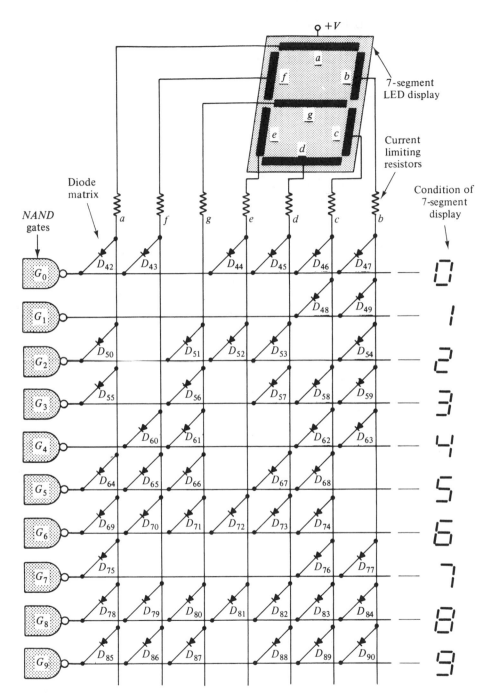

Figure 14-17 Diode matrix for decimal-to-seven-segment conversion. When the output of *NAND* gate G_0 is *low*, segments *a*, *b*, *c*, *d*, *e*, and *f* are energized via diodes D_{42} through D_{47}. Similarly, when another *NAND* gate output is *low*, the appropriate segments are energized to display the correct numeral.

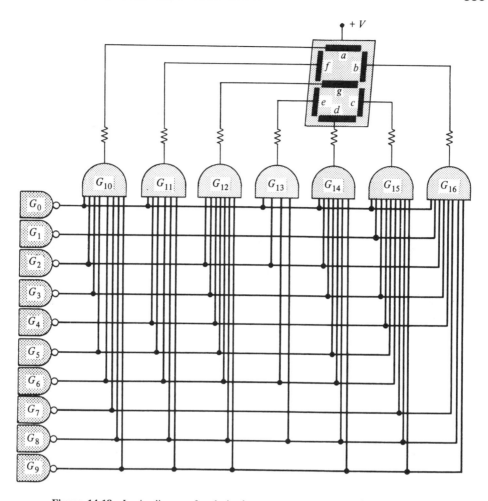

Figure 14-18 Logic diagram for decimal-to-seven-segment conversion. This corresponds to the diode matrix circuit in Figure 14-17, where, for example, diodes D_{44}, D_{52}, D_{72}, and D_{81} constitute an *AND* gate. In Figure 14-18, they are replaced by *AND* gate G_{13}.

Integrated circuits which convert directly from BCD to seven-segment displays are, of course, available. Such ICs are usually listed as *BCD-to-seven-segment decoder/drivers* by their manufacturers. The input/output waveforms for a BCD-to-seven-segment decoder/driver are shown in Figure 14-19. The output waveforms identified as *a*, *b*, *c*, *d*, *e*, *f*, and *g* refer to each of the segments of the display device. (See the seven-segment device in Figure 14-17). For the waveforms in Figure 14-19, the segments are energized when the outputs are *high*. Thus, at the count of zero, outputs *a*, *b*, *c*, *d*, *e*, and *f* are *high*, and output *g* is *low*. This would display the numeral *0*. Similarly, at the count of *3*, only outputs *a*, *b*, *c*, *d*, and *g* are *high*. Examination of the output

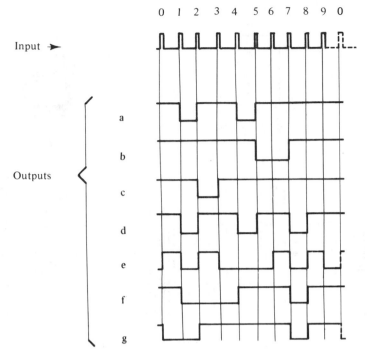

Figure 14-19 Input and output waveforms for BCD-to-seven-segment decoder/driver. To display the numeral 3, segments *a*, *b*, *c*, *d*, and *g* are energized. (See Figure 14-18). The waveforms show that the outputs to those segments are *high* at the count of 3.

waveforms at each stage of the input count shows that the displayed numerals change consecutively from *0* to *9*.

14-7 SCALE-OF-10 000 COUNTER

Counting Circuitry

One decade counter together with a seven-segment display and the necessary binary-to-seven-segment conversion circuitry can be employed to count from 0 to 9. Each time the tenth input pulse is applied, the display goes from 9 to 0 again. When this occurs, the output of the final transistor in the decade counter goes from *1* to *0*. (See Figure 14-5.) This is the only time that the final transistor produces a negative-going output, and this output can be used to trigger another decade counter.

Consider the block diagram of the scale-of-10 000 counter shown in Figure 14-20. The system consists of four complete sets of decade counters, decoders, and displays. Starting from *0*, all four counters can be set at their normal starting conditions.

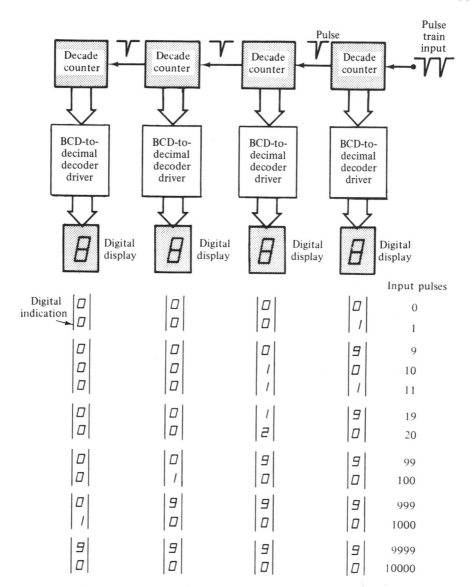

Figure 14-20 Scale-of-10 000 counter constructed of four decade counters, BCD decoders, and numerical displays. The first (right-hand) stage is triggered by the pulse train input, and each successive stage is triggered from the previous stage. Maximum count is 9999.

This gives an indication of *0000*. The first nine input pulses register only on the first (right-hand-side) display. On the tenth input pulse, the first display goes to *0*, and a negative-going pulse output from the first decade counter triggers the second decade counter. The display of the second counter now registers *1*, so that the complete display reads *0010*. The counter has counted to ten, and has also registered *10* on the display system.

The next nine input pulses cause the first counter to go from *0* to *9*, so that the display reads *0019* on the nineteenth pulse. The twentieth pulse causes the first display to go to *0* again. At this time, the final transistor in the first decade counter puts out another negative pulse, which again triggers the second decade counter. The total display now reads *0020*, which indicates that 20 pulses have been applied to the input of the counter. It is seen that the second decade counter and display is counting *tens* of input pulses.

Counting continues as described until the display indicates *0099* after the ninety-ninth input pulse. The one-hundredth input pulse causes the first two displays (from the right) to go to *0*. The second decade counter emits a negative pulse at this time, which triggers the third decade counter. Therefore, the count reads *100*. The system shown in Figure 14-20 can count to a maximum of *9999*. One more pulse causes the display to return to its initial *0000* condition. To increase the maximum count to *99 999*, it is necessary to add one more decade counter, together with a display and BCD-to seven-segment conversion circuitry.

4½ Digit Display

The addition of one flip-flop and a LED display which indicates only the numeral *1* can extend the range of the counter in Figure 14-20 to *19 999*. As illustrated in Figure 14-21, the flip-flop energizes the numeral *1* on receipt of a negative-going pulse from the final (left-hand) decade counter. This occurs after the count has reached *9999*. The next input pulse sets the four right-hand numerals to *0000*, and produces the output pulse from the final decade counter to trigger the flip-flop, giving a display of *10 000*. Another *9999* input pulses can now be applied, to give a total count of *19 999*. Pulse

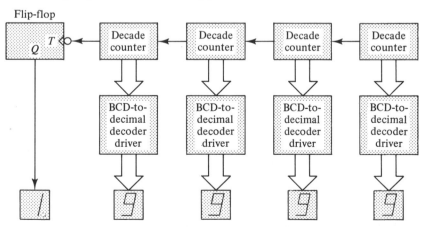

Figure 14-21 Counter with a *4½ digit display*. The *half-unit* at the left side can display only the numeral *1*. It is controlled by a single flip-flop which is triggered by the previous decade counter. The maximum count is *19 999*.

20 000 sets the four seven-segment displays to *0000* once again and produces a pulse which triggers the flip-flop to turn the numeral *1* display *off*. As mentioned earlier, the numeral *1* display is termed a *half-unit*. The complete display is referred to as a *4½ digit display*.

Counter Controls

The type of counter just described can be used directly to count pulses over a given time period, provided that a suitable control circuit is employed. A simple system for switching the counter input pulses *on* and *off* is shown in Figure 14-22. The pulses to be counted are applied to one input of an *AND* gate. The voltage level at the other input of the gate is controlled by the output of a flip-flop. The input triggering pulses pass through the gate to the counter only when the flip-flop $\overline{Q}$ output is at its logic *1* level. The $\overline{Q}$ output can be reset to *1* by switching a negative input voltage to the flip-flop *reset* (*R*) terminal. The *manual start* control shown in the figure is provided for this purpose. A connection to the *reset* input of each decade counter (see Figure 14-7) ensures that the counting circuits return to the *0* condition before counting begins. The *manual stop* control provides a negative voltage which returns the flip-flop to its original *set* condition. This applies to a *0* to the *AND* gate input, thus stopping the pulses from passing through.

The flip-flop can also be triggered, and consequently counting can be started, by means of a negative *start-counting* pulse applied to the *reset* terminal as shown in Figure 14-22. Similarly, a negative pulse applied to the *set* terminal of the flip-flop interrupts the passage of pulses to the counter. The waveforms in the figure show that counting pulses pass through the *AND* gate only during the time interval between application of *start-counting* and *stop-counting* pulses.

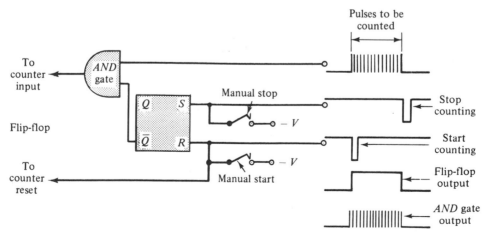

Figure 14-22 Manual and automatic start and stop controls allow a digital counter to count input pulses over a desired time period.

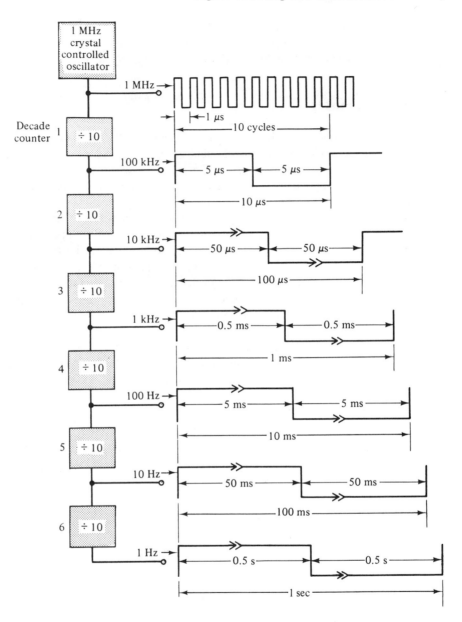

Figure 14-23 The time base for a digital frequency meter uses a very accurate clock source and divides the frequency in stages to produce a precise time period over which the cycles of a waveform can be counted.

14-8 DIGITAL FREQUENCY METER

Timing System

A block diagram and voltage waveforms for a typical timing circuit for a digital frequency meter are shown in Figure 14-23. The source of the time interval over which input pulses are counted is a very accurate crystal-controlled oscillator usually referred to as a *clock source*.

As already discussed, the output frequency from the final flip-flop of a decade counter is exactly one-tenth of the input triggering frequency. This means that the time period of the output waveform is exactly ten times the time period of the input waveform. The 1 MHz output from the crystal oscillator in Figure 14-23 has a time period of 1 μs, and the output waveform from the first decade counter has a time period of 10 μs. The time period of the output from the second decade counter is 100 μs, and that from the third is 1 ms, etc. With all six decade counters shown, the available time periods are 1 μs, 10 μs, 100 μs, 1 ms, 10 ms, 100 ms, and 1 second.

When the counting circuits in a digital frequency meter are triggered for a period of 1 second, the display registers the input frequency directly. A count of 1000 cycles over the 1 second period represents a frequency of 1000 Hz, or 1 kHz; a *5000* display indicates 5000 Hz, etc. These figures are more easily read when a decimal point is placed after the first numeral and the output is identified in kilohertz. Thus a display of *1.000* is 1 kHz, *5.000* is 5 kHz, and *5.473* is 5.473 kHz. In an LED display, the decimal point is created by the use of a single suitably placed light-emitting diode. Also, a *kHz*-indication usually is displayed.

Now consider the effect of using the 100 ms time period to control the counting circuits. A display of *1000* now means 1000 cycles per 100 ms. This is 10 000 cycles per second, or *10 kHz*. When the time period is switched from 1 second to 100 ms, the decimal point also is switched from the first numeral to a position after the second numeral. The *10.00* display is now read as *10.00 kHz*; *50.00* is read as *50.00 kHz*, etc.

When the time period is switched to 10 ms, the decimal point is moved to a new position after the third numeral on the display. A *100.0* display now becomes *100.0 kHz*. Since this is the result of 1000 cycles of input counted over a period of 10 ms, the actual input frequency is 1000/10 ms, that is, 100 kHz. With a 1 ms time period, a display of *1000* indicates 1000 cycles during 1 ms, or 1 MHz. The decimal point is now moved back to its original position after the first numeral, and a *MHz* indication is displayed. Therefore, the display of *1.000* with a 1 ms time base is read as *1.000 MHz*. If the 100 μs and 10 μs time periods are used, the decimal point is again moved so that the *1000* indication becomes 10.00 MHz and 100.0 MHz, respectively.

EXAMPLE 14-5

A 3.5 kHz sine wave is applied to a digital frequency meter. The time base is derived from a 1 MHz clock generator frequency-divided by decade counters.

Determine the meter indication when the time base uses (a) six decade counters, and (b) four decade counters.

Solution   (a) When six decade counters are used,

$$\text{Time base frequency} = f_1 = \frac{1 \text{ MHz}}{10^6} = 1 \text{ Hz}$$

$$\text{Time base} = t_1 = \frac{1}{f_1} = \frac{1}{1 \text{ Hz}} = 1 \text{ sec}$$

$$\text{Cycles of input counted during } t_1 = \text{Input frequency} \times t_1$$

$$= 3.5 \text{ kHz} \times 1 \text{ sec}$$

$$= 3500$$

The display indication is *3500*.

(b) When four decade counters are used,

$$\text{Time base frequency} = f_2 = \frac{1 \text{ MHz}}{10^4} = 100 \text{ Hz}$$

$$\text{Time base} = t_2 = \frac{1}{f_2} = \frac{1}{100 \text{ Hz}} = 10 \text{ ms}$$

$$\text{Cycles of input counted during } t_2 = \text{Input frequency} \times t_2$$

$$= 3.5 \text{ kHz} \times 10 \text{ ms}$$

$$= 35$$

The display indication is 0035.

Latch

If the display devices in a digital frequency meter are controlled directly from the counting circuits, the display changes rapidly as the count progresses from zero. Suppose the input pulses are counted over a period of 1 second, and then the count is held constant for 1 second. The display alternates between being constant for one second and continuously changing for the next second. Therefore, the display is quite difficult to read, and the difficulty is increased when shorter time periods are employed for counting. To overcome this problem, *latch* circuits are employed.

A *latch* isolates each display device from its counting circuits while counting is in progress. At the end of the counting time, a signal to the latch causes the display to change to the decimal equivalent of the final condition of the counting circuits. If the input frequency is constant, as is usually the case, the displayed count remains constant.

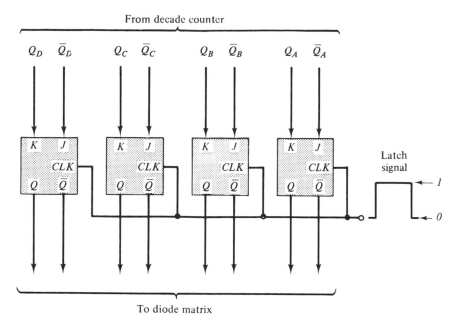

Figure 14-24 *JK* flip-flops used as a latch circuit to isolate the displayed number from the counting circuits in a digital counter while counting is in progress.

The counting circuits continue to sample the input frequency, and the *latches* are repeatedly triggered to check the displays against each final state of the counting circuits.

Figure 14-24 illustrates the use of *JK* flip-flops as a latch circuit. The *J* and *K* input terminals of the flip-flops are connected to the outputs of the decade counters (Figure 14-20), and the latch outputs are fed to the BCD decoders. The latch control signal is applied to the clock input terminal of each *JK* flip-flop.

The theory of the *JK* flip-flop is explained in Section 13-8. When triggered, the flip-flop outputs assume the level of the *J* and *K* input terminals. The output then remains constant until another trigger (i.e., latch) input is applied. If the input levels have changed, the output will change. If the input levels are the same as before, the flip-flop outputs remain unaltered. Thus, the latch circuits sample the output levels of the decade counters and pass these levels to the BCD decoders. Since sampling occurs only when a latch signal is applied, the inputs to the decoders remain constant during the time interval between latch signals. Therefore, the numerical display devices also remain in a constant state during this time, and the counting circuits can go from zero to maximum count without affecting the display.

A *display enable* control which open-circuits the supply voltage to the display devices can sometimes be used instead of a latch. The display is simply switched *off* during the counting time and *on* during the noncounting time. The (normally constant) displayed numerals are thus switched *on* and *off* continuously, with no display occurring during the counting time. When the display time and counting time are brief enough, the *on/off* frequency of the display is so high that the human eye sees only a constant display.

Frequency Meter System

The block diagram of a digital frequency meter is shown in Figure 14-25, and the voltage waveforms for the system are illustrated in Figure 14-26. The input signal which is to have its frequency measured is first amplified and then fed to a Schmitt trigger circuit. Amplification ensures that the signal amplitude is large enough to trigger the Schmitt circuit, and the Schmitt circuit produces a square wave output of the same frequency as the input. A square wave is required for triggering the counting circuits. Before it gets to the counting circuits, however, the square wave must pass through an *AND* gate.

The square wave passes to the counting circuits only when the Q output from the flip-flop is at logic *1* (i.e., positive). The flip-flop changes state each time a negative-going output is received from the *timer*. Therefore, when $T = 1$ sec (see Figure 14-26), the flip-flop output is alternately at level *1* for one second and at level *0* for one second. Consequently, the *AND* gate is alternately *enabled* for one second and *disabled* for one second. During the time that the *AND* gate is *enabled*, the Schmitt output triggers the counting circuits. The exact number of input pulses is counted during that time, and, as already discussed, when $T = 1$ second, the count is a measure of the input frequency. The timer has six (or more) available output time periods over which counting can take place. The desired time period is selected by means of a switch, as illustrated in Figure 14-25. A separate decimal point selector switch moves with the time base selector.

Output $\overline{Q}$ from the flip-flop is the complement of output Q. (See Figure 14-26.) This waveform is employed for resetting the counting circuits, and for opening and closing the latches. At the beginning of the counting time, output $\overline{Q}$ from the flip-flop is a negative-going voltage. This triggers the reset circuitry in each decade counter. Since flip-flop output $\overline{Q}$ is at logic *0* during the counting time, its application to the latching circuits ensures that each latch is *off*. That is, during the counting time, nothing passes through the latching circuits. At the end of the counting time, the waveform fed to the latch control goes to logic *1*. This triggers each latch *on*, so that the conditions of the displays are corrected, if necessary, to reflect the states of the counting circuits. During the latch *on* time, the *AND* gate is *off*, and no counting occurs. Therefore, once corrected, the displays remain constant. The displays also remain constant during the counting time, since the latch circuits are *off*.

A digital frequency meter can be used to measure the frequency of a signal with almost any waveform. Also, it can usually be employed for accurate measurement of time periods, and for determining the ratio of two frequencies.

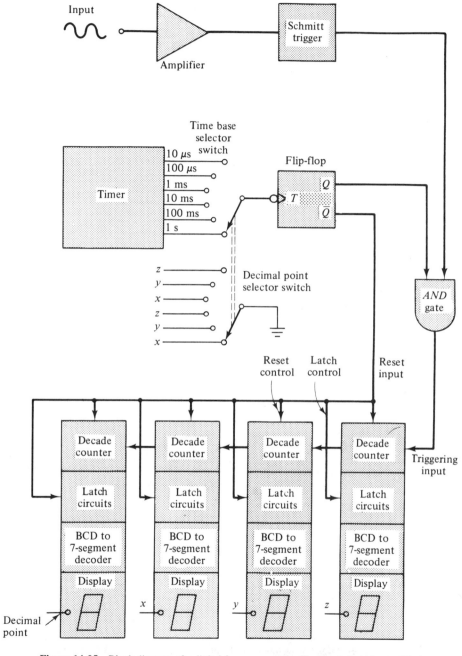

Figure 14-25 Block diagram of a digital frequency meter. The input signal is amplified and converted into a pulse waveform for counting. The pulses are counted over a precise time period generated by the timer. The resultant display can be read as a frequency.

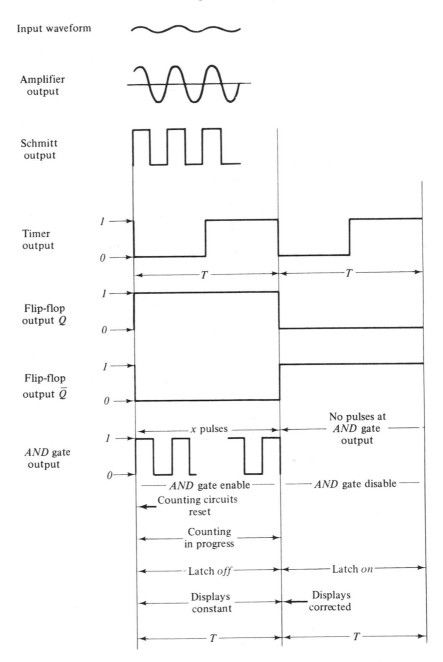

Figure 14-26 Waveforms for the digital frequency meter in Figure 14-25. The input signal is amplified and converted into a pulse waveform, as illustrated, for counting over a precise time period T. While counting is in progress, the latch remains *off* to keep the display constant.

14-9 DIGITAL VOLTMETER

Dual-Slope Integrator

In the dual-slope integration type of digital voltmeter, an integrating circuit is used to generate a ramp over a timed interval. The slope of the ramp is proportional to the dc voltage to be measured. Consequently, the ramp peak amplitude is also proportional to the input voltage. The integrator is then reversed and discharged at a constant rate proportional to an accurate reference voltage. When the time for the ramp to go to zero is measured digitally, the numerical display of the number of input pulses can be read as the dc input voltage.

A dual-slope integrating circuit is shown in Figure 14-27, and the circuit waveforms are illustrated in Figure 14-28. The voltage V_i to be measured is applied to the input terminal of a voltage follower in order to present a high input impedance. The voltage follower output is switched via a FET (Q_1) to the input of a Miller integrator. An accurate current source is also connected to the input of the Miller integrator, and the integrator output level is monitored by a *zero-crossing detector*. The zero-crossing detector (see Section 5-4) is merely a high-gain operational amplifier, which gives a large positive output when the input voltage is slightly above ground, and a large negative output when the input is slightly below ground.

The square wave (control) input to Q_1 provides the time interval over which integration occurs. This square wave is generated by using decade counters to divide the output frequency of a *clock source*, as explained in Section 14-8. When the input to Q_1 is negative, the FET is biased *off*, and V_i is isolated from the integrator. During this time, the reference current is fed into the Miller integrator circuit. The level of this reference current is

$$I_R = \frac{-V_{z1}}{R_3 + R_4 + R_5}$$

The direction of the current is such that C_1 tends to charge positively on the right side, so that the output of the Miller integrator tends to be positive. However, when the output of the Miller circuit becomes slightly positive, the zero-crossing detector generates a large positive output. This biases FET Q_2 *on*, and Q_2 short-circuits C_1. Therefore, at the end of the negative half of the square wave input to Q_1, capacitor C_1 is short-circuited, and the Miller circuit output is held close to ground level.

Transistor Q_1 switches *on* when the square wave input becomes positive. This action connects voltage V_i to resistance R_5 and provides an input current $I_i = V_i/R_5$ to the Miller circuit. Capacitor C_1 now charges with negative polarity on the right side, and this produces a negative-going output from the Miller circuit (Figure 14-28). Consequently, the zero-crossing detector has a large negative output, which biases transistor Q_2 *off*, thus permitting C_1 to charge. The output from the Miller circuit is a linear negative ramp voltage, which continues during the positive portion of the square wave input to Q_1. Since I_i is directly proportional to V_i, the slope of the ramp is also proportional

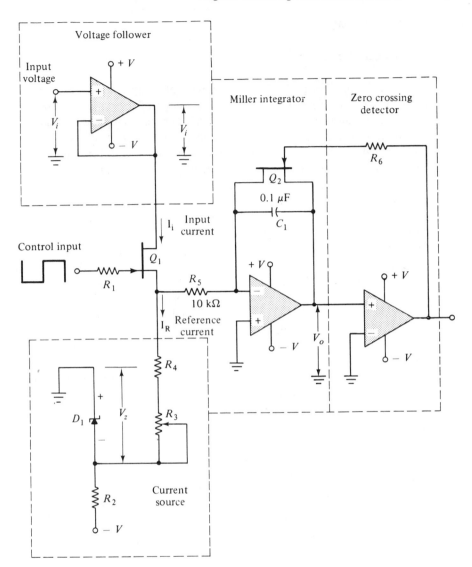

Figure 14-27 Circuit diagram of a dual-slope integrator for use with a digital voltmeter. When Q_1 is *on*, input voltage V_i charges capacitor C_1 to a level proportional to V_i. When Q_1 is *off*, C_1 is discharged to zero by the current source, giving a time period directly proportional to V_i.

to V_i. Also, the time duration t_1 of the positive input voltage is a constant. This means that the ramp amplitude V_o is directly proportional to V_i.

When the square wave input again becomes negative, Q_1 switches *off*, and the reference current I_R commences to flow once more. I_R discharges C_1, so that the Miller circuit output now becomes a positive ramp (Figure 14-28). The positive ramp continues

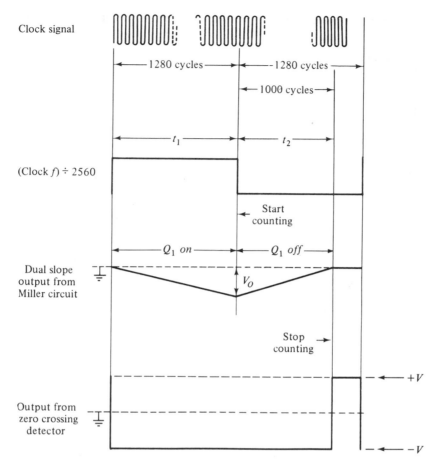

Figure 14-28 Waveforms for the dual-slope integrator in Figure 14-27. Capacitor C_1 (in Fig. 14-27) is charged from V_i over time period t_1, determined by the clock signal. C_1 is then discharged to zero, and the time required (t_2) is measured by counting clock pulses. The displayed count is a digital indication of V_i.

until it arrives at ground level. Then the zero-crossing detector provides an output which switches Q_2 on, discharges C_1, and holds it in short circuit once again.

The time t_2 for the ramp to discharge to zero is now directly proportional to the input voltage. Time t_2 is measured by starting the counting circuits at the negative-going edge of the square wave input to Q_1 and stopping them at the positive-going edge of the output from the zero-crossing detector.

EXAMPLE 14-6

The dual-slope integrator in Figure 14-27 has a square wave input with each half-cycle equivalent to 1280 clock pulses. (See Figure 14-28.) The output frequency from the clock is 200 kHz. If 1000 pulses during time t_2 are to represent an input of $V_i = 1$ V, determine the required level of reference current.

Solution

$$I_i = \frac{V_i}{R_5}$$

For $V_i = 1$ V,

$$I_i = \frac{1 \text{ V}}{10 \text{ k}\Omega}$$

$$= 100 \text{ } \mu\text{A}$$

$$\text{Clock frequency} = 200 \text{ kHz}$$

$$T = \frac{1}{f} = \frac{1}{200 \text{ kHz}}$$

$$= 5 \text{ } \mu\text{s}$$

If t_1 is the time duration of 1280 clock pulses, then

$$t_1 = 5 \text{ } \mu\text{s} \times 1280$$

$$= 6.4 \text{ ms}$$

I_i is applied to the integrator input for a time period t_1. Since

$$C = \frac{It}{V}$$

$$\text{Ramp voltage } V_o = \frac{I_i t_1}{C_1} = \frac{100 \text{ } \mu\text{A} \times 6.4 \text{ ms}}{0.1 \text{ } \mu\text{F}} = 6.4 \text{ V}$$

If t_2 is the time duration of 1000 clock pulses, then

$$t_2 = 5 \text{ } \mu\text{s} \times 1000$$

$$= 5 \text{ ms}$$

and I_R must discharge C_1 in time period t_2.

Therefore,

$$I_R = \frac{C_1 V_o}{t_2}$$

$$= \frac{0.1 \text{ } \mu\text{F} \times 6.4 \text{ V}}{5 \text{ ms}}$$

$$= 128 \text{ } \mu\text{A}$$

One of the most important advantages of the dual-slope integration method is that small drifts in the clock frequency have little or no effect on the accuracy of the measurements. Consider the following example. Let the clock frequency $= f$; then the time period of one cycle of the clock frequency $= T = 1/f$, and the time duration t_1 of 1280 clock pulses is $1280 \times T$. Then,

$$V_o = \frac{I_i t_1}{C_1}$$

$$= \frac{100 \ \mu A \times 1280 \ T}{C_1}$$

and

$$t_2 = \frac{C_1 V_o}{I_R} = \frac{C_1}{128 \ \mu A} \times \frac{100 \ \mu A \times 1280 \ T}{C_1}$$

$$= 1000 \ T$$

The number of clock pulses during t_2 is given by

$$\frac{t_2}{\text{Time period of clock pulses}} = \frac{1000 \ T}{T} = 1000$$

It is seen that when the clock frequency drifts, the digital measurement of voltage is unaffected.

DVM System

Figure 14-29 shows a block diagram of a DVM system employing dual-slope integration. In this particular system, the clock generator has a frequency of 200 kHz. The 200 kHz is divided by a decade counter, two divide-by-16 counters, and one divide-by-2, as illustrated, giving a frequency of approximately 39 Hz. This (39 Hz) is the square wave which controls the integrator, as already explained. The 200 kHz clock signal, the 39 Hz square wave, and the integrator output are all fed to input terminals of an *AND* gate.

The 200 kHz clock output acts as a triggering signal to the counting circuitry when the other two inputs to the *AND* gate are *high*. This occurs during time t_2, as illustrated in Figure 14-28. Note that the output of the zero-crossing detector and the 39 Hz square wave must be inverted before being applied to the *AND* gate. The integrator output (i.e., zero-crossing detector output) is also used to reset the counting circuits and to control the latch. The counting circuits are reset at the beginning of time period t_1. Counting commences at the start of t_2. The latch is switched *on* at the end of t_2 in order to set the displays according to the counting circuits.

The range selector is adjusted to suit the input voltage. An input of less than 2 V is applied directly to the integrator, and a decimal point is selected so that the display can indicate a maximum of *1.999* V. An input voltage between 2 V and 20 V is first potentially divided by 10 and applied to the integrator again as a voltage less than 2

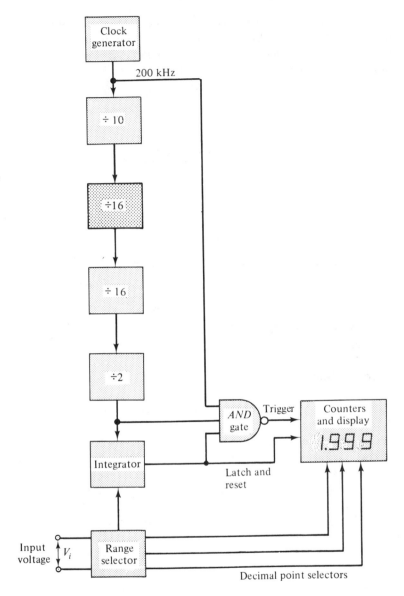

Figure 14-29 Block diagram of a digital voltmeter which uses the dual-slope integrator of Figures 14-27 and 14-28. When the (inverted) output of the integrator and time base are positive (during t_2 in Figure 14-28), the clock pulses pass through the *AND* gate to be counted. Because t_2 is proportional to V_i (in Figure 14-27), the display is read as measured voltage.

V. In this case, the decimal point is selected so that the display can indicate a maximum of *19.99* V. An input voltage between 20 V and 200 V is reduced by a factor of 100 before passing to the integrator. Decimal point selection now allows the meter to indicate a maximum of *199.9*.

REVIEW QUESTIONS AND PROBLEMS

14-1. Sketch the complete circuitry for four cascaded flip-flops. Briefly explain the triggering process.

14-2. Sketch a block diagram for a scale-of-16 counter. Reading only the left-hand transistors in each flip-flop, prepare a truth table showing the state of the counter after each input pulse from 0 to 16. Explain the procedure.

14-3. Sketch the waveforms of collector voltages for every transistor in a scale-of-16 counter for input pulses from 0 to 16. Briefly explain each waveform change.

14-4. Briefly explain how a scale-of-16 counter can be converted to a decade counter. Identify the flip-flops which must be reset in the process, and show which of 16 states of the counter are eliminated.

14-5. Sketch circuitry employed for resetting flip-flops when a scale-of-16 is converted to a decade counter. Briefly explain.

14-6. Sketch the block diagram of a decade counter. Prepare a table that shows the state of the counter after each input pulse from 0 to 10. Also, show the collector waveforms for the even-numbered transistors. Explain the waveforms and the states of the counter.

14-7. Identify the transistors in a decade counter that should be reset to *0* before counting commences. Sketch suitable circuitry for (a) manual resetting, (b) automatic resetting, (c) resetting by pulse input. Briefly explain each circuit.

14-8. Sketch the logic block diagram for a 7493 integrated circuit divide-by-16 counter. Sketch waveforms showing the counter output voltage levels after each input pulse. Explain the operation of the counter.

14-9. Refer to the block diagram for the 7492A IC decade counter, and to the count sequence table for this IC in Appendix 1-24. Sketch waveforms showing the relationship between triggering and signals and the counter outputs.

14-10. Explain the operation of a light-emitting diode. Sketch the cross section of an LED, and identify and explain all component parts of the device.

14-11. Discuss the current and voltage requirements of an LED. Sketch the circuit symbol for the device, and show how it is employed in a seven-segment numerical display.

14-12. Three series-connected LEDs are to have 15 mA passed through them from a −9 V supply. A *pnp* transistor with $h_{FE(min)} = 75$ is to be used as a switch. The base input voltage to the transistor is −9 V. Sketch an appropriate circuit and determine the resistor values required.

14-13. Explain the operation of a liquid crystal cell. Sketch the cross section of a liquid crystal cell, and identify and explain each component part.

14-14. Explain *transmittive-type* and *reflective-type* liquid crystal cells. Discuss the voltage and current requirements for liquid crystal displays, and show how square waves are employed to drive a seven-segment liquid crystal display.

14-15. Explain the construction of a seven-segment fluorescent display, and discuss its characteristics.

14-16. Using sketches, explain the construction of a digital indicator tube. Also, sketch the schematic symbol for the device and explain its operation.

14-17. Using sketches, explain the operation of a seven-segment gas discharge display. Discuss the current and voltage requirements for the display, and explain the function of the *keep-alive cathode*.

14-18. Show how a diode matrix may be employed for BCD-to-decimal conversion. Sketch the complete circuit of the diode matrix, transistor gates, and digital indicator tube.

14-19. For the diode matrix in the BCD-to-decimal conversion circuitry sketched for Problem 14-18, identify the diodes that are forward biased and those that are reverse biased at a decimal count of 6 and at a decimal count of 3.

14-20. Sketch a complete logic diagram for BCD-to-decimal conversion. Briefly explain how the system functions.

14-21. Sketch a complete diode matrix for driving a seven-segment display from a decimal input. Explain the operation of the circuitry, and identify the segments of the display that are energized for each decimal input.

14-22. Sketch a complete logic diagram for decimal-to-seven-segment display conversion. Briefly explain how the system functions.

14-23. Draw the block diagram of a scale-of-1000 counter. Explain the operation of the system.

14-24. Draw the block diagram of a control system that will start and stop a counter manually, and by means of input pulses. Show the waveforms of input and the control voltages. Explain the operation of the system.

14-25. Sketch the block diagram of a timing system for a digital frequency meter. Also, sketch the output waveforms and carefully explain the operation of the system.

14-26. A crystal-controlled oscillator with an output frequency of 100 MHz is available for use in the timing circuit of a digital voltmeter. Draw a block diagram to show how time intervals of 100 μs, 1 ms, 10 ms, and 100 ms can be obtained.

14-27. Discuss the numerical display obtained with a digital frequency meter when the time base is 1 second and the input frequency is 3 kHz. Explain how the time base and the display must be altered when the frequency goes to 30 kHz, 300 kHz, and 3 MHz.

14-28. A digital frequency meter uses a time base derived by decade counters from a 10 MHz source. Determine the display indication produced by a 1.5 kHz input when the time base uses five decade counters. Also, determine the number of decade counters required for the display to read *1500*.

14-29. Sketch a block diagram showing *JK* flip-flops employed as latching circuits. Carefully explain the operation of the latch, and the effect that it has on the numerical display.

14-30. Sketch the complete block diagram of a digital frequency meter. Also, sketch the voltage waveforms that occur throughout the system. Explain the operation of the system.

14-31. Sketch the circuit of a *dual-slope integrator*. Also, sketch the circuit waveforms and carefully explain the operation of the integrator.

14-32. Show that the accuracy of the dual-slope integration method of voltage measurement is not affected by small drifts in clock frequency.

14-33. The Miller circuit in a dual-slope integrator (as in Figure 14-27) has an input resistance $R_5 = 15$ kΩ and a capacitor $C_1 = 0.1$ μF. The input voltage is 1 V, and each half-cycle of the square wave input is equivalent to 1500 cycles of the clock frequency. The clock frequency is 400 kHz. Determine the required reference current if the 1 V input is to be represented by a count of 1000.

14-34. Sketch the block diagram of a digital voltmeter using dual-slope integration. Carefully explain the operation of the system.

Chapter 15

Sampling, Conversion, Modulation, and Multiplexing

INTRODUCTION

A *sampling gate* is a switch which is usually employed to periodically sample the amplitude of an analog voltage. *Sample-and-hold* circuits are essentially sampling gates which hold the voltage sample constant for a specified time. In *analog-to-digital conversion*, amplitude samples of an analog voltage are converted to a number represented by the condition of flip-flops. A *digital-to-analog converter* accepts a digital input and converts it to an equivalent analog output.

Information may be transmitted, recorded, or otherwise processed in the form of pulses. The technique employed may be *pulse amplitude modulation, pulse duration modulation, pulse position modulation*, or *pulse code modulation*. Several separate pulse-modulated signals can be transmitted or recorded on one channel by *time division multiplexing*, which is a technique for inserting pulses in the spaces between other pulses.

15-1 SAMPLING GATES

Series Gate

A *sampling gate* samples the amplitude of a dc or low-frequency ac voltage (or current) signal. Figure 15-1 shows the circuit of a JFET *series sampling gate*. The FET is in series with load resistance R_L, signal voltage V_S, and source resistance R_S. The FET gate terminal has a control voltage V_1 applied to it in the form of a pulse train. When V_1 is at a small positive level (V_1^+), FET Q_1 is *on*. When V_1 is sufficiently negative (V_1^-), the device is biased *off*. It is seen that a sampling gate is essentially a switch

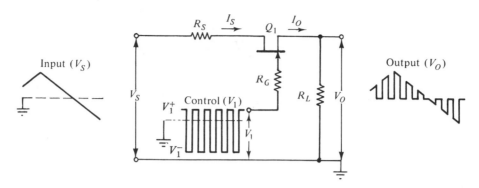

Figure 15-1 A series sampling gate uses a field effect transistor Q_1 as a switch to sample the instantaneous amplitude of a signal voltage V_S several times during each cycle of the signal. Q_1 is switched *on* and *off* by the control voltage V_1 applied to the FET gate terminal.

that is repeatedly opened and closed several times during each cycle of the waveform to be sampled.

In Figure 15-1, the signal voltage has a triangular waveform. The (sampled) output waveform is a series of pulses with amplitudes that vary according to the signal voltage amplitude. These voltage samples may be converted to equivalent digital quantities or to pulse widths, for transmitting, recording, or processing in some other way.

Ideally, a sampling gate should have zero voltage drop across it when *on*, and zero leakage current through it when *off*. Unlike bipolar transistors, which typically have a 0.2 V drop across them when saturated, FETs and MOSFETs can have a very small voltage drop from drain to source. This requires that the channel resistance $R_{D(on)}$ be low, and that the drain current be kept to a minimum. (See Sections 4-7 and 4-8.)

Figure 15-2(a) illustrates the errors introduced by R_S and $R_{D(on)}$. The output voltage, which should ideally be equal to V_S, is

$$V_O = V_S - I_D R_S - I_D R_{D(on)} \tag{15-1}$$

When R_S is very much smaller than R_L, $I_D R_S$ can be neglected. Also, when $R_{D(on)}$ is very much smaller than R_L, $I_D R_{D(on)}$ can be neglected.

The output error due to the FET gate-source leakage current I_{GSS} is illustrated in Figure 15-2(b). When Q_1 is off, an output voltage of $-I_{GSS} R_L$ occurs. Usually, this produces a very small error that can be completely neglected.

Design of a series sampling gate normally involves selection of a suitable FET and determination of the control voltage amplitude. All errors should then be assessed, to determine the accuracy of the output voltage samples. The control frequency should normally be several times greater than the highest frequency of the signal to be sampled, so that several samples are obtained during each cycle of signal voltage.

For Q_1 to switch *on*, its gate voltage should be equal to (or slightly larger than) the voltage level at the source terminal. Since $V_O \approx V_S$, the FET gate voltage should be made equal to the positive peak of V_S to switch Q_1 *on*:

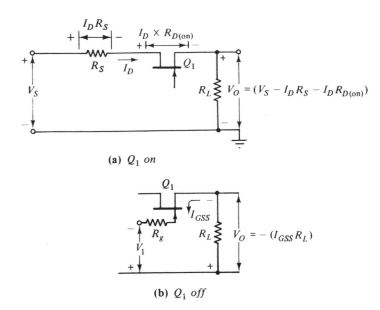

(a) Q_1 *on*

(b) Q_1 *off*

Figure 15-2 Errors occur in a series sampling gate due to the voltage drop $I_D R_{D(\mathrm{on})}$ across the FET when it is *on*, and due to the gate-source leakage current I_{GSS} when the FET is *off*.

$$V_1^+ \approx V_{S(\mathrm{peak})} \tag{15-2}$$

Resistance R_G, in series with the FET gate terminal, is usually selected as 1 MΩ, to ensure that no significant level of gate current flows.

To switch Q_1 *off*, the gate-source voltage should be reverse biased by approximately 1 V greater than the maximum gate-source pinch-off voltage $V_{GS(\mathrm{off})}$. Again, because $V_O \approx V_S$, the FET source terminal goes to the negative peak of V_S. Therefore, to ensure switchoff of Q_1,

$$V_1^- \approx -[V_{S(\mathrm{peak})} + V_{GS(\mathrm{off})(\mathrm{max})} + 1\ \mathrm{V}] \tag{15-3}$$

EXAMPLE 15-1

A sampling gate, as in Figure 15-1, has $V_S = \pm 1$ V and $R_S = 100\ \Omega$. The load resistance is $R_L = 10\ \mathrm{k}\Omega$, and the FET is a 2N4391. (See Appendix 1-10.) Select a suitable control voltage amplitude, and determine the errors due to R_S and $R_{D(\mathrm{on})}$.

Solution   For the 2N4391,

$$V_{GS(\mathrm{off})(\mathrm{max})} = -10\ \mathrm{V} \text{ and } R_{D(\mathrm{on})} = 30\ \Omega$$

From Eq. 15-2, $V_1^+ \approx V_{S(\mathrm{speak})} = +1$ V

From Eq. 15-3, $\quad V_1^- \approx -[V_{S(\text{speak})} + V_{GS(\text{off})(\text{max})} + 1 \text{ V}]$

$$= -[1 \text{ V} + 10 \text{ V} + 1 \text{ V}]$$

$$= -12 \text{ V}$$

$$I_D = \frac{V_S}{R_S + R_{D(\text{on})} + R_L} = \frac{1 \text{ V}}{100 \ \Omega + 30 \ \Omega + 10 \text{ k}\Omega}$$

$$= 98.7 \ \mu\text{A}$$

$$I_D \, R_S = 98.7 \ \mu\text{A} \times 100 \ \Omega$$

$$\approx 9.9 \text{ mV}$$

$$\text{Error due to } R_S = \frac{9.9 \text{ mV}}{1 \text{ V}} \times 100\%$$

$$= 0.99\%$$

$$I_D \, R_{D(\text{on})} = 98.7 \ \mu\text{A} \times 30 \ \Omega$$

$$\approx 3 \text{ mV}$$

$$\text{Error due to } R_{D(\text{on})} = \frac{3 \text{ mV}}{1 \text{ V}} \times 100\%$$

$$= 0.3\%$$

Shunt Gate

The series sampling gate is suitable for signals having a low source resistance. For signals with a very high source resistance, the series gate requirement that R_L be much larger than R_S is difficult to fulfill. In this case, a *shunt sampling gate* is more suitable.

In the shunt gate shown in Figure 15-3, FET Q_1 shorts the input to ground when

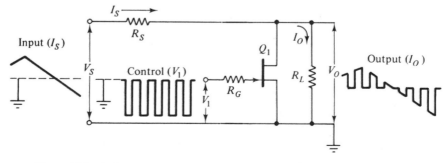

Figure 15-3 A shunt sampling gate samples the instantaneous amplitude of a signal current I_S several times during each cycle of the signal. When Q_1 is switched *on* by the control voltage V_1, I_S is shorted to ground. When Q_1 is *off*, I_S passes to the load.

it is switched *on*. When Q_1 is *off*, current flows from the signal source to the load. Therefore, the shunt sampling gate is essentially a *current switch*, whereas the series sampling gate is a *voltage switch*.

The errors that occur in the shunt gate are illustrated in Figure 15-4. Ideally, as in the series gate, Q_1 in the shunt gate should be a short circuit when *on*, and an open circuit when *off*. The voltage drop across Q_1, when *on*, is $I_D R_{D(on)}$, and this produces some current through R_L when the output level is supposed to be zero [Figure 15-4(a)]. Also, when Q_1 is *off*, there is some leakage current $I_{D(off)}$ through the FET when all of the source current should be passing through the load [Figure 15-4(b)]. This is another source of error.

Operational Amplifier Sampling Gate

An operational amplifier may be employed with a FET, as illustrated in Figure 15-5, to construct a sampling gate which amplifies the signal. The inverting amplifier circuit is designed as discussed in Section 5-2, and the FET is selected to have $R_{D(on)}$ very much smaller than R_1. From Eq. 5-2, when Q_1 is *off*,

$$V_O = V_S \times \frac{R_2}{R_1}$$

and when Q_1 is *on*,

$$V_O = V_S \times \frac{R_2 \parallel R_{D(on)}}{R_1}$$

With $R_{D(on)}$ very much smaller than R_1, $V_O \approx 0$ when Q_1 is *on*.

In addition to amplifying the sampled signal, the operational amplifier provides a low output resistance. Its input resistance is equal to R_1.

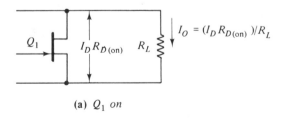

(a) Q_1 *on*

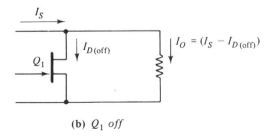

(b) Q_1 *off*

Figure 15-4   Errors occur in a shunt sampling gate due to the voltage drop $I_D R_{D(on)}$ across the FET when it is *on*, and due to the drain-source leakage current $I_{D(off)}$ when the FET is *off*.

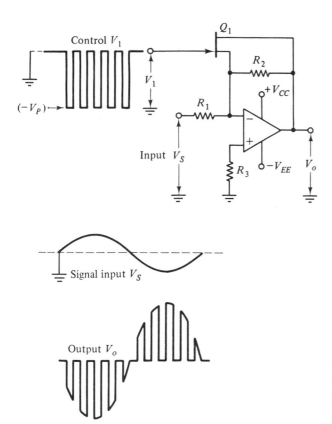

Figure 15-5   An operational amplifier and a FET may be employed as a sampling gate. The signal samples may be amplified to any desired level, and the output resistance is low.

15-2 SAMPLE-AND-HOLD CIRCUIT

A *sample-and-hold circuit*, as its name implies, samples the instantaneous amplitude of a signal and then holds the output voltage constant until the next sampling instant. The circuit (see Figure 15-6) is simply a series gate with a capacitor C_1 to perform the holding function. Operational amplifiers A_1 and A_2 are connected as voltage followers (see Section 5-2) to provide high input impedance and low output impedance.

The waveforms in Figure 15-6 illustrate the relationship between input and output. At time t_a the instantaneous amplitude of the input is V_a. The output holds at the V_a level until time t_b, when it jumps to the input amplitude V_b. Similarly, when the input is falling, the output amplitude remains constant at V_b from t_b to t_c.

During the *sampling time* (also called the *acquisition time*) t_1, Q_1 is *on* and C_1 is charged via the FET channel resistance $R_{D(on)}$, as illustrated in Figure 15-7(a). If the sampling time is

$$t_1 = 5\ CR$$

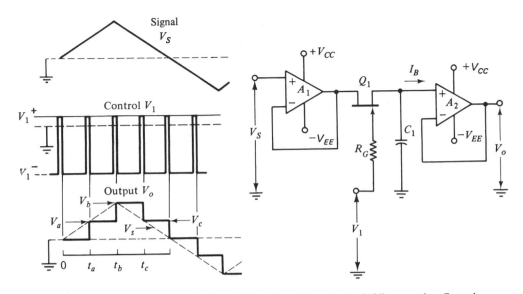

Figure 15-6   Sample-and-hold circuit consisting of sampling gate Q_1, holding capacitor C_1, and two operational amplifiers to provide high input impedance and low output impedance. Each voltage sample is held constant by C_1 until the next sample is taken.

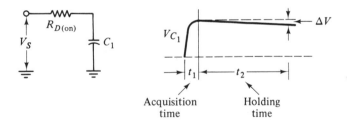

(a) Charging circuit and capacitor voltage

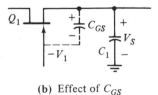

(b) Effect of C_{GS}

Figure 15-7   An error occurs in a sample-and-hold circuit due to C_1 partially discharging during the holding time. Another error is produced by the FET gate-source capacitance C_{GS}.

where R is $R_{D(on)}$, the capacitor is charged to 0.993 of the input voltage. (This comes from Section 2-3.) Allowing the capacitor to charge to 0.993 of V_s results in a 0.7% error in the sampled amplitude. If $t = 7\ CR$, $V_o = 0.999\ V_s$, i.e., a 0.1% error.

During the holding time t_2, C_1 is partially discharged by the bias current I_{B2} flowing into A_2. The FET source-gate leakage current $I_{GS(off)}$ also causes some discharge of C_1. However, $I_{GS(off)}$ is normally very much smaller than I_{B2}, so it can usually be neglected. The capacitance of C_1 is calculated from the knowledge of I_{B2}, the holding time t_2, and the acceptable error due to C_1 discharge. After the value of C_1 is established, the sampling time t_1 is calculated from C_1 and the acceptable charging error.

The FET gate-source capacitance C_{GS} is one more source of error in the output voltage. When the control voltage on the gate goes to its lowest level, C_{GS} is charged to $e_c = (V_s + V_1)$. [This is illustrated in Figure 15-7(b).] The charge on C_{GS} is removed from C_1 and thus reduces V_O. Example 15-2 demonstrates how a sample-and-hold circuit is designed and how the various error sources affect the accuracy of the sample.

EXAMPLE 15-2

A sample-and-hold circuit is to use 741 operational amplifiers and a 2N4391 FET. The signal voltage amplitude $V_s = \pm 1$ V is to be sampled with an accuracy of approximately 0.25%. The holding time is 500 μs. Determine the capacitor value and the minimum acquisition time. Also, calculate the effect of $C_{GS} = 10.5$ pF.

Solution

Capacitor

For the 741, $I_{B(max)} = 500$ nA. Allow $I_{B(max)}$ to discharge C_1 by 0.1% during t_2.

$$\Delta V = 0.1\%\text{ of }1\text{ V}$$

$$= 1\text{ mV}$$

$$C_1 = \frac{I_B \times t_2}{\Delta V} = \frac{500\text{ nA} \times 500\text{ μs}}{1\text{ mV}}$$

$$= 0.25\text{ μF}$$

For the 2N4391, $R_{D(on)} = 30\ \Omega$. Allow another 0.1% error in V_o due to the sampling time t_1.

Minimum acquisition time

$$\text{For }0.1\%\text{ error, }t_1 = 7\ CR_{D(on)}$$

$$= 7 \times 0.25\text{ μF} \times 30\ \Omega$$

$$= 52.5\text{ μs}$$

Effect of C_{GS}

For the 2N4391, $V_{GS(off)(max)} = 10$ V

From Eq. 15-3, $V_1^- \approx -[V_{S(peak)} + V_{GS(off)(max)} + 1\ V]$

$$= -[1\ V + 10\ V + 1\ V]$$

$$= -12\ V$$

$$V_{GS(max)} = +V_S - V_1$$

$$= 1\ V - (-12\ V)$$

$$= 13\ V$$

The charge on C_{GS} is

$$Q = C_{CS} \times V_{GS(max)}$$

$$= 10.5\ pF \times 13\ V$$

$$= 136.5\ pC$$

When Q is removed from C_1,

$$\Delta V_O = \frac{Q}{C_1} = \frac{136.5\ pC}{0.25\ \mu F}$$

$$= 546\ \mu V$$

$$\text{Error due to } C_{GS} = \frac{\Delta V_O}{V_S} \times 100\%$$

$$= \frac{546\ \mu V}{1\ V} \times 100\%$$

$$\approx 0.05\%$$

Sample-and-hold circuits are available in integrated circuit form, which normally require that the hold capacitor be an additional component connected externally. A typical specification for an IC sampling gate has $V_{CC} = \pm(5\ V$ to $18\ V)$, acquisition time $= 10\ \mu s$, and accuracy $= 0.002\%$.

15-3 DIGITAL-TO-ANALOG CONVERSION

DAC Circuit

The circuit for BCD-to-decimal conversion described in Section 14-5 is a form of *digital-to-analog converter (DAC)* with an output which is a numerical display. Normally, however, a DAC converts a digital input into an equivalent analog voltage.

Figure 15-8(a) shows a method known as *digital-to-analog conversion by weighted resistors*. The circuit is also termed a *summing amplifier*. This particular circuit has a

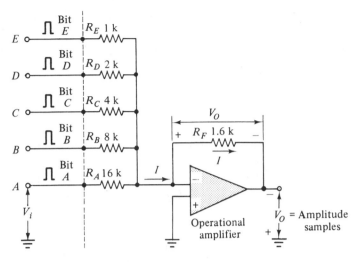

(a) Digital-to-analog conversion circuit

	Binary number represented by bits	Decimal equivalent
Bit E	1 0 0 0 0	16
Bit D	1 0 0 0	8
Bit C	1 0 0	4
Bit B	1 0	2
Bit A	1	1

(b) Binary and decimal equivalents
of bits in 5-bit code

Figure 15-8 Digital-to-analog conversion by weighted resistors. Resistors R_A through R_E have resistances inversely proportional to the digital bit applied to them. The analog output is directly proportional to the decimal number represented by the combination of input bits.

five-bit digital input (bits A through E). Each bit is assumed to have an amplitude of V_i. The digital inputs would typically be derived from the outputs of a cascade of five flip-flops.

The decimal equivalent of each bit in the five-bit input is shown in Figure 15-8(b). The decimal equivalent of bit E is 16, and that for bit D is 8, etc. Each resistor (R_A through R_E) is *weighted* (or selected) to have a resistance inversely proportional to the decimal equivalent of the bit number applied to it. Thus, if $R_A = 16$ kΩ for bit A, which has a decimal equivalent of 1, then $R_E = 1$ kΩ for bit E, with a decimal equivalent of 16. Similarly, $R_B = 8$ kΩ for bit B (decimal equivalent of 2), and $R_D = 2$ kΩ for bit D (decimal equivalent of 8). Note also that the feedback resistor value is $R_F = 1.6$ kΩ.

Consider the effect on the summing amplifier when bit E is present and all other bits are absent. Let the amplitude of all bits be $V_i = 1$ V. Recall, from the previous study of operational amplifiers in Chapter 5, that, with the noninverting terminal grounded, the inverting terminal is always very close to ground potential. Therefore, with the input voltage applied at one end of R_E and a *virtual ground* at the other end, the amplifier input current is

$$I = \frac{V_i}{R_E}$$

Also, in previous studies it was shown that the actual current flow into the amplifier terminal is near zero. Consequently, all of the input current effectively flows through R_F. Since one end of R_F is at the virtual ground input terminal and the other end is at the output terminal, the output voltage is the voltage drop across R_F. Thus, the output voltage is

$$V_o = IR_F = V_i \times \frac{R_F}{R_E}$$

$$= 1 \text{ V} \times (1.6 \text{ k}\Omega/1 \text{ k}\Omega)$$

$$= 1.6 \text{ V}$$

Bit E is converted to 1.6 V, and since the decimal equivalent of bit E is 16, this seems to make sense.

When only bit A is present, the output becomes

$$V_o = V_i \times \frac{R_F}{R_A}$$

$$= 1 \text{ V} \times (1.6 \text{ k}\Omega/16 \text{ k}\Omega)$$

$$= 0.1 \text{ V}$$

This is one-sixteenth of the output produced from bit E. Thus, the two outputs are in correct proportion, since bit E has a decimal equivalent of 16 and bit A has a decimal equivalent of 1.

The combined output produced when bits E and A are present is

$$V_o = V_i \left(\frac{R_F}{R_E}\right) + V_i \left(\frac{R_F}{R_A}\right)$$

$$= V_i R_F \times \left(\frac{1}{R_E} + \frac{1}{R_A}\right)$$

$$= 1 \text{ V} \times 1.6 \text{ k}\Omega \times (1/1 \text{ k}\Omega + 1/16 \text{ k}\Omega)$$

$$= 1.7 \text{ V}$$

The decimal equivalent of binary 10001 (i.e., bit E and bit A) is 17, so that V_O calculated above is once again the analog equivalent of the input. The complete equation for the output is

$$V_O = V_i R_F \left[\frac{1}{R_A} + \frac{1}{R_B} + \frac{1}{R_C} + \frac{1}{R_D} + \frac{1}{R_E} \right] \qquad (15\text{-}4)$$

Any combination of input bits can be substituted into Eq. 15-4, and the analog output calculated. It is always found that the analog output is directly proportional to the decimal equivalent of the binary number represented by the bits. Note that the summing amplifier is an inverting amplifier, so that the output voltage levels are negative. If required, these can easily be converted to positive quantities by the use of another inverting amplifier.

EXAMPLE 15-3

Calculate the output voltage from the DAC in Figure 15-8(a) when the input is 10110.

Solution From Eq. 15-4,

$$V_O = V_i R_F \left[\frac{1}{R_A} + \frac{1}{R_B} + \frac{1}{R_C} + \frac{1}{R_D} + \frac{1}{R_E} \right]$$

$$= 1 \text{ V} \times 1.6 \text{ k}\Omega \left[\frac{1}{1 \text{ k}\Omega} + 0 + \frac{1}{4 \text{ k}\Omega} + \frac{1}{8 \text{ k}\Omega} + 0 \right]$$

$$= 2.2 \text{ V}$$

From Figure 15-8(b),

$$\text{E D C B A}$$

$$1 \ 0 \ 1 \ 1 \ 0 = 16 + 4 + 2$$

$$= 22$$

DAC Performance

Consideration of the DAC circuit in Figure 15-8 shows that the maximum output voltage (when all bits are present) is 3.1 V. Also, the smallest output voltage change occurs when bit A changes from 1 to 0, and *vice versa*. This (smallest output voltage change) is 0.1 V. If the input signal was gradually increased from 0 to its maximum level, the output would increase in steps of 0.1 V. The number of output steps, or the *resolution* of the circuit, is 0.1 V in 3.1 V, or 1 in 31.

The resolution of any DAC can be determined from the number of input bits.

$$\text{Number of steps} = 2^n - 1$$

where n is the number of input bits.

Therefore,

$$\text{Resolution} = \frac{100\%}{\text{number of steps}}$$

or
$$\text{Resolution} = \frac{100\%}{2^n - 1} \qquad (15\text{-}5)$$

For an 8-bit DAC,

$$\text{Resolution} = \frac{100\%}{2^8 - 1} = 0.39\%$$

Obviously, the resolution affects the accuracy of the analog output voltage. Other errors are due to resistor tolerance, input voltage precision, and operational amplifier performance. If these are very much smaller than the resolution, then the accuracy can be defined as the resolution.

15-4 ANALOG-TO-DIGITAL CONVERSION

Ramp Generator Methods

The digital voltmeter (DVM) circuit described in Section 14-9 converts an analog input to a BCD number before converting it into a numerical display. The conversion process involves generating a time period proportional to the input voltage and toggling counting circuits for the duration of that time period. Figure 15-9 shows an *analog-to-digital converter (ADC)*, which also functions by toggling counting circuits over a time period proportional to the analog input voltage.

The analog input V_i is applied to the noninverting input terminal of a voltage comparator as illustrated. A ramp generator is connected to the inverting input terminal. While the ramp voltage is below the level of V_i, the comparator output is *high*. (See the waveforms in Figure 15-9.) When the ramp amplitude becomes equal to V_i, the comparator output switches *low*. The time duration t_1 of the comparator high output is directly proportional to V_i. If V_i were halved, the ramp would become equal to V_i in half the time illustrated, and t_1 would also be halved.

The pulse output from the comparator (V_1) is fed to one input of an *AND* gate which has a clock signal at its other input. While V_1 is *high*, clock pulses pass through the *AND* gate for the time t_1. The *shift register*, which is basically a cascade of flip-flops, is toggled during this time. At the end of t_1, the outputs of the shift register offer a binary number which is the digital equivalent of the analog input voltage. This number is held in the shift register for the remaining duration of the ramp time t_2. The negative-going edge at the end of the ramp resets the shift register to its zero condition once again, preparing it for another cycle of analog-to-digital conversion.

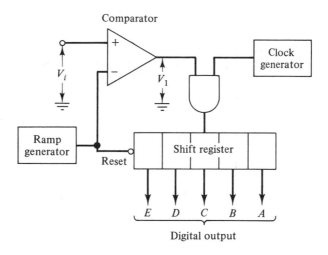

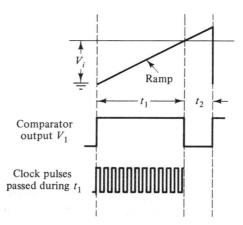

Figure 15-9   Analog-to-digital converter (ADC), using a ramp generator and voltage comparator to generate a time period proportional to the analog input voltage. The shift register is toggled during this time period, to produce a binary number which is equivalent to the analog input.

Digital Ramp ADC

Figure 15-10 shows a modified form of the ramp generator ADC just discussed. The ramp generator in Figure 15-9 is replaced by a DAC which converts the digital output back to analog. As illustrated, the output waveform from the DAC is a *staircase*, which is incremented by each input pulse to the shift register. The shift register resets to zero, and counting is started at the commencement of the *control input* pulse, which is

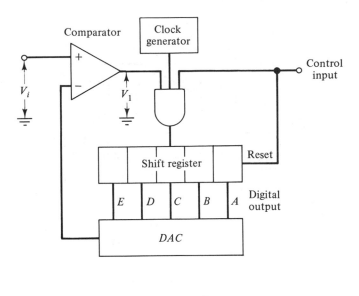

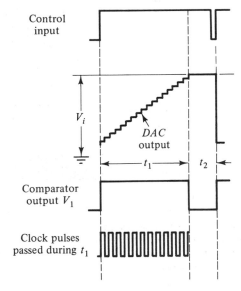

Figure 15-10 Digital-ramp-type analog-to-digital converter using a DAC to convert the digital output back to analog form. This is compared to the analog input, and the counting circuits are toggled until equality is obtained.

applied to the *AND* gate and the shift register. At this time, V_i is greater than the DAC output. Consequently, the clock pulses pass through the *AND* gate to toggle the shift register.

When the output of the shift register is the digital equivalent of the analog input, the DAC output equals V_i. This causes the comparator output to switch to *low*. With one *low* input to the *AND* gate, no clock pulses are passed to the shift register, and counting ceases. The shift register output is now the digital equivalent of the analog input. This condition is maintained for time t_2 until the control input goes positive once again.

Successive Approximation ADC

The operation of a *successive approximation converter* (*SAC*) is similar to that of the digital ramp type described above, except that the shift register bits are toggled in reverse order. That is, instead of being toggled in the sequence A B C D E, they are toggled in the sequence E D C B A. The most significant bit (bit E) is toggled first, instead of the least significant bit (bit A). If this produces a DAC output that exceeds V_i, the register is reset and the second most significant bit is toggled instead. The result of this is that the DAC output is initially incremented in large steps. Consequently, equality between V_i and the DAC output is reached much more quickly than with the usual toggling sequence, and the conversion time is reduced.

Parallel ADC

A *parallel ADC* circuit, also known as a *flash converter*, is shown in Figure 15-11. It uses a resistive potential divider to set up several reference voltage levels, V_{r1} through V_{r7}. The number of reference levels is $2^n - 1$, where n is the required number of bits in the digital output. Each reference level is connected to the noninverting input of a voltage comparator, as illustrated, and the analog signal V_i is applied to the inverting input terminals of all comparators. The comparator outputs are fed into a logic circuit known as a *priority encoder*, which converts them into an n-bit digital output. If V_i is greater than V_{r4}, but less than V_{r5}, for example, comparators *1* through *4* would have *low* output levels, while the outputs of comparators *5* through *7* remain *high*.

With a parallel ADC, analog-to-digital conversion is instantaneous. There is no counting of pulses involved; consequently, there is virtually no conversion time. A disadvantage of parallel converters is that they require a large amount of circuitry.

ADC Performance

To convert an analog voltage into a digital representation, the voltage must be divided up into several discrete levels. The number of discrete levels used determines the precision with which the voltage is represented. For example, if 100 levels are employed, the possible error in the digital representation is 1%, or $\pm\frac{1}{2}\%$. If 1000 levels are used, the error is 0.1%, or $\pm 0.05\%$. The process of dividing the analog input into discrete levels is termed *quantizing*, and the error involved is referred to as the *quantizing error*. If the digital output is represented by a binary number with n bits, the quantizing error is 1 in $(2^n - 1)$, or $\pm\frac{1}{2}$ in $(2^n - 1)$. Thus, a *four-bit* digital output has an error of 1 in 15, while the error in a *six-bit* output is 1 in 63. Great precision requires a large amount of circuitry in an analog-to-digital converter.

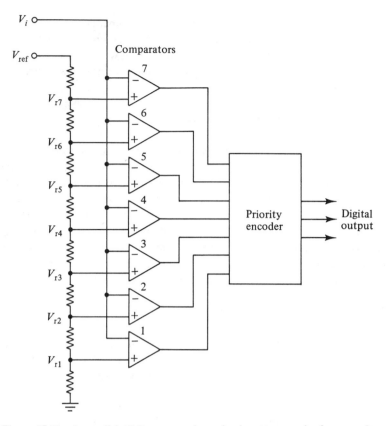

Figure 15-11 A *parallel ADC* compares the analog input to several reference voltage levels provided by a resistive potential divider. The level of V_i is measured by the comparators and converted to a digital number by the priority encoder.

15-5 PULSE MODULATION

Types of Pulse Modulation

The instantaneous amplitude of a signal may be measured (sampled) at regular intervals, and the measured amplitudes converted to pulses. The pulses may then be transmitted, recorded, or otherwise processed. A low-frequency alternating signal and four types of pulse modulation by which the signal may be represented are illustrated in Figure 15-12.

In *pulse amplitude modulation* (PAM), the amplitude of each pulse is made proportional to the instantaneous amplitude of the modulating signal [Figure 15-12(b)]. The largest pulse represents the greatest positive signal amplitude sampled, while the smallest pulse represents the largest negative sample. The time duration of each pulse may be quite short, and the time interval between pulses may be relatively long. If a radio

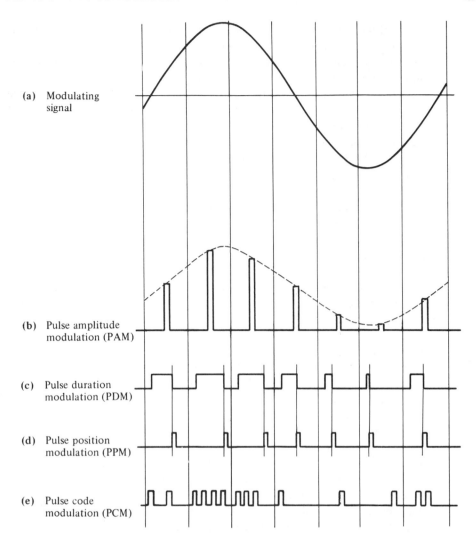

(a) Modulating signal

(b) Pulse amplitude modulation (PAM)

(c) Pulse duration modulation (PDM)

(d) Pulse position modulation (PPM)

(e) Pulse code modulation (PCM)

Figure 15-12   A signal may be represented by a train of pulses with (a) varying amplitudes—*pulse amplitude modulation (PAM)*, (b) varying widths or time durations—*pulse duration modulation (PDM)*, (c) varying positions—*pulse position modulation (PPM)*, or (d) coded groups of pulses—*pulse code modulation (PCM)*.

frequency is pulse-amplitude modulated instead of simply being amplitude modulated, much less power is required for the transmission of information because the transmitter is actually switched *off* between pulses. This is one advantage of pulse modulation.

In *pulse duration modulation* (PDM), also termed *pulse width modulation*, the pulses have a constant amplitude and a variable time duration. The time duration (or width) of each pulse is proportional to the instantaneous amplitude of the modulating signal [Figure 15-12(c)]. In this case, the narrowest pulse represents the most negative sample of the original signal, and the widest pulse represents the largest positive sample. When PDM is applied to radio transmission, the carrier frequency has a constant amplitude, and the transmitter *on*-time is carefully controlled. In some circumstances, PDM can be more accurate than PAM. One example of this is in magnetic tape recording, where pulse widths can be recorded and reproduced with less error than pulse amplitudes.

Pulse position modulation (PPM) [Figure 15-12(d)] is more efficient than PAM or PDM for radio transmission. In PPM, all pulses have the same constant amplitude and narrow pulse width. The position in time of the pulses is made to vary in proportion to the amplitude of the modulating signal. Note that in Figure 15-12(d) each PPM pulse occurs just at the end of a PDM pulse in Figure 15-12(c). Thus, the pulses near the right side of the sampling time period represent the largest positive signal sample, and those toward the left side correspond to the most negative samples of the original signal. PPM uses less power than PDM and, essentially, has all the advantages of PDM. One disadvantage of PPM is that the demodulation process to recover the original signal is more difficult than with PDM.

Pulse code modulation (PCM), illustrated in Figure 15-12(e), is the most complicated type of pulse modulation. However, it can be the most accurate and the most efficient of the four methods. In certain circumstances, it may be the only type of pulse modulation that can be employed. In PCM, each amplitude sample of the signal is converted from an analog to a digital signal. The sample amplitude is then represented by a group of pulses, the presence of a pulse indicating *1* and the absence of a pulse indicating *0*. The *four-bit* code illustrated in Figure 15-12(e) can represent only 16 discrete levels of signal amplitude, consequently, it is far from accurate. As discussed in Section 15-4, the accuracy can be improved by increasing the number of *bits* (i.e., pulses) employed. A seven-bit code, for example, can represent 128 discrete levels of signal amplitude; an accuracy of better than 1%.

For all four pulse modulation methods, the sampling frequency is determined by the highest signal frequency that must be processed. It can be shown that if samples are taken at a rate greater than twice the signal frequency, then the original signal can be recovered. In practice, it is normal to sample at a minimum rate of about three times the highest signal frequency. For audio—voice transmission, for example, with a "high" frequency of 3 kH—the sampling frequency might be 8 kHz.

Another major advantage of pulse modulation, other than pulse amplitude modulation, is illustrated in Figure 15-13. When radio signals are very weak, they may be almost completely lost in circuit or atmospheric noise. If the modulation method is PDM, PPM, or PCM, the signals can be recovered simply by clipping off the noise. In this case, PCM gives the best results, since it is only necessary to determine whether each pulse is present or absent.

Pulse Amplitude Modulation

A process for producing a pulse-amplitude-modulated waveform is illustrated in Figure 15-14. The block diagram of Figure 15-14(a) shows a pulse generator triggering a monostable multivibrator at an appropriate sampling frequency. The output pulses from

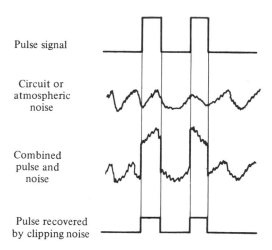

Pulse signal

Circuit or
atmospheric
noise

Combined
pulse and
noise

Pulse recovered
by clipping noise

Figure 15-13 A major advantage of pulse modulation, other than pulse amplitude modulation, is that the signal can be recovered in extremely noisy conditions.

the multivibrator are made to increase and decrease in amplitude by the modulating signal. Figure 15-14(b) shows the circuit of a monostable multivibrator (Chapter 7) with its load resistance R_5 supplied from the modulating signal source. When Q_3 is *on*, the output voltage is the saturation level of Q_3 collector. When Q_3 is switched *off* for the pulse time, the output voltage (i.e., the pulse amplitude) is equal to the modulating signal level. The actual voltage applied as a supply to R_5 must have a dc component as well as the ac modulating signal.

Demodulation of PAM is accomplished simply by passing the amplitude-modulated pulses through a low-pass filter. This process is illustrated in Figure 15-15. The PAM waveform consists of the fundamental modulating frequency and a number of high-frequency components which give the pulses their shape. The resistance R_1 and capacitance C_1 shown in Figure 15-15 form a potential divider. At low frequencies, the impedance of C_1 is very much larger than R_1. Consequently, low-frequency signals (i.e., the fundamental) are passed with very little attenuation. At high frequencies, the impedance of C_1 becomes quite small, and the signals experience severe attenuation. Thus, the filter output is the signal frequency, with perhaps a very small pulse frequency component. If necessary, more than one stage of filtering can be employed to sufficiently attenuate the pulse frequency.

Pulse Duration Modulation

In pulse duration modulation, the signal samples must be converted to pulses which have a time duration directly proportional to the amplitude of the samples. One method of producing PDM, shown in Figure 15-16, uses a free-running ramp generator and a *voltage comparator*. The modulating signal is capacitor coupled via C_1 to the noninverting input terminal of the comparator. Therefore, the voltage at that terminal is a dc level (provided by R_1 and R_2) with the ac signal superimposed. The free-running ramp generator produces a ramp waveform output at the required sampling frequency. This is directly coupled to the inverting input terminal of the comparator. When the ramp is at its zero

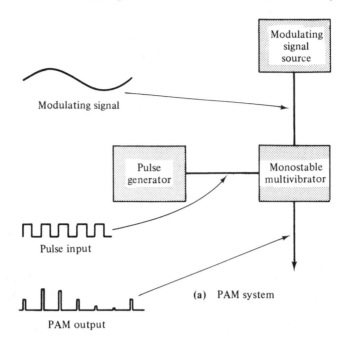

(a)   PAM system

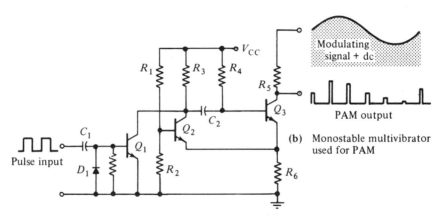

(b)   Monostable multivibrator used for PAM

Figure 15-14   Pulse amplitude modulation system, showing a monostable multivibrator triggered at a constant frequency by a square wave generator. The modulating signal is superimposed upon the dc supply voltage to the output stage of the multivibrator.

level, the comparator inverting input terminal voltage is below the level at the noninverting input terminal. In this condition, the comparator output is a positive voltage level. The ramp voltage grows linearly, and eventually becomes equal to the voltage at the noninverting terminal of the comparator. When this occurs, the comparator output switches rapidly from positive to a negative voltage level. When the ramp voltage returns to zero, the

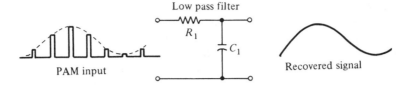

Figure 15-15   Basic method of demodulating a PAM waveform, using a low-pass filter to pass the signal frequency and to block all high-frequency components of the waveform.

inverting input voltage is once again below the level of the noninverting input, and the comparator output returns to its positive voltage level.

It is seen that the output from the comparator is a series of positive-going pulses. Each pulse commences at the instant the ramp waveform returns to zero volts and ends when the ramp level coincides with the signal voltage. When the signal is at its highest level, the ramp takes its longest time to reach equality with the noninverting input voltage. Therefore, the output pulses at this time are of the longest duration. At the instant that the signal is at is lowest level, the ramp takes the shortest time to arrive at

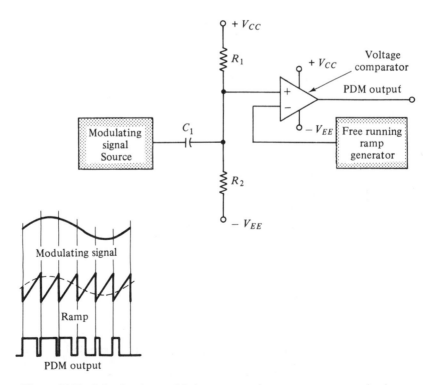

Figure 15-16   Pulse duration modulation system, using a ramp generator and voltage comparator to convert the instantaneous amplitudes of the signal to time periods.

the same voltage as that at the noninverting terminal. Consequently, the width of the output pulses is a minimum at this instant.

Demodulation of PDM waves can be accomplished by first converting each duration-modulated pulse into an amplitude-modulated pulse. Then, filtering can be employed to recover the original modulating signal. The block diagram and waveforms for such a PDM demodulation system are shown in Figure 15-17. The PDM wave is applied to an integrator which generates a ramp-type output. The integrator output ramp always increases linearly at the same rate for constant amplitude pulses. The ramp commences at the start of each input pulse and finishes at the end of the pulse. Consequently, the ramp peak value is proportional to the pulse width. Since the pulse widths are made proportional to the instantaneous sample of the original signal voltage, the ramp peaks represent amplitude samples of the original signal. The integrator output waveform is now fed to the low-pass filter, which removes the high-frequency components and passes the low-frequency signal waveform.

Pulse Position Modulation

The simplest modulation process for pulse position modulation is a PDM system with the addition of a monostable multivibrator. (See Figure 15-18.) The monostable is arranged so that it is triggered by the trailing edges of the PDM pulses. Thus, the monostable output is a series of constant-width, constant-amplitude pulses which vary in position according to the original signal amplitude.

For demodulation of PPM, a PDM waveform is first constructed by triggering an *RS flip-flop*, as shown in Figure 15-19. The flip-flop is triggered into its *set* condition by the leading edges of a square wave which must be synchronized with the original signal source. Synchronization is necessary so that the leading edges of the square

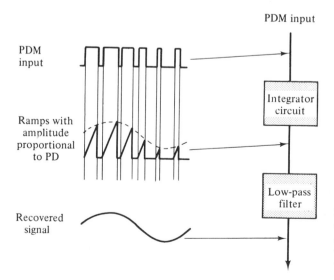

Figure 15-17 PDM demodulating system, using an integrator to generate ramps with amplitudes proportional to the pulse durations. The ramp waveform is then filtered to recover the signal.

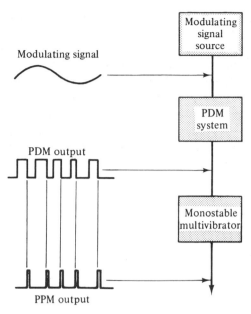

Figure 15-18 Pulse position modulation system. A PDM waveform is first generated, and then a monostable multivibrator is triggered at the lagging edges of the (PDM) pulses to create a PPM waveform.

wave coincide with the leading edges of the PDM wave that was employed to generate the PPM pulses (Figure 15-18). The flip-flop in Figure 15-19 is *reset* by the leading edge of the PPM pulses. The output of the flip-flop is now a PDM wave which may be demodulated by the process illustrated in Figure 15-17.

One inportant requirement for PPM demodulation is synchronization of the square wave for triggering the flip-flop. Several alternatives are available. PPM pulses may be

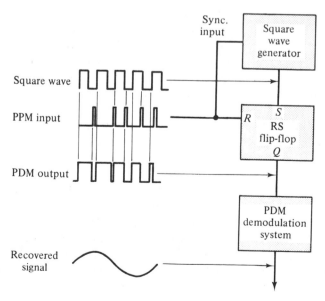

Figure 15-19 To demodulate PPM, a square wave, synchronized to the original modulating frequency, is used to *set* an RS flip-flop. The flip-flop is then *reset* by the PPM pulses. This produces a PDM waveform, which can be demodulated in the usual way.

generated at both the leading and lagging edges of the PDM waveform in Figure 15-18. For demodulation, the leading edge pulses are applied to the *set* terminal of the RS flip-flop. In this case, a square wave generator is not required. However, some method of identifying the leading edge pulses must be employed. A more efficient system is to include periodic synchronizing pulses. For example, every fiftieth pulse might be a synchronizing pulse which corrects any frequency drift in the square wave generator. Again, some means of identifying the synchronizing pulses is essential. This could be done, for example, by making all synchronizing pulses negative while the other pulses are positive. Alternatively, the synchronizing pulses could have a greater amplitude or longer duration than the other PPM pulses. Perhaps the best method of identifying a sychronizing pulse is to precede it with a longer-than-normal space. Methods for generating such a space during the modulation process and for identifying it during demodulation are similar to those employed in time division multiplexing.

15-6 TIME DIVISION MULTIPLEXING

TDM Waveforms

Suppose that a 2.5 kHz signal is sampled ten times in every cycle (at 25 kHz). The samples occur at time intervals of one-tenth of the time period of the waveform, that is, at 40 μs intervals. If PDM is employed, and the maximum pulse width is made less than 10 μs, a 30 μs time interval (or space) is left between pulses. If other signals are sampled at the same rate, the additional pulse samples might be included in the 30 μs space. This process, known as *time division multiplexing* (TDM), is illustrated in Figure 15-20.

Channel 1 in Figure 15-20(a) is a series of PDM samples, with the first pulse shown commencing at time t_1. Channel 2, in Figure 15-20(b), is another series of PDM pulses with the first pulse shown commencing at t_2, 10 μs after t_1. The second pulse in channel 2 occurs 40 μs after t_2 (10 μs after t_5). For channel 3, [Figure 15-20(c)], the first pulse shown starts at t_3, which is 20 μs after t_1 and 10 μs after t_2. As in the case of the other channels, there is a 40 μs time interval between commencement of each pulse in channel 3. When the pulses are time multiplexed, as shown in Figure 15-20(d), three channels of information are contained in the waveform. This waveform may now be recorded on a single magnetic track, transmitted on a single radio frequency, or otherwise processed.

Each channel in the time-multiplexed waveform [Figure 15-20(d)] is allocated a 10 μs time period. Therefore, the maximum pulse width must be less than 10 μs, so that the end of one pulse can be distinguished from the beginning of the next. At each sampling, the three channels occupy a total time period of 30 μs. This leaves a 10 μs time interval (one unoccupied channel space) between the end of the three pulses and commencement of the next three. This time period is deliberately left clear so that it may be used for synchronization in the demodulation process. Channel 1 can easily be identified during demodulation as the first information pulse after the *sync space*. Channels

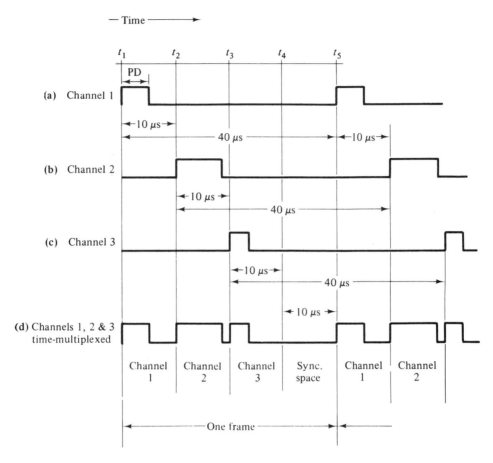

Figure 15-20 *Time division multiplexing (TDM) of three PDM channels. Channel 1* pulses are generated only during time t_1 to t_2, channel 2 pulses occupy only the period t_2 to t_3, and channel 3 pulses occur only during t_3 to t_4. When the three channels are put together, each pulse occupies a space between the other pulses.

2 and 3 can then be identified in sequence after channel 1. The series of three or more information pulses together with the synchronizing space is usually termed *one frame* of the TDM waveform.

PDM is not the only pulse modulation process that lends itself to time division multiplexing. Figure 15-21 shows typical waveforms of PAM, PPM, and PCM signals in time-multiplexed form. Note that the PPM synchronization problem discussed earlier is solved by the TDM synchronization arrangements. Instead of using a space between pulses, synchronization may be accomplished by means of a negative pulse, or perhaps a wider or taller pulse than normal. The number of channels that may be time multiplexed is by no means limited to three. The maximum number of channels is determined by the highest frequency to be sampled and the time period that must be allocated to each channel.

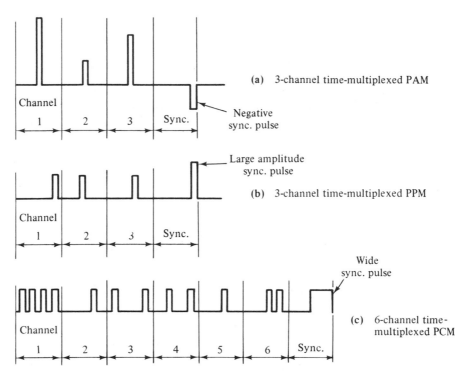

(a)   3-channel time-multiplexed PAM

Channel
1   2   3   Sync.

Negative
sync. pulse

Large amplitude
sync. pulse

(b)   3-channel time-multiplexed PPM

Channel
1   2   3   Sync.

Wide
sync. pulse

(c)   6-channel time-
multiplexed PCM

Channel
1   2   3   4   5   6   Sync.

Figure 15-21   TDM waveforms for PAM, PPM, and PCM, and various synchronizing methods for identifying the end of each group of pulses.

Ring Counter

A *ring counter* is the circuit employed in a TDM coding system to select the signals to be sampled in the correct repetitive sequence. Triggered by input pulses, the circuit switches through a number of states equal to one more than the number of TDM channels. For a three-channel system, the ring counter must have four states, that is, three channels plus the synchronizing space. Ring counters are usually constructed of flip-flops and logic gates. The BCD-to-decimal conversion system shown in Figure 14-16 could be employed as a ten-state ring counter. In this case, the outputs would control nine signal channels and a synchronizing space.

Figure 15-22 shows a four-state ring counter made up of two flip-flops and four *AND* gates. The operation of the circuit is similar to the BCD-to-decimal conversion system explained in Chapter 14. Assume an initial condition of $\overline{Q_A} = 1$, $Q_A = 0$, $\overline{Q_B} = 1$, and $Q_B = 0$, as illustrated by the truth table. This means that both inputs to gate 1 are at *logic 1*, and consequently, the output of gate 1 is *high*. All other gates have at least one *logic 0* input, which renders their outputs *low*. (See the gate output waveforms.) When the first toggle pulse is applied, the state of the flip-flops change to $\overline{Q_A} = 0$, $Q_A = 1$, $\overline{Q_B} = 1$, and $Q_B = 0$. Gate 1 now has one *logic 0* input, so its

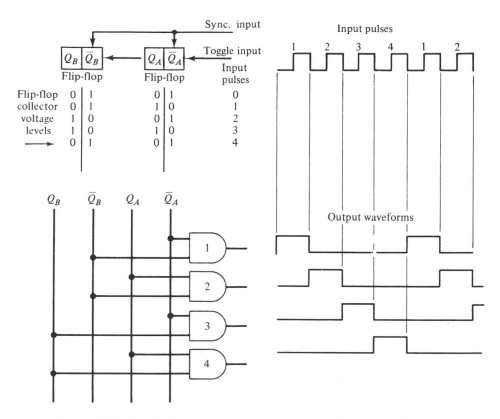

Figure 15-22   Two flip-flops and a logic circuit connected to function as a four-state *ring counter*. The ring counter can be toggled continuously through each of the four states.

output is *low*. Gate 2 has two *logic 1* inputs, giving it a *high* output. Gates 3 and 4 have at least one *logic 0* input, making their outputs *low*. Continuing through the second and third input pulses, it is seen that the outputs of gates 3 and 4 switch *high* in turn, while all other outputs become *low*. After the fourth input pulse, the next pulse returns the circuit to *high* output at gate 1 once again. Then the sequence is repeated.

The ring counter derives its name from the fact that it toggles in a repetitive, or *ring*, sequence, through each of its states. A synchronizing input can also be included, as explained in Section 14-2, so that the circuit can be set into its initial state.

TDM Coding System

The block diagram and waveforms for the time multiplexing of three signals are shown in Figure 15-23. A time division multiplexing system is usually referred to as a TDM *coding system*. The system for separating the waveform into individual channels is

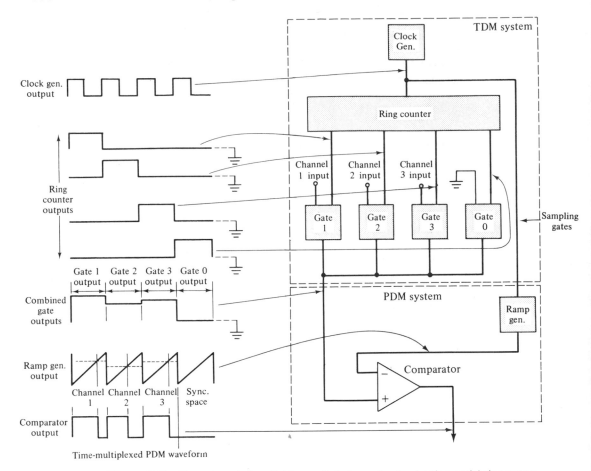

Figure 15-23 Three-channel time division multiplexing and pulse duration modulation system. The clock generator toggles the ring counter, to switch the sampling gates *on* in sequence. The amplitude samples are fed to a PDM system to create a TDM waveform consisting of three PDM channels and a synchronizing space.

termed a TDM *decoding system*. If pulse modulation is performed concurrently with time multiplexing, a single modulation system can be used for all channels.

In Figure 15-23, a clock generator produces a square waveform to toggle the ring counter at the desired frequency. The ring counter outputs switch the sampling gates *on* and *off* in the correct sequence. When one output from the ring counter is *high*, all others are *low*. Thus, one sampling gate is *on*, and all others are *off*. The time periods during which a sampling gate is in the *on* and *off* states are determined by the clock generator frequency and the number of channels. For example, if the clock generator output is a 1 kHz square wave, each gate is *on* for 1 ms. For the three-channel system shown, each gate is *off* for three time periods of the clock generator (i.e., for 3 ms).

Each signal to be sampled is applied to the input terminal of a sampling gate. The input of gate 0, the synchronizing gate, is grounded, so that during the sync time its output is zero volts. The outputs of all four gates are connected in common. Thus, as each gate switches *on* in turn, amplitude samples of the signal waveforms are time multiplexed into a single waveform. The combined output of the gates (see Figure 15-23) is in the form of a pulse-amplitude-modulated waveform with no spaces between pulses, and with a sync space at the end of each set of samples.

The waveform from the gates is converted to PDM by applying it to one input of a voltage comparator. The other input terminal of the comparator has a ramp generator output applied to it, as shown in Figure 15-23. The ramp generator has a linear output and is triggered by the negative-going edges of the clock generator output, so that the ramp commences at the beginning of the output from each sampling gate. The voltage comparator output becomes high when the ramp output becomes zero. When the ramp amplitude is equal to the amplitude of the signal being sampled, the comparator output goes to zero. (The process of converting amplitude samples to PDM has already been explained). During the sync space, the comparator input (from the gates) remains at zero; thus, the output level from the comparator does not switch positively when the ramp goes to zero. The output waveform from the system is seen to be time-multiplexed, duration-modulated pulses with intervening sync spaces.

TDM Decoding System

Before demodulation, time-multiplexed signals must be *decoded*, or separated into individual channels. Figure 15-24 shows the block diagram and waveforms of a system for decoding three-channel, time-multiplexed PDM signals.

The waveform to be decoded is applied simultaneously to an integrator circuit, the toggle input of a ring counter, and one input on each of three *AND* gates. The function of the integrator and the Schmitt trigger circuit which follows it is to detect the sync space in the TDM waveform. A suitable integrator circuit for this situation is the ramp generator shown in Figure 9-3. During the time that an input pulse is applied, the ramp generator output remains at zero. When no pulse is present the ramp output grows linearly, then it rapidly returns to zero at commencement of the next input pulse. Thus, the integrator output is a series of small ramps generated during the space time. During the synchronizing space, the integrator generates a larger-than-usual output. When this output arrives at the upper trigger point of the Schmitt circuit, the Schmitt produces an output pulse which resets the ring counter to its synchronized state. The next input pulse (i.e., from channel 1) toggles the ring counter to its channel 1 state.

In the channel 1 state, the ring counter provides a positive output to *AND* gate 1, and a zero output to gates 2 and 3. At this time, channel 1 pulse is present at one input of all three *AND* gates. Only gate 1 has a positive voltage at both input terminals. Therefore, only gate 1 produces an output during the application of the pulse from channel 1. The output pulse from gate 1 commences at the beginning of channel 1 and ends at the end of the channel 1 PDM pulse. At the commencement of channel 2 (i.e., on receipt of the next positive-going pulse edge), the ring counter toggles to its channel

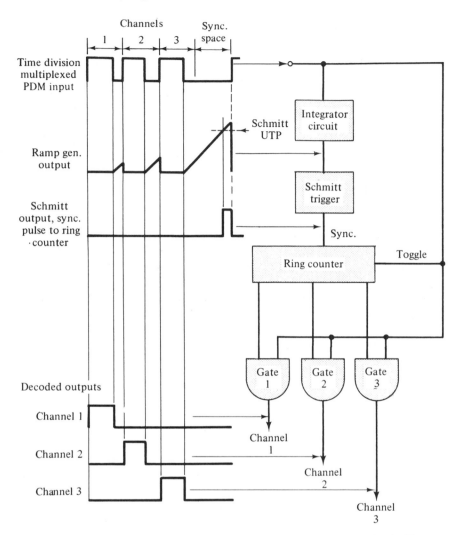

Figure 15-24 Decoding system for three-channel, time-multiplexed PDM signals. The integrator identifies the sync space and uses it to *reset* the ring counter. The ring counter opens the *AND* gates in correct sequence, so that the waveform is separated into three channels.

2 state. Therefore, *AND* gate 2 now has positive voltage levels at both input terminals, while *AND* gates 1 and 3 have zero levels applied from the ring counter. Thus, gate 2 produces the channel 2 PDM pulse at its output terminals. Similarly, at commencement of channel 3, the ring counter toggles to its channel 3 state. *AND* gate 3 output then becomes positive for the duration of the PDM pulse in the channel 3 portion of the input waveform. It is seen that the decoding system separates the TDM wave into the

individual channels. Now, each channel can be demodulated to recover the original modulating signals.

15-7 PULSE CODE MODULATION AND DEMODULATION

PCM Modulation System

PCM signals are essentially time-division-multiplexed pulses; therefore, modulation and demodulation techniques for PCM are similar to TDM methods. The block diagram of a PCM modulation system is shown in Figure 15-25, and the waveforms for the system are illustrated in Figure 15-26. The system is seen to consist of an ADC, similar to that shown in Figure 15-9, and a TDM circuit like the one in Figure 15-23.

The input signal voltage which is to be converted to PCM (V_s in Figure 15-25) is applied to one input of a voltage comparator, and a ramp generator output is applied

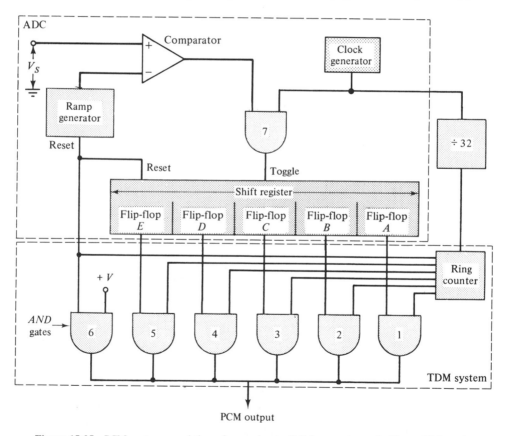

Figure 15-25 PCM system, consisting of an analog-to-digital converter, as in Figure 15-9, and a time-division multiplexing system similar to that in Figure 15-23. The ADC generates a digital number in the shift register, and the TDM system reads the shift register and produces the PCM output.

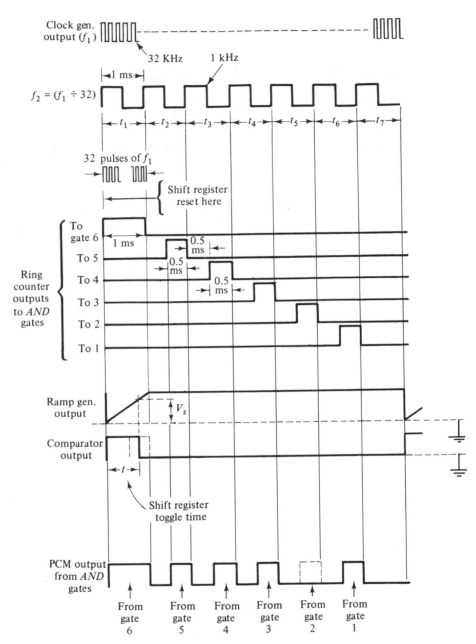

Figure 15-26 Voltage waveforms for the PCM system in Figure 15-25. The clock generator (frequency f_1) toggles the shift register (for time t) until the ramp amplitude equals the signal voltage V_S. The ring counter, which is toggled by frequency f_2, turns *on* the output AND gates in sequence, to generate the PCM waveform.

to the other input of the comparator. This is similar to the PDM method, in which the signal amplitude sample is converted to a time period. In Figure 15-26, the comparator output is shown as having a time duration t. To convert this time period to a five-bit binary number, the shift register is toggled by the clock generator during the time period. Once the shift register has settled at the digital equivalent of the signal sample, the ring counter is triggered to *read* the state of the shift register and produce a PCM output.

Suppose that the clock generator frequency f_1 is 32 kHz. This frequency is divided by a factor of 32 to produce $f_2 = (f_1/32) = 1$ kHz. (See Figure 15-26.) (Frequency-dividing techniques are discussed in Chapter 14.) The time period of f_2 is 1 ms, and this is used to trigger the ring counter. As shown by the waveforms, the ring counter provides a 1 ms pulse to *AND* gate 6. This pulse also goes to the ramp generator and to the *reset* input terminal of the shift register. Therefore, at commencement of time period t_1 (Figure 15-26) the shift register is set to its zero condition, and the ramp generator output is triggered from its maximum output voltage level to zero. When the ramp generator output becomes zero, it causes the output of the voltage comparator to switch from *low* to *high*. Thus, the input (from the comparator) to *AND* gate 7 is positive, and the clock generator pulses (i.e., those *not* divided by 32) pass through to toggle the shift register.

At the commencement of time period t_1, the ramp voltage also starts to grow linearly from the zero level. When the ramp amplitude becomes equal to the instantaneous amplitude of the signal voltage V_s, the comparator output switches to *low* once again. This means that no more clock generator pulses can pass through *AND* gate 7. At this time, the state of the shift register represents the binary equivalent of the signal voltage amplitude. Since no more toggle pulses arrive, the shift register state remains constant.

The 1 ms ring counter output pulse that is applied to the ramp generator during time t_1 is also applied to *AND* gate 6. Since the other input to the gate is at $+V$, the output of gate 6 is a positive 1 ms pulse. During time period t_2, the ring counter provides a 0.5 ms pulse to *AND* gate 5. The other input to *AND* gate 5 is derived from FF_E in the shift register. If the output of FF_E is *high*, gate 5 produces a 0.5 ms output pulse, as shown. If the flip-flop output is *low*, there is no pulse present at this point in the PCM output waveform. Through time periods t_3, t_4, t_5, and t_6, the ring counter switches *on* gates D, C, B, and A, respectively, to sample the shift register voltage levels at each flip-flop. Thus, the PCM output waveform is produced, with pulses that represent the digital equivalent of the signal sample, as measured during t_1. At the end of t_6, a gap (t_7) is generated by the ring counter holding gates 1 to 6 *off*, and then the long synchronizing pulse t_1 commences again, and the sampling and coding process is repeated.

PCM Decoding and Demodulation

For decoding and demodulation, PCM signals are first processed through a TDM decoding system (Figure 15-24) to separate the bits into different channels. Then, a DAC, as described in Section 15-3, is employed to convert the digital signals to analog voltage levels. The DAC output is a train of amplitude-modulated pulses, which can be demodulated by filtering.

REVIEW QUESTIONS AND PROBLEMS

15-1. Sketch the circuit of a FET series sampling gate. Show the voltage waveforms, explain the operation of the circuit, and discuss the error sources.

15-2. A signal of 1.5 V with a very low source resistance is to be sampled and passed to a circuit with $R_i = 20$ kΩ. Design a suitable FET gate circuit, and estimate the output errors.

15-3. Sketch the circuit of a FET shunt sampling gate and the related voltage waveforms. Explain the circuit operation and discuss the error sources.

15-4. A 200 μA signal current is to be sampled and passed to a circuit with $R_i = 15$ kΩ. Design a suitable FET gate and estimate the output errors.

15-5. An operational amplifier and FET are to be employed as a sampling gate. Sketch the circuit and briefly explain its operation.

15-6. The circuit described for question 15-5 is to have an input voltage of $V_s = 750$ mV and a voltage gain of 5. Design the circuit to use an operational amplifier which has $I_{B(max)} = 750$ nA, and a FET with $V_{GS(off)} = 8$ V and $R_{D(on)} = 50$ Ω. Estimate the output error due to $R_{D(on)}$.

15-7. Sketch a sample-and-hold circuit using two operational amplifiers and a FET. Show all waveforms and explain how the circuit operates.

15-8. A sample-and-hold circuit is to use operational amplifiers with $I_{B(max)} = 750$ nA, and a FET with $V_{GS(off)} = 8$ V, $R_{D(on)} = 50$ Ω, and $C_{GS} = 7$ pF. The signal voltage amplitude is $V_s = \pm 3$ V, and the sampled accuracy is to be approximately 1%. The holding time is 750 μs. Calculate the capacitor value and minimum sampling time. Also, calculate the effect of C_{GS}.

15-9. Sketch the circuit of a five-bit DAC and explain its operation. Discuss the relationship between the resistors and the input bits. Write an equation for V_o.

15-10. A DAC, as in Figure 15-8, has $R_F = 10$ kΩ. Determine suitable resistor values of R_A through R_F to give $V_o = -5$ V when $V_i = 2$ V and all inputs are present.

15-11. Explain the term *resolution* in reference to a DAC. Calculate the resolution for a five-bit DAC. Also, determine the number of bits required for a DAC to have a resolution better than 0.1%.

15-12. Sketch an ADC circuit using the ramp generator method. Show the circuit waveforms and explain its operation.

15-13. Sketch the circuit and waveforms for a digital-ramp-type ADC. Explain its operation and discuss the successive approximation ADC.

15-14. Sketch the circuit of a parallel ADC and explain its operation. Determine the number of voltage comparators required for a parallel ADC with a five-bit output.

15-15. Sketch waveforms to illustrate the various types of pulse modulation. Identify and briefly explain each type of modulation.

15-16. Discuss the performance of the various types of pulse modulation with respect to noise.

15-17. Show how a monostable multivibrator may be employed for pulse amplitude modulation. Briefly explain.

15-18. Sketch a block diagram for a PAM modulating system. Show the waveforms and explain the system.

15-19. Sketch and explain the process for demodulating PAM signals.

15-20. Draw a block diagram for a PDM modulating system. Show the system waveforms. Explain the modulation process.

15-21. Sketch a system block diagram and waveform for demodulation of PDM signals. Explain the process.

15-22. Using illustrations, explain the process of producing a PPM waveform.

15-23. Explain the process of demodulating a PPM waveform. Sketch the appropriate block diagram and voltage waveforms. Also, discuss the *synchronizing* problem which occurs with PPM and state the possible solutions.

15-24. Explain *time division multiplexing* and discuss its advantages. Sketch waveforms for time-multiplexed signals using PAM, PDM, PPM, and PCM.

15-25. Draw a block diagram and diode circuit to show how three flip-flops and a diode matrix can be employed as an eight-state ring counter. Also, sketch the input and output waveforms and explain how the ring counter functions.

15-26. Draw a block diagram for a four-channel TDM system which includes pulse duration modulation. Show the waveforms throughout and explain the operation of the system.

15-27. Draw a block diagram for a decoding system for four-channel, time-multiplexed PDM signals. Show the system waveforms and explain the operation of the system.

15-28. A signal having a maximum frequency of 5 kHz is to be pulse-code-modulated with not more than 2% quantizing error. Determine the number of pulses required to represent each sample and the number of samples required per second. Explain.

15-29. Draw a block diagram of a PCM modulating system that would be suitable for the signal described in Problem 15-28. Sketch appropriate waveforms and explain the system process.

15-30. Explain the process of decoding and demodulating a five-bit PCM signal. Sketch the appropriate circuitry and identify the level of output for each input pulse.

Appendix 1

Manufacturers' Data Sheets

APPENDIX 1-1*

TYPES 1N914, 1N914A, 1N914B, 1N915, 1N916, 1N916A, 1N916B and 1N917
DIFFUSED SILICON SWITCHING DIODES

- **Extremely Stable and Reliable High–Speed Diodes**

mechanical data

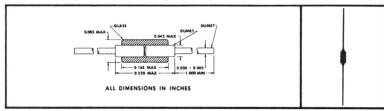

ALL DIMENSIONS IN INCHES

absolute maximum ratings at 25°C ambient temperature (unless otherwise noted)

		1N914	1N914A	1N914B	1N915	1N916	1N916A	1N916B	1N917	Unit
V_R	Reverse Voltage at −65 to +150°C	75	75	75	50	75	75	75	30	v
I_o	Average Rectified Fwd. Current	75	75	75	75	75	75	75	50	ma
I_o	Average Rectified Fwd. Current at +150°C	10	10	10	10	10	10	10	10	ma
i_f	Recurrent Peak Fwd. Current	225	225	225	225	225	225	225	150	ma
$i_{f(surge)}$	Surge Current, 1 sec	500	500	500	500	500	500	500	300	ma
P	Power Dissipation	250	250	250	250	250	250	250	250	mw
T_A	Operating Temperature Range	−65 to +175								°C
T_{stg}	Storage Temperature Range	200								°C

maximum electrical characteristics at 25°C ambient temperature (unless otherwise noted)

		1N914	1N914A	1N914B	1N915	1N916	1N916A	1N916B	1N917	Unit
BV_R	Min Breakdown Voltage at 100 μa	100	100	100	65	100	100	100	40	v
I_R	Reverse Current at V_R	5	5	5	5	5	5	5		μa
I_R	Reverse Current at −20 v	0.025	0.025	0.025		0.025	0.025	0.025		μa
I_R	Reverse Current at −20 v at 100°C	3	3	3	5	3	3	3	25	μa
I_R	Reverse Current at −20 v at +150°C	50	50	50		50	50	50		μa
I_R	Reverse Current at −10 v				0.025				0.05	μa
I_R	Reverse Current at −10 v at 125°C									μa
I_F	Min Fwd Current at $V_F = 1$ v	10	20	100	50	10	20	30	10	ma
V_F	at 250 μa								0.64	v
V_F	at 1.5 ma								0.74	v
V_F	at 3.5 ma								0.83	v
V_F	at 5 ma				0.72	0.73		0.73		v
V_F	Min at 5 ma					0.60				v
C	Capacitance at $V_R = 0$	4	4	4	4	2	2	2	2.5	pf

operating characteristics at 25°C ambient temperature (unless otherwise noted)

		1N914	1N914A	1N914B	1N915	1N916	1N916A	1N916B	1N917	Unit
t_{rr}	Max Reverse Recovery Time	**4	**4	**4	°10	**4	**4	**4	°3	nsec
		°8	°8	°8		°8	°8	°8		nsec
V_f	Fwd Recovery Voltage (50 ma Peak Sq. wave, 0.1 μsec pulse width, 10 nsec rise time, 5 kc to 100 kc rep. rate)	2.5	2.5	2.5	2.5	2.5	2.5	2.5	2.5	v

* Trademark of Texas Instruments
° Lumatron (10 ma I_F, 10 ma I_R, recover to 1 ma)
** EG&G (10 ma I_F, 6v V_R, recover to 1 ma)

*Courtesy of Texas Instruments, Incorporated

APPENDIX 1-2*

1N4001thru 1N4007

$I_O = 1\ A$
V_R — to 1000 V

CATHODE

CASE 59

Low-current, passivated silicon rectifiers in subminiature void-free, flame-proof silicone polymer case. Designed to operate under military environmental conditions.

MAXIMUM RATINGS (At 60 cps Sinusoidal, Input, Resistive or Inductive Load)

Rating	Symbol	1N4001	1N4002	1N4003	1N4004	1N4005	1N4006	1N4007	Unit
Peak Repetitive Reverse Voltage DC Blocking Voltage	$V_{RM(rep)}$ V_R	50	100	200	400	600	800	1000	Volts
RMS Reverse Voltage	V_r	35	70	140	280	420	560	700	Volts
Average Half-Wave Rectified Forward Current (75°C Ambient) (100°C Ambient)	I_O	1000 750	1000 750	1000 750	1000 750	1000 750	1000 750	1000 750	mA mA
Peak Surge Current 25°C (1/2 Cycle Surge, 60 cps) Peak Repetitive Forward Current	$I_{FM(surge)}$ $I_{FM(rep)}$	30 10	30 10	30 10	30 10	30 10	30 10	30 10	Amps Amps
Operating and Storage Temperature Range	T_J, T_{stg}	-65 to + 175							°C

ELECTRICAL CHARACTERISTICS

Characteristic	Symbol	Rating	Unit
Maximum Forward Voltage Drop (1 Amp Continuous DC, 25°C)	V_F	1. 1	Volts
Maximum Full-Cycle Average Forward Voltage Drop (Rated Current @ 25°C)	$V_{F(AV)}$	0. 8	Volts
Maximum Reverse Current @ Rated DC Voltage (25°C) (100°C)	I_R	0. 01 0. 05	mA
Maximum Full-Cycle Average Reverse Current (Max Rated PIV and Current, as Half-Wave Rectifier, Resistive Load, 100°C)	$I_{R(AV)}$	0. 03	mA

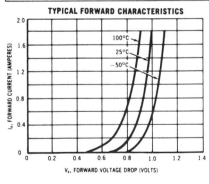

TYPICAL FORWARD CHARACTERISTICS

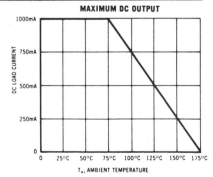

MAXIMUM DC OUTPUT

*Courtesy of Motorola, Inc.

APPENDIX 1-3*

TYPES 1N746 THRU 1N759, 1N746A THRU 1N759A
SILICON VOLTAGE REGULATOR DIODES

3.3 TO 12 VOLTS • 400 mw

GUARANTEED DYNAMIC ZENER IMPEDANCE

Available in 5% and 10% tolerances

-65 to 175°C operation & storage

*electrical characteristics at 25°C free-air temperature (unless otherwise noted)

PARAMETER	V_Z Zener Breakdown Voltage					α_Z Temperature Coefficient of Breakdown Voltage	Z_Z Small-Signal Breakdown Impedance	I_R Static Reverse Current	
TEST CONDITIONS	$I_{ZT} = 20$ ma					$I_{ZT} = 20$ ma	$I_{ZT} = 20$ ma, $I_{z\dagger} = 1$ ma	$V_R = 1$ v	$V_R = 1$ v, $T_A = 150°C$
LIMIT →	NOM	1N746 – 1N759		1N746A–1N759A		TYP	MAX	MAX	MAX
		MIN	MAX	MIN	MAX				
UNIT →	v	v	v	v	v	% / °C	Ω	μa	μa
1N746	3 3	2 97	3 63	3 135	3 465	−0 062	28	10	30
1N747	3 6	3 24	3 96	3 420	3 780	−0 055	24	10	30
1N748	3 9	3 51	4 29	3 705	4 095	−0 049	23	10	30
1N749	4 3	3 87	4 73	4 085	4 515	−0 036	22	2	30
1N750	4 7	4 23	5 17	4 465	4 935	−0 018	19	2	30
1N751	5 1	4 59	5 61	4 845	5 355	−0 008	17	1	20
1N752	5 6	5 04	6 16	5 320	5 880	+0 006	11	1	20
1N753	6 2	5 58	6 82	5 890	6 510	+0 022	7	0 1	20
1N754	6 8	6 12	7 48	6 460	7 140	+0 035	5	0 1	20
1N755	7 5	6 75	8 25	7 125	7 875	+0 045	6	0 1	20
1N756	8 2	7 38	9 02	7 790	8 610	+0 052	8	0 1	20
1N757	9 1	8 19	10 01	8 645	9 555	+0 056	10	0 1	20
1N758	10 0	9 00	11 00	9 500	10 500	+0 060	17	0 1	20
1N759	12 0	10 80	13 20	11 400	12 000	+0 060	30	0 1	20

*absolute maximum ratings

Average Rectified Forward Current at (or below) 25°C Free-Air Temperature . . . . 230 ma

Average Rectified Forward Current at 150°C Free-Air Temperature . . . . 85 ma

Continuous Power Dissipation at (or below) 50°C Free-Air Temperature . . . . 400 mw

Continuous Power Dissipation at 150°C Free-Air Temperature . . . . 100 mw

Operating Free-Air Temperature Range . . . . −65°C to 175°C

Storage Temperature Range . . . . −65°C to 175°C

*Indicates JEDEC registered data

*Courtesy of Texas Instruments, Incorporated

APPENDIX 1-4*

2N3903 (SILICON)
2N3904

$V_{CB} = 60$ V
$I_C = 200$ mA
$C_{ob} = 4.0$ pf (max)

CASE 29
(TO-92)

NPN silicon annular transistors, designed for general purpose switching and amplifier applications, features one-piece, injection-molded plastic package for high reliability. The 2N3903 and 2N3904 are complementary with types 2N3905 and 2N3906, respectively.

MAXIMUM RATINGS (T$_A$ = 25°C unless otherwise noted)

Characteristic	Symbol	Rating	Unit
Collector-Base Voltage	V_{CB}	60	Vdc
Collector-Emitter Voltage	V_{CEO}	40	Vdc
Emitter-Base Voltage	V_{EB}	6	Vdc
Collector Current	I_C	200	mAdc
Total Device Dissipation @ T_A = 60°C	P_D	210	mW
Total Device Dissipation @ T_A = 25°C Derate above 25°C	P_D	310 2.81	mW mW/°C
Thermal Resistance, Junction to Ambient	θ_{JA}	0.357	°C/mW
Junction Operating Temperature	T_J	135	°C
Storage Temperature Range	T_{stg}	-55 to +135	°C

ELECTRICAL CHARACTERISTICS (T$_A$ = 25°C unless otherwise noted)

Characteristic	Symbol	Min	Max	Unit
OFF CHARACTERISTICS				
Collector-Base Breakdown Voltage (I_C = 10 μAdc, I_E = 0)	BV_{CBO}	60	—	Vdc
Collector-Emitter Breakdown Voltage* (I_C = 1 mAdc)	BV_{CEO}*	40	—	Vdc
Emitter-Base Breakdown Voltage (I_E = 10 μAdc, I_C = 0)	BV_{EBO}	6	—	Vdc
Collector Cutoff Current (V_{CE} = 40 Vdc, V_{OB} = 3 Vdc)	I_{CEX}	—	50	nAdc
Base Cutoff Current (V_{CE} = 40 Vdc, V_{OB} = 3 Vdc)	I_{BL}	—	50	nAdc

*Pulse Test: Pulse Width = 300 μsec, Duty Cycle = 2%. V_{OB} = Base Emitter Reverse Bias

*Courtesy of Motorola, Inc.

2N3903, 2N3904 (continued)

ELECTRICAL CHARACTERISTICS (continued)

Characteristic		Symbol	Min	Max	Unit
ON CHARACTERISTICS					
DC Current Gain *		h_{FE}*		—	—
(I_C = 0.1 mAdc, V_{CE} = 1 Vdc) 2N3903			20	—	
2N3904			40	—	
(I_C = 1.0 mAdc, V_{CE} = 1 Vdc) 2N3903			35	—	
2N3904			70	—	
(I_C = 10 mAdc, V_{CE} = 1 Vdc) 2N3903			50	150	
2N3904			100	300	
(I_C = 50 mAdc, V_{CE} = 1 Vdc) 2N3903			30	—	
2N3904			60	—	
(I_C = 100 mAdc, V_{CE} = 1 Vdc) 2N3903			15	—	
2N3904			30	—	
Collector-Emitter Saturation Voltage*		$V_{CE(sat)}$*			Vdc
(I_C = 10 mAdc, I_B = 1 mAdc)			—	0.2	
(I_C = 50 mAdc, I_B = 5 mAdc)			—	0.3	
Base-Emitter Saturation Voltage*		$V_{BE(sat)}$*			Vdc
(I_C = 10 mAdc, I_B = 1 mAdc)			0.65	0.85	
(I_C = 50 mAdc, I_B = 5 mAdc)			—	0.95	
SMALL SIGNAL CHARACTERISTICS					
High Frequency Current Gain 2N3903		$|h_{fe}|$	2.5	—	—
(I_C = 10 mA, V_{CE} = 20 V, f = 100 mc) 2N3904			3.0	—	
Current-Gain—Bandwidth Product 2N3903		f_T	250	—	mc
(I_C = 10 mA, V_{CE} = 20 V, f = 100 mc) 2N3904			300	—	
Output Capacitance		C_{ob}	—	4	pf
(V_{CB} = 5 Vdc, I_E = 0, f = 100 kc)					
Input Capacitance		C_{ib}	—	8	pf
(V_{OB} = 0.5 Vdc, I_C = 0, f = 100 kc)					
Small Signal Current Gain 2N3903		h_{fe}	50	200	—
(I_C = 1.0 mA, V_{CE} = 10 V, f = 1 kc) 2N3904			100	400	
Voltage Feedback Ratio 2N3903		h_{re}	0.1	5.0	×10⁻⁴
(I_C = 1.0 mA, V_{CE} = 10 V, f = 1 kc) 2N3904			0.5	8.0	
Input Impedance 2N3903		h_{ie}	0.5	8	Kohms
(I_C = 1.0 mA, V_{CE} = 10 V, f = 1 kc) 2N3904			1.0	10	
Output Admittance		h_{oe}	1.0	40	μmhos
(I_C = 1.0 mA, V_{CE} = 10 V, f = 1 kc) Both Types					
Noise Figure		NF			db
(I_C = 100 μA, V_{CE} = 5 V, R_g = 1 Kohms, 2N3903			—	6	
Noise Bandwidth = 10 cps to 15.7 kc) 2N3904			—	5	
SWITCHING CHARACTERISTICS					
Delay Time	V_{CC} = 3 Vdc, V_{OB} = 0.5 Vdc,	t_d	—	35	nsec
Rise Time	I_C = 10 mAdc, I_{B1} = 1 mA	t_r	—	35	nsec
Storage Time	V_{CC} = 3 Vdc, I_C = 10 mAdc, 2N3903	t_s	—	175	nsec
	I_{B1} = I_{B2} = 1 mAdc 2N3904		—	200	
Fall Time		t_f	—	50	nsec

*Pulse Test: Pulse Width = 300 μsec, Duty Cycle = 2% V_{OB} = Base Emitter Reverse Bias

APPENDIX 1-5*

2N3905 (SILICON)
2N3906

$V_{CB} = 40$ V
$I_C = 200$ mA
$C_{ob} = 4.5$ pf (max)

CASE 29
(TO-92)

PNP silicon annular transistor, designed for general purpose switching and amplifier applications, features one-piece, injection-molded plastic package for high reliability. The 2N3905 and 2N3906 are complementary with types 2N3903 and 2N3904, respectively.

MAXIMUM RATINGS (T$_A$ = 25°C unless otherwise noted)

Characteristic	Symbol	Rating	Unit
Collector-Base Voltage	V_{CB}	40	Vdc
Collector-Emitter Voltage	V_{CEO}	40	Vdc
Emitter-Base Voltage	V_{EB}	5	Vdc
Collector Current	I_C	200	mAdc
Total Device Dissipation @ T$_A$ = 60°C	P_D	210	mW
Total Device Dissipation @ T$_A$ = 25°C Derate above 25°C	P_D	310 2.81	mW mW/°C
Thermal Resistance, Junction to Ambient	θ_{JA}	0.357	°C/mW
Junction Operating Temperature	T_J	135	°C
Storage Temperature Range	T_{stg}	-55 to +135	°C

ELECTRICAL CHARACTERISTICS (T$_A$ = 25°C unless otherwise noted)

Characteristic	Symbol	Min	Max	Unit
OFF CHARACTERISTICS				
Collector-Base Breakdown Voltage (I_C = 10 μAdc, I_E = 0)	BV_{CBO}	40	—	Vdc
Collector-Emitter Breakdown Voltage* (I_C = 1 mAdc)	BV_{CEO}*	40	—	Vdc
Emitter-Base Breakdown Voltage (I_E = 10 μAdc, I_C = 0)	BV_{EBO}	5	—	Vdc
Collector Cutoff Current (V_{CE} = 40 Vdc, V_{OB} = 3 Vdc)	I_{CEX}	—	50	nAdc
Base Cutoff Current (V_{CE} = 40 Vdc, V_{OB} = 3 Vdc)	I_{BL}	—	50	nAdc

*Pulse Test: Pulse Width = 300 μsec, Duty Cycle = 2% V_{OB} = Base Emitter Reverse Bias

*Courtesy of Motorola, Inc.

2N3905, 2N3906 (continued)

ELECTRICAL CHARACTERISTICS   (continued)

Characteristic		Symbol	Min	Max	Unit
ON CHARACTERISTICS					
DC Current Gain*		h_{FE}*		—	—
(I_C = 0.1 mAdc, V_{CE} = 1 Vdc)	2N3905		30	—	
	2N3906		60	—	
(I_C = 1.0 mAdc, V_{CE} = 1 Vdc)	2N3905		40	—	
	2N3906		80	—	
(I_C = 10 mAdc, V_{CE} = 1 Vdc)	2N3905		50	150	
	2N3906		100	300	
(I_C = 50 mAdc, V_{CE} = 1 Vdc)	2N3905		30	—	
	2N3906		60	—	
(I_C = 100 mAdc, V_{CE} = 1 Vdc)	2N3905		15	—	
	2N3906		30	—	
Collector-Emitter Saturation Voltage*		$V_{CE(sat)}$*			Vdc
(I_C = 10 mAdc, I_B = 1 mAdc)			—	0.25	
(I_C = 50 mAdc, I_B = 5 mAdc)			—	0.4	
Base-Emitter Saturation Voltage*		$V_{BE(sat)}$*			Vdc
(I_C = 10 mAdc, I_B = 1 mAdc)			0.65	0.85	
(I_C = 50 mAdc, I_B = 5 mAdc)			—	0.95	
SMALL SIGNAL CHARACTERISTICS					
High-Frequency Current Gain	2N3905	$|h_{fe}|$	2.0	—	—
(I_C = 10 mAdc, V_{CE} = 20 Vdc, f = 100 mc)	2N3906		2.5	—	
Current-Gain—Bandwidth Product	2N3905	f_T	200	—	mc
(I_C = 10 mAdc, V_{CE} = 20 Vdc, f = 100 mc)	2N3906		250	—	
Output Capacitance		C_{ob}			pf
(V_{CB} = 5 Vdc, I_E = 0, f = 100 kc)			—	4.5	
Input Capacitance		C_{ib}			pf
(V_{OB} = 0.5 Vdc, I_C = 0, f = 100 kc)			—	10	
Small Signal Current Gain	2N3905	h_{fe}	50	200	—
(I_C = 1.0 mA, V_{CE} = 10 V, f = 1 kc)	2N3906		100	400	
Voltage Feedback Ratio	2N3905	h_{re}	0.1	5	$X10^{-4}$
(I_C = 1.0 mA, V_{CE} = 10 V, f = 1 kc)	2N3906		1.0	10	
Input Impedance	2N3905	h_{ie}	0.5	8	Kohms
(I_C = 1.0 mA, V_{CE} = 10 V, f = 1 kc)	2N3906		2.0	12	
Output Admittance	2N3905	h_{oe}	1.0	40	μ mhos
(I_C = 1.0 mA, V_{CE} = 10 V, f = 1 kc)	2N3906		3.0	60	
Noise Figure		NF			db
(I_C = 100 μA, V_{CE} = 5 V, Rg = 1 Kohms,	2N3905		—	5.0	
Noise Bandwidth = 10 cps to 15.7 kc)	2N3906		—	4.0	

SWITCHING CHARACTERISTICS

			Symbol	Min	Max	Unit
Delay Time	V_{CC} = 3 Vdc, V_{OB} = 0.5 Vdc,		t_d	—	35	nsec
Rise Time	I_C = 10 mAdc, I_{B1} = 1 mA		t_r	—	35	nsec
Storage Time		2N3905	t_s	—	200	nsec
	V_{CC} = 3 Vdc, I_C = 10 mAdc,	2N3906		—	225	
Fall Time	$I_{B1} = I_{B2}$ = 1 mAdc	2N3905	t_f	—	60	nsec
		2N3906		—	75	

*Pulse Test: PW = 300 μsec, Duty Cycle = 2% V_{OB} = Base-Emitter Reverse Bias

APPENDIX 1-6*

TYPES 2N929, 2N930
N-P-N PLANAR SILICON TRANSISTORS

FOR EXTREMELY LOW-LEVEL, LOW-NOISE, HIGH-GAIN,
SMALL-SIGNAL AMPLIFIER APPLICATIONS

- Guaranteed h_{FE} at 10 μa, $T_A = -55°C$ and 25°C
- Guaranteed Low-Noise Characteristics at 10 μa
- Usable at Collector Currents as Low as 1 μa
- Very High Reliability
- 2N929 and 2N930 Also Are Available to MIL-S-19500/253 (Sig C)

*mechanical data

THE COLLECTOR IS IN ELECTRICAL
CONTACT WITH THE CASE

ALL JEDEC TO-18 DIMENSIONS
AND NOTES ARE APPLICABLE

ALL DIMENSIONS ARE
IN INCHES
UNLESS OTHERWISE
SPECIFIED

*absolute maximum ratings at 25°C free-air temperature (unless otherwise noted)

Collector-Base Voltage	45 v
Collector-Emitter Voltage (See Note 1)	45 v
Emitter-Base Voltage	5 v
Collector Current	30 ma
Total Device Dissipation at (or below) 25°C Free-Air Temperature (See Note 2)	300 mw
Total Device Dissipation at (or below) 25°C Case Temperature (See Note 3)	600 mw
Operating Collector Junction Temperature	175°C
Storage Temperature Range	−65°C to + 200°C

NOTES: 1. This value applies when the base-emitter diode is open-circuited.
    2. Derate linearly to 175°C free-air temperature at the rate of 2.0 mw/C°.
    3. Derate linearly to 175°C case temperature at the rate of 4.0 mw/C°.

*Indicates JEDEC registered data

*Courtesy of Texas Instruments, Incorporated

TYPES 2N929, 2N930
N-P-N PLANAR SILICON TRANSISTOR

*electrical characteristics at 25°C free-air temperature (unless otherwise noted)

PARAMETER		TEST CONDITIONS	2N929 MIN	2N929 MAX	2N930 MIN	2N930 MAX	UNIT
BV_{CEO}	Collector-Emitter Breakdown Voltage	$I_C = 10$ ma, $I_B = 0$, (See Note 4)	45		45		v
BV_{EBO}	Emitter-Base Breakdown Voltage	$I_E = 10$ na $I_C = 0$	5		5		v
I_{CBO}	Collector Cutoff Current	$V_{CB} = 45$ v, $I_E = 0$		10		10	na
I_{CES}	Collector Cutoff Current (See Note 5)	$V_{CE} = 45$ v, $V_{BE} = 0$		10		10	na
		$V_{CE} = 45$ v, $V_{BE} = 0$, $T_A = 170°C$		10		10	μa
I_{CEO}	Collector Cutoff Current	$V_{CE} = 5$ v, $I_B = 0$		2		2	na
I_{EBO}	Emitter Cutoff Current	$V_{EB} = 5$ v, $I_C = 0$		10		10	na
h_{FE}	Static Forward Current Transfer Ratio	$V_{CE} = 5$ v, $I_C = 10$ μa	40	120	100	300	
		$V_{CE} = 5$ v, $I_C = 10$ μa, $T_A = -55°C$	10		20		
		$V_{CE} = 5$ v, $I_C = 500$ μa	60		150		
		$V_{CE} = 5$ v, $I_C = 10$ ma, (See Note 4)		350		600	
V_{BE}	Base-Emitter Voltage	$I_B = 0.5$ ma, $I_C = 10$ ma, (See Note 4)	0.6	1.0	0.6	1.0	v
$V_{CE(sat)}$	Collector-Emitter Saturation Voltage	$I_B = 0.5$ ma, $I_C = 10$ ma, (See Note 4)		1.0		1.0	v
h_{ib}	Small-Signal Common-Base Input Impedance	$V_{CB} = 5$ v, $I_E = -1$ ma, f = 1 kc	25	32	25	32	ohm
h_{rb}	Small-Signal Common-Base Reverse Voltage Transfer Ratio	$V_{CB} = 5$ v, $I_E = -1$ ma, f = 1 kc	0	6.0×10^{-4}	0	6.0×10^{-4}	
h_{ob}	Small-Signal Common-Base Output Admittance	$V_{CB} = 5$ v, $I_E = -1$ ma, f = 1 kc	0	1.0	0	1.0	μmho
h_{fe}	Small-Signal Common-Emitter Forward Current Transfer Ratio	$V_{CE} = 5$ v, $I_C = 1$ ma, f = 1 kc	60	350	150	600	
$|h_{fe}|$	Small-Signal Common-Emitter Forward Current Transfer Ratio	$V_{CE} = 5$ v, $I_C = 500$ μa, f = 30 mc	1.0		1.0		
C_{ob}	Common-Base Open-Circuit Output Capacitance	$V_{CB} = 5$ v, $I_E = 0$, f = 1 mc		8		8	pf

*operating characteristics at 25°C free-air temperature

PARAMETER		TEST CONDITIONS	2N929 MAX	2N930 MAX	UNIT
$\overline{NF}$	Average Noise Figure	$V_{CE} = 5$ v, $I_C = 10$ μa, $R_G = 10$ kΩ Noise Bandwidth 10 cps to 15.7 kc	4	3	db

NOTES: 4. These parameters must be measured using pulse techniques. PW = 300 μsec, Duty Cycle $\leq$ 2%.
     5. I_{CES} may be used in place of I_{CBO} for circuit stability calculations.
*Indicates JEDEC registered data

APPENDIX 1-7*

TYPES 2N4418 AND 2N4419
N-P-N EPITAXIAL PLANAR SILICON TRANSISTORS

SILECT† TRANSISTORS FOR HIGH-SPEED SWITCHING APPLICATIONS
- 2N4418 Electrically Similar to the 2N2369
- Rugged, One-Piece Construction with Standard TO-18 100-mil Pin Circle

*switching characteristics at 25°C free-air temperature

PARAMETER		TEST CONDITIONS†	2N4418 MAX	2N4419 MAX	UNIT
t_d	Delay Time	$I_C = 10$ mA, $I_{B(1)} = 1$ mA, $V_{BE(off)} = 0$, $R_L = 280\ \Omega$, See Figure 1	10	10	ns
t_r	Rise Time		12	14	ns
t_{on}	Turn-on Time		20	22	ns
t_s	Storage Time	$I_C = 10$ mA, $I_{B(1)} = 1$ mA, $I_{B(2)} = -1$ mA, $R_L = 280\ \Omega$, See Figure 2	12	14	ns
t_f	Fall Time		14	16	ns
t_{off}	Turn-off Time		22	28	ns
t_s	Storage Time	$I_C = 10$ mA, $I_{B(1)} = 10$ mA, $I_{B(2)} = -10$ mA, See Figure 3	18	20	ns

†Voltage and current values shown are nominal; exact values vary slightly with transistor parameters.

*absolute maximum ratings at 25°C free-air temperature (unless otherwise noted)

	2N4418	2N4419
Collector-Base Voltage	40 V	30 V
Collector-Emitter Voltage (See Note 1)	40 V	30 V
Collector-Emitter Voltage (See Note 2)	15 V	12 V
Emitter-Base Voltage	4.5 V	4.5 V
Continuous Collector Current	← 200 mA →	
Peak Collector Current (See Note 3)	← 500 mA →	
Continuous Device Dissipation at (or below) 25°C Free-Air Temperature (See Note 4) .	← 360 mW→	
Continuous Device Dissipation at (or below) 25°C Lead Temperature (See Note 5) . .	← 500 mW→	
Storage Temperature Range	−65°C to 150°C	
Lead Temperature 1/16 Inch from Case for 10 Seconds	← 260°C --→	

*electrical characteristics at 25°C free-air temperature (unless otherwise noted)

PARAMETER		TEST CONDITIONS	2N4418 MIN	2N4418 MAX	2N4419 MIN	2N4419 MAX	UNIT
$V_{(BR)CBO}$	Collector-Base Breakdown Voltage	$I_C = 10\ \mu A$, $I_E = 0$	40		30		V
$V_{(BR)CEO}$	Collector-Emitter Breakdown Voltage	$I_C = 10$ mA, $I_B = 0$, See Note 6	15		12		V
$V_{(BR)CES}$	Collector-Emitter Breakdown Voltage	$I_C = 10\ \mu A$, $V_{BE} = 0$	40		30		V
$V_{(BR)EBO}$	Emitter-Base Breakdown Voltage	$I_E = 10\ \mu A$, $I_C = 0$	4.5		4.5		V
I_{CBO}	Collector Cutoff Current	$V_{CB} = 20$ V, $I_E = 0$		0.4		0.4	μA
		$V_{CB} = 20$ V, $I_E = 0$, $T_A = 70°C$		3		3	μA
I_{EBO}	Emitter Cutoff Current	$V_{EB} = 3$ V, $I_C = 0$		20		25	nA
h_{FE}	Static Forward Current Transfer Ratio	$V_{CE} = 1$ V, $I_C = 10$ mA, See Note 6	40	120	30		
		$V_{CE} = 2$ V, $I_C = 100$ mA, See Note 6	20				
V_{BE}	Base-Emitter Voltage	$I_B = 1$ mA, $I_C = 10$ mA	0.72	0.87	0.72	0.87	V
$V_{CE(sat)}$	Collector-Emitter Saturation Voltage	$I_B = 1$ mA, $I_C = 10$ mA		0.25		0.25	V
$\|h_{fe}\|$	Small-Signal Common-Emitter Forward Current Transfer Ratio	$V_{CE} = 10$ V, $I_C = 10$ mA, $f = 100$ MHz	5		4		
C_{cb}	Collector-Base Capacitance	$V_{CB} = 5$ V, $I_E = 0$, $f = 1$ MHz, See Note 7		4		4	pF

NOTES:
1. This value applies when the base-emitter diode is short-circuited.
2. These values apply between 0 and 200 mA collector current when the base-emitter diode is open-circuited. Maximum rated voltage and 200 mA collector current may be simultaneously applied provided the time of application is 10 μs or less and the duty cycle is 2% or less.
3. This value applies for $t_p \leq 10\ \mu s$ and duty cycle ≤ 2%.
4. Derate linearly to 150°C free-air temperature at the rate of 2.88 mW/deg.
5. Derate linearly to 150°C lead temperature at the rate of 4 mW/deg. Lead temperature is measured on the collector lead 1/16 inch from the case.
6. These parameters must be measured using pulse techniques. $t_p = 300\ \mu s$, duty cycle ≤ 2%.
7. C_{cb} is measured using three-terminal measurement techniques with the emitter guarded.

*Indicates JEDEC registered data
†Trademark of Texas Instruments
‡Patent Pending

*Courtesy of Texas Instruments, Incorporated

APPENDIX 1-8*

TYPES 2N4856, 2N4857, 2N4858, 2N4859, 2N4860, 2N4861
N-CHANNEL EPITAXIAL PLANAR SILICON FIELD-EFFECT TRANSISTORS

**SYMMETRICAL N-CHANNEL FIELD-EFFECT TRANSISTORS
FOR HIGH-SPEED COMMUTATOR AND CHOPPER APPLICATIONS**
2N4859 Formerly TIXS41
- Low $r_{ds(on)}$: 25 Ω Max (2N4856, 2N4859)
- Low $I_{D(off)}$: 0.25 nA Max

*mechanical data

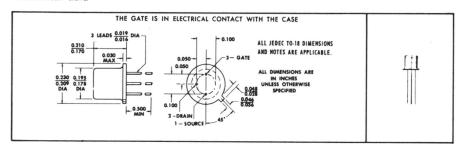

*absolute maximum ratings at 25°C free-air temperature (unless otherwise noted)

	2N4856 2N4857 2N4858	2N4859 2N4860 2N4861
Drain-Gate Voltage	40 V	30 V
Drain-Source Voltage	40 V	30 V
Reverse Gate-Source Voltage	–40 V	–30 V
Forward Gate Current	←— 50 mA —→	
Continuous Device Dissipation at (or below) 25°C Free-Air Temperature (See Note 1)	←— 360 mW —→	
Storage Temperature Range	–65°C to 200°C	
Lead Temperature ¹⁄₁₆ Inch from Case for 10 Seconds	←—300°C—→	

NOTE 1: Derate linearly to 175°C free-air temperature at the rate of 2.4 mW/deg.
*Indicates JEDEC registered data

---

*Courtesy of Texas Instruments, Incorporated

TYPES 2N4856 THRU 2N4861
N-CHANNEL EPITAXIAL PLANAR SILICON FIELD-EFFECT TRANSISTORS

*electrical characteristics at 25°C free-air temperature (unless otherwise noted)

PARAMETER		TEST CONDITIONS	2N4856 MIN	2N4856 MAX	2N4857 MIN	2N4857 MAX	2N4858 MIN	2N4858 MAX	2N4859 MIN	2N4859 MAX	2N4860 MIN	2N4860 MAX	2N4861 MIN	2N4861 MAX	UNIT
$V_{(BR)GSS}$	Gate-Source Breakdown Voltage	$I_G = -1\ \mu A$, $V_{DS} = 0$	−40		−40		−40		−30		−30		−30		V
I_{GSS}	Gate Reverse Current	$V_{GS} = -20$ V, $V_{DS} = 0$		−0.25		−0.25		−0.25							nA
		$V_{GS} = -20$ V, $V_{DS} = 0$, $T_A = 150°C$		−0.5		−0.5		−0.5							μA
		$V_{GS} = -15$ V, $V_{DS} = 0$								−0.25		−0.25		−0.25	nA
		$V_{GS} = -15$ V, $V_{DS} = 0$, $T_A = 150°C$								−0.5		−0.5		−0.5	μA
$I_{D(off)}$	Drain Cutoff Current	$V_{DS} = 15$ V, $V_{GS} = -10$ V		0.25		0.25		0.25		0.25		0.25		0.25	nA
		$V_{DS} = 15$ V, $V_{GS} = -10$ V, $T_A = 150°C$		0.5		0.5		0.5		0.5		0.5		0.5	μA
$V_{GS(off)}$	Gate-Source Cutoff Voltage	$V_{DS} = 15$ V, $I_D = 0.5$ nA	−4	−10	−2	−6	−0.8	−4	−4	−10	−2	−6	−0.8	−4	V
I_{DSS}	Zero-Gate-Voltage Drain Current	$V_{DS} = 15$ V, $V_{GS} = 0$, See Note 2	50		20	100	8	80	50		20	100	8	80	mA
$V_{DS(on)}$	Drain-Source On-State Voltage	$I_D = 20$ mA, $V_{GS} = 0$		0.75						0.75					V
		$I_D = 10$ mA, $V_{GS} = 0$				0.50						0.50			V
		$I_D = 5$ mA, $V_{GS} = 0$						0.50						0.50	V
$r_{ds(on)}$	Small-Signal Drain-Source On-State Resistance	$V_{GS} = 0$, $I_D = 0$, $f = 1$ kHz		25		40		60		25		40		60	Ω
C_{iss}	Common-Source Short-Circuit Input Capacitance	$V_{GS} = -10$ V, $V_{DS} = 0$, $f = 1$ MHz		18		18		18		18		18		18	pF
C_{rss}	Common-Source Short-Circuit Reverse Transfer Capacitance	$V_{GS} = -10$ V, $V_{DS} = 0$, $f = 1$ MHz		8		8		8		8		8		8	pF

*switching characteristics at 25°C free-air temperature

PARAMETER		TEST CONDITIONS		2N4856 2N4859 MAX	2N4857 2N4860 MAX	2N4858 2N4861 MAX	UNIT
$t_{d(on)}$	Turn-On Delay Time	$V_{DD} = 10$ V,	$I_{D(on)}† = \begin{cases} 20 \text{ mA (2N4856, 2N4859)} \\ 10 \text{ mA (2N4857, 2N4860)} \\ 5 \text{ mA (2N4858, 2N4861)} \end{cases}$	6	6	10	ns
t_r	Rise Time	$V_{GS(on)} = 0$,		3	4	10	ns
t_{off}	Turn-Off Time	See Figure 1	$V_{GS(off)} = \begin{cases} -10 \text{ V (2N4856, 2N4859)} \\ -6 \text{ V (2N4857, 2N4860)} \\ -4 \text{ V (2N4858, 2N4861)} \end{cases}$	25	50	100	ns

NOTE 2: This parameter must be measured using pulse techniques. $t_p \approx 100$ ms, duty cycle $\leq 10\%$.

*Indicates JEDEC registered data

†These are nominal values; exact values vary slightly with transistor parameters.

APPENDIX 1-9*

2N5457 (SILICON)
2N5458
2N5459

Silicon N-channel junction field-effect transistors depletion mode (Type A) designed for general-purpose audio and switching applications.

CASE 29 (5)
(TO-92)

Drain and source may be interchanged.

MAXIMUM RATINGS

Rating	Symbol	Value	Unit
Drain-Source Voltage	V_{DS}	25	Vdc
Drain-Gate Voltage	V_{DG}	25	Vdc
Reverse Gate-Source Voltage	$V_{GS(r)}$	25	Vdc
Gate Current	I_G	10	mAdc
Total Device Dissipation @ T_A= 25°C Derate above 25°C	P_D [2]	310 2.82	mW mW/°C
Operating Junction Temperature	T_J [2]	135	°C
Storage Temperature Range	T_{stg} [2]	-65 to +150	°C

ELECTRICAL CHARACTERISTICS (T_A = 25°C unless otherwise noted)

Characteristic		Symbol	Min	Typ	Max	Unit
OFF CHARACTERISTICS						
Gate-Source Breakdown Voltage (I_G = -10μAdc, V_{DS} = 0)		BV_{GSS}	25	—	—	Vdc
Gate Reverse Current (V_{GS} = -15 Vdc, V_{DS} = 0)		I_{GSS}	—	—	1.0	nAdc
(V_{GS} = -15 Vdc, V_{DS} = 0, T_A = 100°C)			—	—	200	
Gate-Source Cutoff Voltage (V_{DS} = 15 Vdc, I_D = 10 nAdc)	2N5457	$V_{GS(off)}$	0.5	—	6.0	Vdc
	2N5458		1.0	—	7.0	
	2N5459		2.0	—	8.0	
Gate-Source Voltage (V_{DS} = 15 Vdc, I_D = 100 μAdc)	2N5457	V_{GS}	—	2.5	—	Vdc
(V_{DS} = 15 Vdc, I_D = 200 μAdc)	2N5458		—	3.5	—	
(V_{DS} = 15 Vdc, I_D = 400 μAdc)	2N5459		—	4.5	—	
ON CHARACTERISTICS						
Zero-Gate-Voltage Drain Current [1] (V_{DS} = 15 Vdc, V_{GS} = 0)	2N5457	I_{DSS}	1.0	3.0	5.0	mAdc
	2N5458		2.0	6.0	9.0	
	2N5459		4.0	9.0	16	
DYNAMIC CHARACTERISTICS						
Forward Transfer Admittance [1] (V_{DS} = 15 Vdc, V_{GS} = 0, f = 1 kHz)	2N5457	$\|y_{fs}\|$	1000	3000	5000	μmhos
	2N5458		1500	4000	5500	
	2N5459		2000	4500	6000	
Output Admittance [1] (V_{DS} = 15 Vdc, V_{GS} = 0, f = 1 kHz)		$\|y_{os}\|$	—	10	50	μmhos
Input Capacitance (V_{DS} = 15 Vdc, V_{GS} = 0, f = 1 MHz)		C_{iss}	—	4.5	7.0	pF
Reverse Transfer Capacitance (V_{DS} = 15 Vdc, V_{GS} = 0, f = 1 MHz)		C_{rss}	—	1.5	3.0	pF

[1] Pulse Test: Pulse Width ≤ 630 ms; Duty Cycle ≤ 10%

[2] Continuous package improvements have enhanced these guaranteed Maximum Ratings as follows P_D = 1.0 W @ T_C = 25°C. Derate above 25°C − 8.0 mW/°C, T_J = -65 to +150°C, θ_{JC} = 125° C/W

*Courtesy of Motorola, Inc.

APPENDIX 1-10*

2N**4391** (SILICON)
2N**4392**
2N**4393**

SILICON N-CHANNEL
JUNCTION FIELD-EFFECT TRANSISTORS

Depletion Mode (Type A) Junction Field-Effect Transistors designed for chopper and high-speed switching applications.

- Low Drain-Source "On" Resistance —
  $r_{ds(on)}$ = 30 Ohms (Max) @ f = 1.0 kHz (2N4391)
- Low Gate Reverse Current —
  I_{GSS} = 0.1 nAdc (Max) @ V_{GS} = 20 Vdc
- Guaranteed Switching Characteristics

N-CHANNEL
JUNCTION FIELD-EFFECT
TRANSISTORS
(Type A)

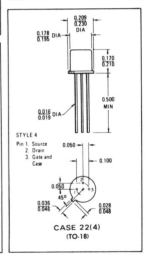

CASE 22(4)
(TO-18)

MAXIMUM RATINGS

Rating	Symbol	Value	Unit
Drain-Source Voltage	V_{DS}	40	Vdc
Drain-Gate Voltage	V_{DG}	40	Vdc
Gate-Source Voltage	V_{GS}	40	Vdc
Forward Gate Current	$I_{G(f)}$	50	mAdc
Total Device Dissipation @ T_C = 25°C Derate above 25°C	P_D	1.8 10	Watts mW/°C
Operating Junction Temperature Range	T_J	-65 to +175	°C
Storage Temperature Range	T_{stg}	-65 to +200	°C

FIGURE 1   SWITCHING TIMES TEST CIRCUIT

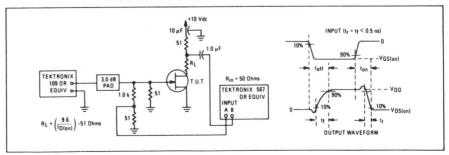

*Courtesy of Motorola, Inc.

2N4391, 2N4392, 2N4393 (continued)

ELECTRICAL CHARACTERISTICS ($T_A = 25°C$ unless otherwise noted)

Characteristic		Symbol	Min	Max	Unit
OFF CHARACTERISTICS					
Gate-Source Breakdown Voltage ($I_G = 1.0$ μAdc, $V_{DS} = 0$)		$V_{(BR)GSS}$	40	–	Vdc
Gate-Source Forward Voltage ($I_G = 1.0$ mAdc, $V_{DS} = 0$)		$V_{GS(f)}$	–	1.0	Vdc
Gate-Source Voltage ($V_{DS} = 20$ Vdc, $I_D = 1.0$ nAdc)	2N4391	V_{GS}	4.0	10	Vdc
	2N4392		2.0	5.0	
	2N4393		0.5	3.0	
Gate Reverse Current ($V_{GS} = 20$ Vdc, $V_{DS} = 0$)		I_{GSS}	–	0.1	nAdc
($V_{GS} = 20$ Vdc, $V_{DS} = 0$, $T_A = 150°C$)			–	0.2	μAdc
Drain-Cutoff Current ($V_{DS} = 20$ Vdc, $V_{GS} = 12$ Vdc)	2N4391	$I_{D(off)}$	–	0.1	nAdc
($V_{DS} = 20$ Vdc, $V_{GS} = 7.0$ Vdc)	2N4392		–	0.1	
($V_{DS} = 20$ Vdc, $V_{GS} = 5.0$ Vdc)	2N4393		–	0.1	
($V_{DS} = 20$ Vdc, $V_{GS} = 12$ Vdc, $T_A = 150°C$)	2N4391		–	0.2	μAdc
($V_{DS} = 20$ Vdc, $V_{GS} = 7.0$ Vdc, $T_A = 150°C$)	2N4392		–	0.2	
($V_{DS} = 20$ Vdc, $V_{GS} = 5.0$ Vdc, $T_A = 150°C$)	2N4393		–	0.2	
ON CHARACTERISTICS					
Zero-Gate Voltage Drain Current (1) ($V_{DS} = 20$ Vdc, $V_{GS} = 0$)	2N4391	I_{DSS}	50	150	mAdc
	2N4392		25	75	
	2N4393		5.0	30	
Drain-Source "ON" Voltage ($I_D = 12$ mAdc, $V_{GS} = 0$)	2N4391	$V_{DS(on)}$	–	0.4	Vdc
($I_D = 6.0$ mAdc, $V_{GS} = 0$)	2N4392		–	0.4	
($I_D = 3.0$ mAdc, $V_{GS} = 0$)	2N4393		–	0.4	
Static Drain-Source "ON" Resistance ($I_D = 1.0$ mAdc, $V_{GS} = 0$)	2N4391	$r_{DS(on)}$	–	30	Ohms
	2N4392		–	60	
	2N4393		–	100	
SMALL-SIGNAL CHARACTERISTICS					
Drain-Source "ON" Resistance ($V_{GS} = 0$, $I_D = 0$, $f = 1.0$ kHz)	2N4391	$r_{ds(on)}$	–	30	Ohms
	2N4392		–	60	
	2N4393		–	100	
Input Capacitance ($V_{DS} = 20$ Vdc, $V_{GS} = 0$, $f = 1.0$ MHz)		C_{iss}	–	14	pF
Reverse Transfer Capacitance ($V_{DS} = 0$, $V_{GS} = 12$ Vdc, $f = 1.0$ MHz)	2N4391	C_{rss}	–	3.5	pF
($V_{DS} = 0$, $V_{GS} = 7.0$ Vdc, $f = 1.0$ MHz)	2N4392		–	3.5	
($V_{DS} = 0$, $V_{GS} = 5.0$ Vdc, $f = 1.0$ MHz)	2N4393		–	3.5	
SWITCHING CHARACTERISTICS					
Turn-On Time (See Figure 1) ($I_{D(on)} = 12$ mAdc)	2N4391	t_{on}	–	15	ns
($I_{D(on)} = 6.0$ mAdc)	2N4392		–	15	
($I_{D(on)} = 3.0$ mAdc)	2N4393		–	15	
Rise Time (See Figure 1) ($I_{D(on)} = 12$ mAdc)	2N4391	t_r	–	5.0	ns
($I_{D(on)} = 6.0$ mAdc)	2N4392		–	5.0	
($I_{D(on)} = 3.0$ mAdc)	2N4393		–	5.0	
Turn-Off Time (See Figure 1) ($V_{GS(off)} = 12$ Vdc)	2N4391	t_{off}	–	20	ns
($V_{GS(off)} = 7.0$ Vdc)	2N4392		–	35	
($V_{GS(off)} = 5.0$ Vdc)	2N4393		–	50	
Fall Time (See Figure 1) ($V_{GS(off)} = 12$ Vdc)	2N4391	t_f	–	15	ns
($V_{GS(off)} = 7.0$ Vdc)	2N4392		–	20	
($V_{GS(off)} = 5.0$ Vdc)	2N4393		–	30	

(1) Pulse Test: Pulse Width ⩽ 100 μs, Duty Cycle ⩽ 1.0%.

APPENDIX 1-11*

μA741
FREQUENCY-COMPENSATED OPERATIONAL AMPLIFIER
FAIRCHILD LINEAR INTEGRATED CIRCUITS

GENERAL DESCRIPTION — The μA741 is a high performance monolithic Operational Amplifier constructed using the Fairchild Planar* epitaxial process. It is intended for a wide range of analog applications. High common mode voltage range and absence of "latch-up" tendencies make the μA741 ideal for use as a voltage follower. The high gain and wide range of operating voltage provides superior performance in integrator, summing amplifier, and general feedback applications.

- **NO FREQUENCY COMPENSATION REQUIRED**
- **SHORT CIRCUIT PROTECTION**
- **OFFSET VOLTAGE NULL CAPABILITY**
- **LARGE COMMON-MODE AND DIFFERENTIAL VOLTAGE RANGES**
- **LOW POWER CONSUMPTION**
- **NO LATCH UP**

ABSOLUTE MAXIMUM RATINGS

Supply Voltage	
Military (741)	±22 V
Commercial (741C)	±18 V
Internal Power Dissipation (Note 1)	
Metal Can	500 mW
DIP	670 mW
Mini DIP	310 mW
Flatpak	570 mW
Differential Input Voltage	±30 V
Input Voltage (Note 2)	±15 V
Storage Temperature Range	
Metal Can, DIP, and Flatpak	−65°C to +150°C
Mini DIP	−55°C to +125°C
Operating Temperature Range	
Military (741)	−55°C to +125°C
Commercial (741C)	0°C to +70°C
Lead Temperature (Soldering)	
Metal Can, DIP, and Flatpak (60 seconds)	300°C
Mini DIP (10 seconds)	260°C
Output Short Circuit Duration (Note 3)	Indefinite

CONNECTION DIAGRAMS

8-LEAD METAL CAN
(TOP VIEW)
PACKAGE OUTLINE 5B

Note: Pin 4 connected to case

ORDER INFORMATION

TYPE	PART NO.
741	741HM
741C	741HC

14-LEAD DIP
(TOP VIEW)
PACKAGE OUTLINE 6A

ORDER INFORMATION

TYPE	PART NO.
741	741DM
741C	741DC

10-LEAD FLATPAK
(TOP VIEW)
PACKAGE OUTLINE 3F

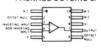

ORDER INFORMATION

TYPE	PART NO.
741	741FM

8-LEAD MINIDIP
(TOP VIEW)
PACKAGE OUTLINE 9T

ORDER INFORMATION

TYPE	PART NO.
741C	741TC

EQUIVALENT CIRCUIT

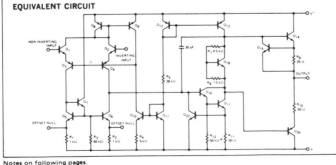

Notes on following pages.

*Planar is a patented Fairchild process.

*Courtesy of Fairchild Semiconductors

741

ELECTRICAL CHARACTERISTICS ($V_S = \pm 15$ V, $T_A = 25^\circ$C unless otherwise specified)

PARAMETERS (see definitions)		CONDITIONS	MIN.	TYP.	MAX.	UNITS
Input Offset Voltage		$R_S \leq 10$ kΩ		1.0	5.0	mV
Input Offset Current				20	200	nA
Input Bias Current				80	500	nA
Input Resistance			0.3	2.0		MΩ
Input Capacitance				1.4		pF
Offset Voltage Adjustment Range				± 15		mV
Large Signal Voltage Gain		$R_L \geq 2$ kΩ, $V_{OUT} = \pm 10$ V	50,000	200,000		
Output Resistance				75		Ω
Output Short Circuit Current				25		mA
Supply Current				1.7	2.8	mA
Power Consumption				50	85	mW
Transient Response (Unity Gain)	Risetime	$V_{IN} = 20$ mV, $R_L = 2$ kΩ, $C_L \leq 100$ pF		0.3		μs
	Overshoot			5.0		%
Slew Rate		$R_L \geq 2$ kΩ		0.5		V/μs

The following specifications apply for -55°C $\leq T_A \leq +125^\circ$C:

Input Offset Voltage		$R_S \leq 10$ kΩ		1.0	6.0	mV
Input Offset Current	$T_A = +125^\circ$C			7.0	200	nA
	$T_A = -55^\circ$C			85	500	nA
Input Bias Current	$T_A = +125^\circ$C			0.03	0.5	μA
	$T_A = -55^\circ$C			0.3	1.5	μA
Input Voltage Range			± 12	± 13		V
Common Mode Rejection Ratio		$R_S \leq 10$ kΩ	70	90		dB
Supply Voltage Rejection Ratio		$R_S \leq 10$ kΩ		30	150	μV/V
Large Signal Voltage Gain		$R_L \geq 2$ kΩ, $V_{OUT} = \pm 10$ V	25,000			
Output Voltage Swing	$R_L \geq 10$ kΩ		± 12	± 14		V
	$R_L \geq 2$ kΩ		± 10	± 13		V
Supply Current	$T_A = +125^\circ$C			1.5	2.5	mA
	$T_A = -55^\circ$C			2.0	3.3	mA
Power Consumption	$T_A = +125^\circ$C			45	75	mW
	$T_A = -55^\circ$C			60	100	mW

TYPICAL PERFORMANCE CURVES FOR 741

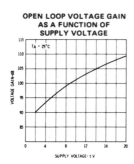

OPEN LOOP VOLTAGE GAIN
AS A FUNCTION OF
SUPPLY VOLTAGE

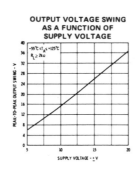

OUTPUT VOLTAGE SWING
AS A FUNCTION OF
SUPPLY VOLTAGE

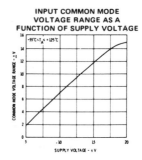

INPUT COMMON MODE
VOLTAGE RANGE AS A
FUNCTION OF SUPPLY VOLTAGE

APPENDIX 1-12

FAIRCHILD LINEAR INTEGRATED CIRCUITS • µA709

709A

ELECTRICAL CHARACTERISTICS (T_A = +25°C, ±9 V $\leq$ V_S $\leq$ ±15 V unless otherwise specified)

PARAMETER (see definitions)	CONDITIONS	MIN.	TYP.	MAX.	UNITS
Input Offset Voltage	$R_S \leq 10\ k\Omega$		0.6	2.0	mV
Input Offset Current			10	50	nA
Input Bias Current			100	200	nA
Input Resistance		350	700		kΩ
Output Resistance			150		Ω
Supply Current	V_S = ±15 V		2.5	3.6	mA
Power Consumption	V_S = ±15 V		75	108	mW
Transient Response — Risetime	V_S = ±15 V, V_{IN} = 20 mV, R_L = 2 kΩ, C_1 = 5 nF, R_1 = 1.5 kΩ, C_2 = 200 pF, R_2 = 50Ω $C_L \leq$ 100 pF			1.5	µs
Transient Response — Overshoot				30	%

The following specifications apply for $-55°C \leq T_A \leq +125°C$:

PARAMETER (see definitions)	CONDITIONS	MIN.	TYP.	MAX.	UNITS
Input Offset Voltage	$R_S \leq 10\ k\Omega$			3.0	mV
Average Temperature Coefficient of Input Offset Voltage	R_S = 50Ω, T_A = +25°C to +125°C		1.8	10	µV/°C
	R_S = 50Ω, T_A = +25°C to −55°C		1.8	10	µV/°C
	R_S = 10 kΩ, T_A = +25°C to +125°C		2.0	15	µV/°C
	R_S = 10 kΩ, T_A = +25°C to −55°C		4.8	25	µV/°C
Input Offset Current	T_A = +125°C		3.5	50	nA
	T_A = −55°C		40	250	nA
Average Temperature Coefficient of Input Offset Current	T_A = +25°C to +125°C		0.08	0.5	nA/°C
	T_A = +25°C to −55°C		0.45	2.8	nA/°C
Input Bias Current	T_A = −55°C		300	600	nA
Input Resistance	T_A = −55°C	85	170		kΩ
Input Voltage Range	V_S = ±15 V	±8.0			V
Common Mode Rejection Ratio	$R_S \leq 10\ k\Omega$	80	110		dB
Supply Voltage Rejection Ratio	$R_S \leq 10\ k\Omega$		40	100	µV/V
Large Signal Voltage Gain	V_S = ±15 V, $R_L \geq$ 2 kΩ, V_{OUT} = ±10 V	25,000		70,000	V/V
Output Voltage Swing	V_S = ±15 V, $R_L \geq$ 10 kΩ	±12	±14		V
	V_S = ±15 V, $R_L \geq$ 2 kΩ	±10	±13		V
Supply Current	T_A = +125°C, V_S = ±15 V		2.1	3.0	mA
	T_A = −55°C, V_S = ±15 V		2.7	4.5	mA
Power Consumption	T_A = +125°C, V_S = ±15 V		63	90	mW
	T_A = −55°C, V_S = ±15 V		81	135	mW

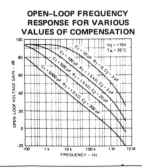

OPEN-LOOP FREQUENCY RESPONSE FOR VARIOUS VALUES OF COMPENSATION

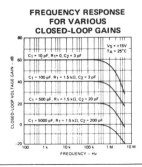

FREQUENCY RESPONSE FOR VARIOUS CLOSED-LOOP GAINS

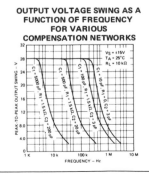

OUTPUT VOLTAGE SWING AS A FUNCTION OF FREQUENCY FOR VARIOUS COMPENSATION NETWORKS

*Courtesy of Fairchild Camera and Instrument Corporation © 1982

APPENDIX 1-13*

TYPES 2N3980, 2N4947 THRU 2N4949
P-N PLANAR UNIJUNCTION SILICON TRANSISTORS

PLANAR UNIJUNCTION TRANSISTORS SPECIFICALLY CHARACTERIZED FOR A WIDE RANGE OF MILITARY, SPACE, AND INDUSTRIAL APPLICATIONS:

2N3980 for General-Purpose UJT Applications
2N4947 for High-Frequency Relaxation-Oscillator Circuits
2N4948 for Thyristor (SCR) Trigger Circuits
2N4949 for Long-Time-Delay Circuits

- Planar Process Ensures Extremely Low Leakage, High Performance for Low Driving Currents, and Greatly Improved Reliability

*mechanical data

Package outline is same as JEDEC TO-18 except for lead position. All TO-18 registration notes also apply to this outline.

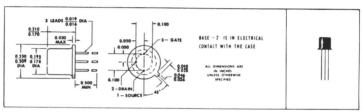

*absolute maximum ratings at 25°C free-air temperature (unless otherwise noted)

Emitter--Base-Two Reverse Voltage	. . . . −30 V
Interbase Voltage	. . . . See Note 1
Continuous Emitter Current	. . . . 50 mA
Peak Emitter Current (See Note 2)	. . . . 1 A
Continuous Device Dissipation at (or below) 25°C Free-Air Temperature (See Note 3)	. . . . 360 mW
Storage Temperature Range	. . . . −65°C to 200°C
Lead Temperature 1/16 Inch from Case for 10 Seconds	. . . . 260°C

*electrical characteristics at 25°C free-air temperature (unless otherwise noted)

PARAMETER		TEST CONDITIONS	2N3980		2N4947		2N4948		2N4949		UNIT
			MIN	MAX	MIN	MAX	MIN	MAX	MIN	MAX	
r_{BB}	Static Interbase Resistance	$V_{B2-B1} = 3$ V, $I_E = 0$	4	8	4	9.1	4	12	4	12	kΩ
α_{rBB}	Interbase Resistance Temperature Coefficient	$V_{B2-B1} = 3$ V, $I_E = 0$, $T_A = -65°C$ to 100°C, See Note 4	0.4	0.9	0.1	0.9	0.1	0.9	0.1	0.9	%/deg
η	Intrinsic Standoff Ratio	$V_{B2-B1} = 10$ V, See Figure 1	0.68	0.82	0.51	0.69	0.55	0.82	0.74	0.86	
$I_{B2(mod)}$	Modulated Interbase Current	$V_{B2-B1} = 10$ V, $I_E = 50$ mA, See Note 5	12		12		12		12		mA
I_{EB2O}	Emitter Reverse Current	$V_{EB2} = -30$ V, $I_{B1} = 0$		−10		−10		−10		−10	nA
		$V_{EB2} = -30$ V, $I_{B1} = 0$, $T_A = 125°C$		−1		−1		−1		−1	μA
I_P	Peak-Point Emitter Current	$V_{B2-B1} = 25$ V		2		2		2		1	μA
$V_{EB1(sat)}$	Emitter — Base-One Saturation Voltage	$V_{B2-B1} = 10$ V, $I_E = 50$ mA, See Note 5		.3		3		3		3	V
I_V	Valley-Point Emitter Current	$V_{B2-B1} = 20$ V	1	10	4		2		2		mA
V_{OB1}	Base-One Peak Pulse Voltage	See Figure 2	6		3		6		3		V

*Courtesy of Texas Instruments, Incorporated

APPENDIX 1-14*

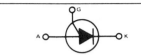

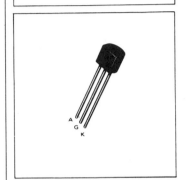

SILICON PROGRAMMABLE UNIJUNCTION TRANSISTORS	SILICON PROGRAMMABLE UNIJUNCTION TRANSISTORS 40 VOLTS 375 mW

SILICON PROGRAMMABLE UNIJUNCTION TRANSISTORS

. . . designed to enable the engineer to "program" unijunction characteristics such as R_{BB}, η, I_V, and I_P by merely selecting two resistor values. Application includes thyristor-trigger, oscillator, pulse and timing circuits. These devices may also be used in special thyristor applications due to the availability of an anode gate. Supplied in an inexpensive TO-92 plastic package for high-volume requirements, this package is readily adaptable for use in automatic insertion equipment.

- Programmable — R_{BB}, η, I_V and I_P.
- Low On-State Voltage — 1.5 Volts Maximum @ I_F = 50 mA
- Low Gate to Anode Leakage Current — 10 nA Maximum
- High Peak Output Voltage — 11 Volts Typical
- Low Offset Voltage — 0.35 Volt Typical (R_G = 10 k ohms)

ELECTRICAL CHARACTERISTICS (T_A = 25°C unless otherwise noted)

Characteristic		Figure	Symbol	Min	Typ	Max	Unit
• Peak Current		2,9,11	I_P				µA
(V_S = 10 Vdc, R_G = 1.0 MΩ)	2N6027			—	1.25	2.0	
	2N6028			—	0.08	0.15	
(V_S = 10 Vdc, R_G = 10 k ohms)	2N6027			—	4.0	5.0	
	2N6028			—	0.70	1.0	
• Offset Voltage		1	V_T				Volts
(V_S = 10 Vdc, R_G = 1.0 MΩ)	2N6027			0.2	0.70	1.6	
	2N6028			0.2	0.50	0.6	
(V_S = 10 Vdc, R_G = 10 k ohms)	(Both Types)			0.2	0.35	0.6	
• Valley Current		1,4,5,	I_V				µA
(V_S = 10 Vdc, R_G = 1.0 MΩ)	2N6027			—	18	50	
	2N6028			—	18	25	
(V_S = 10 Vdc, R_G = 10 k ohms)	2N6027			70	270	—	
	2N6028			25	270	—	
(V_S = 10 Vdc, R_G = 200 Ohms)	2N6027			1.5	—	—	mA
	2N6028			1.0	—	—	
• Gate to Anode Leakage Current		—	I_{GAO}				nAdc
(V_S = 40 Vdc, T_A = 25°C, Cathode Open)				—	1.0	10	
(V_S = 40 Vdc, T_A = 75°C, Cathode Open)				—	3.0	—	
Gate to Cathode Leakage Current		—	I_{GKS}	—	5.0	50	nAdc
(V_S = 40 Vdc, Anode to Cathode Shorted)							
• Forward Voltage (I_F = 50 mA Peak)		1,6	V_F	—	0.8	1.5	Volts
• Peak Output Voltage		3,7	V_O	6.0	11	—	Volts
(V_B = 20 Vdc, C_C = 0.2 µF)							
Pulse Voltage Rise Time		3	t_r	—	40	80	ns
(V_B = 20 Vdc, C_C = 0.2 µF)							

*Indicates JEDEC Registered Data

*Courtesy of Motorola, Inc.

APPENDIX 1-15*

SCHMITT-TRIGGER POSITIVE-NAND GATES AND INVERTERS WITH TOTEM-POLE OUTPUTS

recommended operating conditions

	54 FAMILY 74 FAMILY	SERIES 54 SERIES 74					
		'13			'14		
		MIN	NOM	MAX	MIN	NOM	MAX
Supply voltage, V_{CC}	54 Family	4.5	5	5.5	4.5	5	5.5
	74 Family	4.75	5	5.25	4.75	5	5.25
High-level output current, I_{OH}				−800			−800
Low-level output current, I_{OL}				16			16
Operating free-air temperature, T_A	54 Family	−55		125	−55		125
	74 Family	0		70	0		70

electrical characteristics over recommended operating free-air temperature range (unless otherwise noted)

PARAMETER		TEST FIGURE	TEST CONDITIONS†		SERIES 54 SERIES 74					
					'13			'14		
					MIN	TYP‡	MAX	MIN	TYP‡	MAX
V_{T+}	Positive-going threshold voltage	8	V_{CC} = 5 V		1.5	1.7	2	1.5	1.7	2
V_{T-}	Negative-going threshold voltage	9	V_{CC} = 5 V		0.6	0.9	1.1	0.6	0.9	1.1
	Hysteresis ($V_{T+}-V_{T-}$)	8, 9	V_{CC} = 5 V		0.4	0.8		0.4	0.8	
V_I	Input clamp voltage	3	V_{CC} = MIN, I_I = §				−1.5			−1.5
V_{OH}	High-level output voltage	9	V_{CC} = MIN, V_I = V_{T-}min, I_{OH} = MAX	54 Family	2.4	3.4		2.4	3.4	
				74 Family	2.4	3.4		2.4	3.4	
V_{OL}	Low-level output voltage	8	V_{CC} = MIN, V_I = V_{T+}max, I_{OL} = MAX			0.2	0.4		0.2	0.4
I_{T+}	Input current at positive-going threshold	8	V_{CC} = 5 V, V_I = V_{T+}			−0.65			−0.43	
I_{T-}	Input current at negative-going threshold	9	V_{CC} = 5 V, V_I = V_{T-}			−0.85			−0.56	
I_I	Input current at maximum input voltage	4	V_{CC} = MAX, V_I = 5.5 V				1			1
I_{IH}	High-level input current	4	V_{CC} = MAX, V_I = 2.4 V				40			40
				V_I = 2.7 V						
I_{IL}	Low-level input current	5	V_{CC} = MAX, V_{IL} = 0.4 V			−1	−1.6		−0.8	−1.2
				V_{IL} = 0.5 V						
I_{OS}	Short-circuit output current◆	6	V_{CC} = MAX		−18		−55	−18		−55
I_{CC}	Supply current	Total, output high	7	V_{CC} = MAX		14	23		22	36
		Total, output low		V_{CC} = MAX		20	32		39	60
		Average per gate		V_{CC} = 5 V, 50% duty cycle		8.5			5.1	

† For conditions shown as MIN or MAX, use the appropriate value specified under recommended operating conditions.
‡ All typical values are at V_{CC} = 5 V, T_A = 25°C.
§ I_I = −12 mA for SN54'/SN74' and −18 mA for 'S132.
◆ Not more than one output should be shorted at a time, and for 'S132, duration of output short-circuit should not exceed one second.

switching characteristics, V_{CC} = 5 V, T_A = 25°C

TYPE	TEST CONDITIONS	t_{PLH} (ns) Propagation delay time, low-to-high-level output		t_{PHL} (ns) Propagation delay time, high-to-low-level output	
		TYP	MAX	TYP	MAX
'13	C_L = 15 pF, R_L = 400 Ω	18	27	15	22
'14, '132		15	22	15	22
'S132	C_L = 15 pF, R_L = 280 Ω	7	10.5	8.5	13

*Courtesy of Texas Instruments, Incorporated

APPENDIX 1-16*

LM555/LM555C

absolute maximum ratings

Supply Voltage	+18V
Power Dissipation	600 mW
Operating Temperature Ranges	
LM555C	0°C to +70°C
LM555	−55°C to +125°C
Storage Temperature Range	−65°C to +150°C
Lead Temperature (Soldering, 10 sec)	300°C

electrical characteristics ($T_A = 25°C$, $V_{CC} = +5V$ to $+15V$ unless otherwise specified)

PARAMETER	CONDITIONS	LM555 MIN	LM555 TYP	LM555 MAX	LM555C MIN	LM555C TYP	LM555C MAX	UNITS
Supply Voltage		4.5		18	4.5		16	V
Supply Current	V_{CC} 5V, $R_L = \infty$		3	5		3	6	mA
	$V_{CC} = 15V$, $R_L = \infty$ (Low State) (Note 1)		10	12		10	15	mA
Timing Error, Monostable								
Initial Accuracy	R_A, $R_B = 1k$ to 100k,		.5	2		1		%
Drift with Temperature	$C = .1\mu F$, (Note 2)		30	100		50		ppm/C
Drift with Supply			.005	.02		.01		%/V
Threshold Voltage			.667			.667		× V_{CC}
Trigger Voltage	$V_{CC} = 15V$	4.8	5	5.2		5		V
	$V_{CC} = 5V$	1.45	1.67	1.9		1.67		V
Trigger Current			.5			.5		µA
Reset Voltage		4	.7	1	4	.7	1	V
Reset Current			.1			.1		mA
Threshold Current	(Note 3)		.1	.25		.1	.25	µA
Control Voltage Level	$V_{CC} = 15V$	9.6	10	10.4	9	10	11	V
	$V_{CC} = 5V$	2.9	3.33	3.8	2.6	3.33	4	V
Output Voltage Drop (Low)	$V_{CC} = 15V$							
	$I_{SINK} = 10$ mA		.1	.15		.1	.25	V
	$I_{SINK} = 50$ mA		.4	.5		.4	.75	V
	$I_{SINK} = 100$ mA		2	2.2		2	2.5	V
	$I_{SINK} = 200$ mA		2.5			2.5		V
	$V_{CC} = 5V$							
	$I_{SINK} = 8$ mA		.1	.25				V
	$I_{SINK} = 5$ mA					.25	.35	V
Output Voltage Drop (High)	$I_{SOURCE} = 200$ mA $V_{CC} = 15V$		12.5			12.5		V
	$I_{SOURCE} = 100$ mA $V_{CC} = 15V$	13	13.3		12.75	13.3		V
	$V_{CC} = 5V$	3	3.3		2.75	3.3		V
Rise Time of Output			100			100		ns
Fall Time of Output			100			100		ns

Note 1: Supply current when output high typically 1 mA less at $V_{CC} = 5V$.
Note 2: Tested at $V_{CC} = 5V$ and $V_{CC} = 15V$.
Note 3: This will determine the maximum value of $R_A + R_B$ for 15V operation. The max total $(R_A + R_B) = 20 M\Omega$.

*Courtesy of **National Semiconductor Corp.**

APPENDIX 1-17*

SERIES 54/74 FLIP-FLOPS

recommended operating conditions

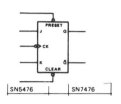

		SERIES 54/74	'70			'72, '73, '76, '107		
			MIN	NOM	MAX	MIN	NOM	MAX
Supply voltage, V_{CC}		Series 54	4.5	5	5.5	4.5	5	5.5
		Series 74	4.75	5	5.25	4.75	5	5.25
High-level output current, I_{OH}					−400			−400
Low-level output current, I_{OL}					16			16
Pulse width, t_W	Clock high		20			20		
	Clock low		30			47		
	Preset or clear low		25			25		
Input setup time, t_{setup}			20↑			0↑		
Input hold time, t_{hold}			5↑			0↓		
Operating free-air temperature, T_A		Series 54	−55		125	−55		125
		Series 74	0		70	0		70

↑↓The arrow indicates the edge of the clock pulse used for reference: ↑ for the rising edge, ↓ for the falling edge.

electrical characteristics over recommended operating free-air temperature range (unless otherwise noted)

PARAMETER		TEST CONDITIONS†	'70			'72, '73, '76, '107		
			MIN	TYP‡	MAX	MIN	TYP‡	MAX
V_{IH}	High-level input voltage		2			2		
V_{IL}	Low-level input voltage				0.8			0.8
V_I	Input clamp voltage	V_{CC} = MIN, I_I = −12 mA		*	−1.5		*	−1.5
V_{OH}	High-level output voltage	V_{CC} = MIN, V_{IH} = 2 V, V_{IL} = 0.8 V, I_{OH} = MAX	2.4	3.4		2.4	3.4	
V_{OL}	Low-level output voltage	V_{CC} = MIN, V_{IH} = 2 V, V_{IL} = 0.8 V, I_{OL} = 16 mA		0.2	0.4		0.2	0.4
I_I	Input current at maximum input voltage	V_{CC} = MAX, V_I = 5.5 V			1			1
I_{IH}	High-level input current	D, J, K, or $\overline{K}$	V_{CC} = MAX, V_I = 2.4 V			40			40
		Clear				80			80
		Preset				80			80
		Clock				40			80
I_{IL}	Low-level input current	D, J, K, or $\overline{K}$	V_{CC} = MAX, V_I = 0.4 V			−1.6			−1.6
		Clear				−3.2			−3.2
		Preset				−3.2			−3.2
		Clock				−1.6			−3.2
I_{OS}	Short-circuit output current♦	Series 54	V_{CC} = MAX	−20		−57	−20		−57
		Series 74		−18		−57	−18		−57
I_{CC}	Supply current (Average per flip-flop)	V_{CC} = MAX, See Note 1		13	26		10	20

switching characteristics, V_{CC} = 5 V, T_A = 25°C

PARAMETER¶	FROM (INPUT)	TO (OUTPUT)	TEST CONDITIONS	'70			'72, '73 '76, '107		
				MIN	TYP	MAX	MIN	TYP	MAX
f_{max}				20	35		15	20	
t_{PLH}	Preset	Q	C_L = 15 pF, R_L = 400 Ω, See Note 2			50		16	25
t_{PHL}	(as applicable)	$\overline{Q}$				50		25	40
t_{PLH}	Clear	$\overline{Q}$				50		16	25
t_{PHL}	(as applicable)	Q				50		25	40
t_{PLH}	Clock	Q or $\overline{Q}$		10	27	50	10	16	25
t_{PHL}				10	18	50	10	25	40

*Courtesy of Texas Instruments, Incorporated

APPENDIX 1-18*

POSITIVE-NAND GATES AND INVERTERS WITH TOTEM-POLE OUTPUTS

recommended operating conditions

		SERIES 54 / SERIES 74 '00, '04, '10, '20, '30			SERIES 54H / SERIES 74H 'H00, 'H04, 'H10, 'H20, 'H30			SERIES 54L / SERIES 74L 'L00, 'L04, 'L10, 'L20, 'L30			SERIES 54LS / SERIES 74LS 'LS00, 'LS04, 'LS10, 'LS20, 'LS30			SERIES 54S / SERIES 74S 'S00, 'S04, 'S10, 'S20, 'S30, 'S133			UNIT
		MIN	NOM	MAX	MIN	NOM	MAX	MIN	NOM	MAX	MIN	NOM	MAX	MIN	NOM	MAX	
Supply voltage, V_{CC}	54 Family	4.5	5	5.5	4.5	5	5.5	4.5	5	5.5	4.5	5	5.5	4.5	5	5.5	V
	74 Family	4.75	5	5.25	4.75	5	5.25	4.75	5	5.25	4.75	5	5.25	4.75	5	5.25	
High-level output current, I_{OH}	54 Family			-400			-500			-100			-400			-1000	µA
	74 Family			-400			-500			-200			-400			-1000	
Low-level output current, I_{OL}	54 Family			16			20			2			4			20	mA
	74 Family			16			20			3.6			8			20	
Operating free-air temperature, T_A	54 Family	-55		125	-55		125	-55		125	-55		125	-55		125	°C
	74 Family	0		70	0		70	0		70	0		70	0		70	

electrical characteristics over recommended operating free-air temperature range (unless otherwise noted)

PARAMETER	TEST FIGURE	TEST CONDITIONS[†]		SERIES 54 / SERIES 74 '00, '04, '10, '20, '30			SERIES 54H / SERIES 74H 'H00, 'H04, 'H10, 'H20, 'H30			SERIES 54L / SERIES 74L 'L00, 'L04, 'L10, 'L20, 'L30			SERIES 54LS / SERIES 74LS 'LS00, 'LS04, 'LS10, 'LS20, 'LS30			SERIES 54S / SERIES 74S 'S00, 'S04, 'S10, 'S20, 'S30, 'S133			UNIT
				MIN	TYP[‡]	MAX	MIN	TYP[‡]	MAX	MIN	TYP[‡]	MAX	MIN	TYP[‡]	MAX	MIN	TYP[‡]	MAX	
V_{IH} High-level input voltage	1, 2			2			2			2			2			2			V
V_{IL} Low-level input voltage	1, 2	54 Family				0.8			0.8			0.7			0.7			0.8	V
		74 Family				0.8			0.8			0.7			0.8			0.8	
V_I Input clamp voltage	3	V_{CC} = MIN, I_I = §				*-1.5			*-1.5						-1.5			-1.2	V
V_{OH} High-level output voltage	1	V_{CC} = MIN, V_{IL} = V_{IL} max, I_{OH} = MAX	54 Family	2.4	3.4		2.4	3.5		2.4	3.3		2.5	3.4		2.5	3.4		V
			74 Family	2.4	3.4		2.4	3.5		2.4	3.2		2.7	3.4		2.7	3.4		
V_{OL} Low-level output voltage	2	V_{CC} = MIN, V_{IH} = 2 V, I_{OL} = MAX	54 Family		0.2	0.4		0.2	0.4		0.15	0.3		0.25	0.4			0.5	V
			74 Family		0.2	0.4		0.2	0.4		0.2	0.4		0.35	0.5			0.5	
I_I Input current at maximum input voltage	4	V_{CC} = MAX, V_I = 5.5 V				1			1			0.1			0.1			1	mA
I_{IH} High-level input current	4	V_{CC} = MAX	V_{IH} = 2.4 V			40			50			10			20				µA
			V_{IH} = 2.7 V															50	
I_{IL} Low-level input current	5	V_{CC} = MAX	V_{IL} = 0.3 V 'LS30												-0.4				mA
			V_{IL} = 0.4 V Others			-1.6			-2			-0.18			-0.36			-2	
			V_{IL} = 0.5 V																
I_{OS} Short-circuit output current◆	6	V_{CC} = MAX	54 Family	-20		-55	-40		-100	-3		-15	-6		-40	-40		-100	mA
			74 Family	-18		-55	-40		-100	-3		-15	-5		-42	-40		-100	
I_{CC} Supply current	7	V_{CC} = MAX							See table on next page											mA

[†] For conditions shown as MIN or MAX, use the appropriate value specified under recommended operating conditions.
[‡] All typical values are at V_{CC} = 5 V, T_A = 25°C.
§ I_I = -12 mA for SN54'/SN74', -8 mA for SN54H'/SN74H', and -18 mA for SN54LS'/SN74LS' and SN54S'/SN74S'.
● I_I = -12 mA for SN54H'/SN74H', duration of short-circuit should not exceed 1 second.
◆ Not more than one output should be shorted at a time, and for SN54S'/SN74S' and SN54S'/SN74S' parts date-coded 7332 or higher.
* The input clamp voltage specification is effective for Series 54/74 and 54H/74H parts date-coded 7332 or higher.

*Courtesy of Texas Instruments, Incorporated

POSITIVE-NAND GATES AND INVERTERS WITH TOTEM-POLE OUTPUTS

supply current¶

schematics (each gate)

TYPE	ICCH (mA) Total with outputs high		ICCL (mA) Total with outputs low		ICC (mA) Average per gate (50% duty cycle)
	TYP	MAX	TYP	MAX	TYP
'00	4	8	12	22	2
'04	6	12	18	33	2
'10	3	6	9	16.5	2
'20	2	4	6	11	2
'30	1	2	3	6	2
'H00	10	16.8	26	40	4.5
'H04	16	26	40	58	4.5
'H10	7.5	12.6	19.5	30	4.5
'H20	5	8.4	13	20	4.5
'H30	2.5	4.2	6.5	10	4.5
'L00	0.44	0.8	1.16	2.04	0.20
'L04	0.66	1.2	1.74	3.06	0.20
'L10	0.33	0.6	0.87	1.53	0.20
'L20	0.22	0.4	0.58	1.02	0.20
SN54L30	0.11	0.33	0.29	0.51	0.20
SN74L30	0.11	0.2	0.29	0.51	0.20
'LS00	0.8	1.6	2.4	4.4	0.4
'LS04	1.2	2.4	3.6	6.6	0.4
'LS10	0.6	1.2	1.8	3.3	0.4
'LS20	0.4	0.8	1.2	2.2	0.4
'LS30	0.35	0.5	0.6	1.1	0.48
'S00	10	16	20	36	3.75
'S04	15	24	30	54	3.75
'S10	7.5	12	15	27	3.75
'S20	5	8	10	18	3.75
'S30	3	5	5.5	10	4.25
'S133	3	5	5.5	10	4.25

¶Maximum values of I_{CC} are over the recommended operating ranges of V_{CC} and T_A; typical values are at $V_{CC} = 5$ V, $T_A = 25°$ C.

switching characteristics at $V_{CC} = 5$ V, $T_A = 25°$ C

TYPE	TEST CONDITIONS#	tPLH (ns) Propagation delay time, low-to-high-level output			tPHL (ns) Propagation delay time, high-to-low-level output		
		MIN	TYP	MAX	MIN	TYP	MAX
'00, '10	$C_L = 15$ pF, $R_L = 400$ Ω		11	22		7	15
'04, '20			12	22		8	15
'30			13	22		8	15
'H00	$C_L = 25$ pF, $R_L = 280$ Ω		5.9	10		6.2	10
'H04			5.9	10		6.5	10
'H10			5.9	10		6.3	10
'H20			6	10		7	10
'H30			6.8	10		8.9	12
'L00, 'L04, 'L10, 'L20	$C_L = 50$ pF, $R_L = 4$ kΩ		35	60		31	60
'L30			35	60		70	100
'LS00, 'LS04	$C_L = 15$ pF, $R_L = 2$ kΩ		9	20		10	20
'LS10, 'LS20			9	20		10	20
'LS30							
'S00, 'S04	$C_L = 15$ pF, $R_L = 280$ Ω	2	9	20	2	25	35
'S10, 'S20	$C_L = 50$ pF, $R_L = 280$ Ω		3	4.5		3	5
'S30	$C_L = 15$ pF, $R_L = 280$ Ω		4.5	6		4.5	6
'S133	$C_L = 50$ pF, $R_L = 280$ Ω	2	5.5		2	6.5	7

#Load circuits and voltage waveforms are shown on pages 148 and 149.

'S00, 'S04, 'S10, 'S20, 'S30, 'S133 CIRCUITS

'H00, 'H04, 'H10, 'H20, 'H30 CIRCUITS

'LS00, 'LS04, 'LS10, 'LS20, 'LS30 CIRCUITS

'00, '04, '10, '20, '30 CIRCUITS

'L00, 'L04, 'L10, 'L20, 'L30 CIRCUITS

CIRCUIT	R1	R2	R3	R4
'00, '04, '10, '20, '30	4 k	1.6 k	130	1 k
'L00, 'L04, 'L10, 'L20, 'L30	40 k	20 k	500	12 k

Input clamp diodes not on SN54L'/SN74L' circuits.

Resistor values shown are nominal and in ohms.

APPENDIX 1-19[*]

| 3-INPUT GATES | MECL MC300 series |

MC306 · MC307

Expandable 3-input gates that provide the positive logic "NOR" function and its complement simultaneously. MC307 omits output pull-down resistors, permitting reduction of power dissipation.

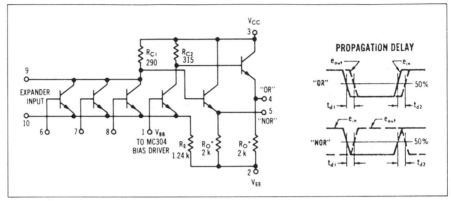

PROPAGATION DELAY

ELECTRICAL CHARACTERISTICS

		Test Conditions Vdc ± 1%									Test Limits						
@ Test Temperature	−55°C	—	−0.945	−1.450	−5.20	−1.25					−55°C		+25°C		+125°C		Unit
	+25°C	0.690	−0.795	−1.350	−5.20	−1.15					Min	Max	Min	Max	Min	Max	
	+125°C		0.655	−1.300	−5.20	−1.00											
Characteristic		V_H Pin No	$V_{1\,max}$ Pin No	V_L Pin No	V_{EE} Pin No	V_{BB} Pin No	dV_n Pin No	I_L Pin No	Ground Pin No	Symbol Pin No in ( )	Min	Max	Min	Max	Min	Max	
Power Supply MC306		—	—	—	2,6,7,8	1	—	—	3	I_E (2)	—	8.85	—	8.85	—	8.15	mAdc
Drain Current MC307		—	—	—	2,6,7,8	1	—	—	3	I_E (2)	—	3.6	—	3.6	—	3.3	mAdc
Input Current		6	—	—	2,7,8	1	—	—	3	I_{in} (6)	—	—	—	100	—	—	µAdc
		7	—	—	2,6,8	1	—	—	3	I_{in} (7)	—	—			—	—	
		8	—	—	2,6,7	1	—	—	3	I_{in} (8)	—	—			—	—	
"NOR" Logical "1" Output Voltage		—	—	6	2,7,8	1	—	—	3	V_1 (5)	−0.825	−0.945	0.690	0.795	−0.525	0.655	Vdc
		—	—	7	2,6,8	1	—	—	3	V_1 (5)							
		—	—	8	2,6,7	1	—	—	3	V_1 (5)							
"NOR" Logical "0" Output Voltage		—	6	—	2,7,8	1	—	—	3	V_4 (5)	−1.560	−1.850	−1.465	−1.750	−1.340	−1.675	Vdc
		—	7	—	2,6,8	1	—	—	3	V_4 (5)							
		—	8	—	2,6,7	1	—	—	3	V_4 (5)							
"OR" Logical "1" Output Voltage		—	6	—	2,7,8	1	—	—	3	V_5 (4)	−0.825	−0.945	0.690	−0.795	−0.525	0.655	Vdc
		—	7	—	2,6,8	1	—	—	3	V_5 (4)							
		—	8	—	2,6,7	1	—	—	3	V_5 (4)							
"OR" Logical "0" Output Voltage		—	—	6	2,7,8	1	—	—	3	V_2 (4)	−1.560	−1.850	−1.465	−1.750	−1.340	−1.675	Vdc
		—	—	7	2,6,8	1	—	—	3	V_2 (4)							
		—	—	8	2,6,7	1	—	—	3	V_2 (4)							
"NOR" Output Voltage Change (No load to full load)		—	—	6	2,7,8	1	—	5①	3	$\triangle V_1$ (5)	—	0.055	—	0.055	—	−0.060	Volts
"OR" Output Voltage Change (No load to full load)		—	6	—	2,7,8	1	—	4①	3	$\triangle V_5$ (4)	—	0.055	—	0.055	—	−0.060	Volts
"NOR" Saturation Breakpoint Voltage		—	—	—	2,7,8	1	6①	—	3	V_3 (5)	—	0.40	—	−0.55	—	−0.68	Vdc
		—	—	—	2,6,8	1	7①	—	3	V_3 (5)							
		—	—	—	2,6,7	1	8①	—	3	V_3 (5)							
Switching Times		Pulse In	Pulse Out								Typ	Max	Typ	Max	Typ	Max	ns
Propagation Delay Time		6	4	—	2,7,8	1	—	—	3	t_{d1} (4)	7.0	11.0	7.0	11.5	9.5	14.5	ns
		6	5	—	2,7,8	1	—	—	3	t_{d1} (5)	5.5	10.0	5.5	10.5	7.0	12.5	
		6	4	—	2,7,8	1	—	—	3	t_{d2} (4)	5.5	10.0	5.5	11.0	7.0	12.5	
		6	5	—	2,7,8	1	—	—	3	t_{d2} (5)	7.0	10.5	7.0	11.0	9.5	14.5	
Rise Time		6	4	—	2,7,8	1	—	—	3	t_r (4)	6.0	8.5	6.0	10.0	8.0	13.0	
		6	5	—	2,7,8	1	—	—	3	t_r (5)	7.5	11.5	7.5	12.5	9.5	15.0	
Fall Time		6	4	—	2,7,8	1	—	—	3	t_f (4)	6.5	10.5	6.5	12.0	9.0	15.0	
		6	5	—	2,7,8	1	—	—	3	t_f (5)	6.5	12.0	6.5	12.5	9.0	15.0	

[*]Courtesy of Motorola, Inc.

APPENDIX 1-20*

µA710
HIGH-SPEED DIFFERENTIAL COMPARATOR
FAIRCHILD LINEAR INTEGRATED CIRCUITS

ABSOLUTE MAXIMUM RATINGS

Positive Supply Voltage	+14.0 V
Negative Supply Voltage	−7.0 V
Peak Output Current	10 mA
Differential Input Voltage	±5.0 V
Input Voltage	±7.0 V
Internal Power Dissipation	
TO-99 [Note 1]	300 mW
Flat Package [Note 2]	200 mW
Operating Temperature Range	−55°C to +125°C
Storage Temperature Range	−65°C to +150°C
Lead Temperature (Soldering, 60 sec.)	300°C

ELECTRICAL CHARACTERISTICS ($T_A = +25°C$, $V^+ = 12.0V$, $V^- = -6.0V$ unless otherwise specified)

PARAMETER (see definitions)	CONDITIONS (Note 4)	MIN.	TYP.	MAX.	UNITS
Input Offset Voltage	$R_s \leq 200\Omega$		0.6	2.0	mV
Input Offset Current			0.75	3.0	µA
Input Bias Current			13	20	µA
Voltage Gain		1250	1700		
Output Resistance			200		Ω
Output Sink Current	$\Delta V_{in} \geq 5$ mV, $V_{out} = 0$	2.0	2.5		mA
Response Time [Note 3]			40		ns

The following specifications apply for $-55°C \leq T_A \leq +125°C$:

Input Offset Voltage	$R_s \leq 200\Omega$			3.0	mV
Average Temperature Coefficient of Input Offset Voltage	$R_s = 50\Omega$, $T_A = 25°C$ to $T_A = +125°C$ $R_s = 50\Omega$, $T_A = 25°C$ to $T_A = -55°C$		3.5 2.7	10 10	µV/°C µV/°C
Input Offset Current	$T_A = +125°C$		0.25	3.0	µA
	$T_A = -55°C$		1.8	7.0	µA
Average Temperature Coefficient of Input Offset Current	$T_A = 25°C$ to $T_A = +125°C$ $T_A = 25°C$ to $T_A = -55°C$		5.0 15	25 75	nA/°C nA/°C
Input Bias Current	$T_A = -55°C$		27	45	µA
Input Voltage Range	$V^- = -7.0V$	±5.0			V
Common Mode Rejection Ratio	$R_s \leq 200\Omega$	80	100		dB
Differential Input Voltage Range		±5.0			V
Voltage Gain		1000			
Positive Output Level	$\Delta V_{in} \geq 5$ mV, $0 \leq I_{out} \leq 5.0$ mA	2.5	3.2	4.0	V
Negative Output Level	$\Delta V_{in} \geq 5$ mV	−1.0	−0.5	0	V
Output Sink Current	$T_A = +125°C$, $\Delta V_{in} \geq 5$ mV, $V_{out} = 0$	0.5	1.7		mA
	$T_A = -55°C$, $\Delta V_{in} \geq 5$ mV, $V_{out} = 0$	1.0	2.3		mA
Positive Supply Current	$V_{out} \leq 0$		5.2	9.0	mA
Negative Supply Current			4.6	7.0	mA
Power Consumption			90	150	mW

*Courtesy of Fairchild Semiconductors

APPENDIX 1-21*

LM311

absolute maximum ratings

Total Supply Voltage (V_{84})	36V
Output to Negative Supply Voltage (V_{74})	40V
Ground to Negative Supply Voltage (V_{14})	30V
Differential Input Voltage	±30V
Input Voltage (Note 1)	±15V
Power Dissipation (Note 2)	500 mW
Output Short Circuit Duration	10 sec
Operating Temperature Range	0°C to 70°C
Storage Temperature Range	−65°C to 150°C
Lead Temperature (soldering, 10 sec)	300°C

electrical characteristics (Note 3)

PARAMETER	CONDITIONS	MIN	TYP	MAX	UNITS
Input Offset Voltage (Note 4)	$T_A = 25°C$, $R_S \le 50K$		2.0	7.5	mV
Input Offset Current (Note 4)	$T_A = 25°C$		6.0	50	nA
Input Bias Current	$T_A = 25°C$		100	250	nA
Voltage Gain	$T_A = 25°C$		200		V/mV
Response Time (Note 5)	$T_A = 25°C$		200		ns
Saturation Voltage	$V_{IN} \le -10$ mV, $I_{OUT} = 50$ mA $T_A = 25°C$		0.75	1.5	V
Strobe On Current	$T_A = 25°C$		3.0		mA
Output Leakage Current	$V_{IN} \ge 10$ mV, $V_{OUT} = 35$ V $T_A = 25°C$		0.2	50	nA
Input Offset Voltage (Note 4)	$R_S \le 50K$			10	mV
Input Offset Current (Note 4)				70	nA
Input Bias Current				300	nA
Input Voltage Range			±14		V
Saturation Voltage	$V^+ \ge 4.5V$, $V^- = 0$ $V_{IN} < -10$ mV, $I_{SINK} \le 8$ mA		0.23	0.4	V
Positive Supply Current	$T_A = 25$ C		5.1	7.5	mA
Negative Supply Current	$T_A = 25$ C		4.1	5.0	mA

Note 1: This rating applies for ±15V supplies. The positive input voltage limit is 30V above the negative supply. The negative input voltage limit is equal to the negative supply voltage or 30V below the positive supply, whichever is less.

Note 2: The maximum junction temperature of the LM311 is 85°C. For operating at elevated temperatures, devices in the TO-5 package must be derated based on a thermal resistance of 150°C/W, junction to ambient, or 45°C/W, junction to case. For the flat package, the derating is based on a thermal resistance of 185°C/W when mounted on a 1/16-inch-thick epoxy glass board with ten, 0.03-inch-wide, 2-ounce copper conductors. The thermal resistance of the dual-in-line package is 100°C/W, junction to ambient.

Note 3: These specifications apply for V_S = ±15V and 0°C < T_A < 70°C, un. otherwise specified. The offset voltage, offset current and bias current specifications apply for any supply voltage from a single 5V supply up to ±15V supplies.

Note 4: The offset voltages and offset currents given are the maximum values required to drive the output within a volt of either supply with 1 mA load. Thus, these parameters define an error band and take into account the worst case effects of voltage gain and input impedance.

Note 5: The response time specified (see definitions) is for a 100 mV input step with 5 mV overdrive.

*Courtesy of National Semiconductor Corp.

APPENDIX 1-22*

HEWLETT **hp** PACKARD

COMPONENTS

0.3" SOLID STATE SEVEN SEGMENT INDICATOR | 5082-7740

TENTATIVE DATA AUGUST 1973

Features

- COMMON CATHODE
- RIGHT HAND DP
- EXCELLENT CHARACTER APPEARANCE
  - Continuous Uniform Segments
  - Wide Viewing Angle
  - High Contrast
- IC COMPATIBLE
  - 1.7V per Segment
- STANDARD 0.3" DIP LEAD CONFIGURATION
  - PC Board or Standard Socket Mountable
- CATEGORIZED FOR LUMINOUS INTENSITY
  - Assures Uniformity of Light Output from Unit to Unit within a Single Category

Description

The HP 5082-7740 is a common cathode LED numeric display with a right hand decimal point. The large 0.3" high character size generates a bright, continuously uniform 7 segment display. Designed for viewing distances of up to 10 feet, this single digit display has been human engineered to provide a high contrast ratio and wide viewing angle.

The 5082-7740 utilizes a standard 0.3" dual-in-line package configuration that allows for quick mounting on PC boards or in standard IC sockets. Requiring a forward voltage of only 1.7V, the display is inherently IC compatible allowing for easy integration into electronic calculators, credit card verifiers, TVs, radios, and digital clocks.

Package Dimensions

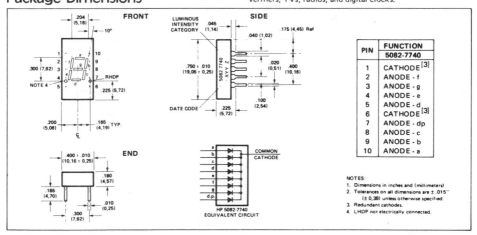

PIN	FUNCTION 5082-7740
1	CATHODE [3]
2	ANODE - f
3	ANODE - g
4	ANODE - e
5	ANODE - d
6	CATHODE [3]
7	ANODE - dp
8	ANODE - c
9	ANODE - b
10	ANODE - a

NOTES:
1. Dimensions in inches and (millimeters)
2. Tolerances on all dimensions are ± .015" (± 0,38) unless otherwise specified.
3. Redundant cathodes.
4. LHDP not electrically connected.

*Courtesy of Hewlett Packard, Inc.

Absolute Maximum Ratings

Power Dissipation T_A = 25°C. . . . . 400 mW
Operating Temperature Range . . . . −20°C to ⊹ 85°C
Storage Temperature Range . . . . −20°C to + 85°C
Average Forward Current/Segment or Decimal Pt. T_A = 25°C [1]. . . . . 25 mA
Peak Forward Current/Segment or Decimal Pt. T_A = 25°C (Pulse Duration ⩽ 500 μs) . . . . 150 mA
Reverse Voltage/Segment or Decimal Pt. 6V
Max. Solder Temperature 1/16'' Below Seating Plane (t ⩽ 5 sec.) [2] . . . . 230°C

NOTES: 1. Derate from 25°C at .25 mA/°C per segment or D.P.   2. Clean only in Freon TF, Isopropanol, or water.

Electrical/Optical Characteristics at T_A=25°C

Description	Symbol	Test Condition	Min.	Typ.	Max.	Units
Luminous Intensity/Segment [1]	I_{ν} AVE	I_{PEAK} = 100 mA 10% Duty Cycle	50	150		μcd
		I_F = 20 mA DC		250		
Peak Wavelength	λ_{PEAK}			655		nm
Forward Voltage/Segment or D.P.	V_F	I_F = 100 mA		1.6	2.3	V
Reverse Current/Segment or D.P.	I_R	V_R = 6V			100	μA
Rise and Fall Time [2]	t_r, t_f			10		ns
Temperature Coefficient of Forward Voltage	ΔV_F / °C			−2.0		mV/°C
Temperature Coefficient of Luminous Intensity	ΔI_{ν}/°C			−1.0		%/°C

NOTES: 1. The digits are categorized for luminous intensity such that the variation from digit to digit within a category is not discernible to
the eye. Intensity categories are designated by a letter located on the right hand side of the package.
2. Time for a 10%-90% change of light intensity for step change in current.

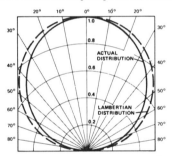

Figure 1. Normalized Angular Distribution of
Luminous Intensity.

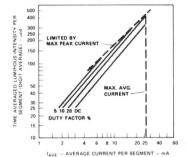

Figure 2. Typical Time Averaged Luminous Intensity
per Segment versus Average Current.

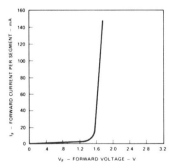

Figure 3. Forward Current versus Forward
Voltage.

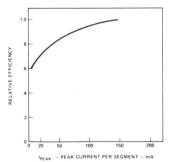

Figure 4. Relative Efficiency (Luminous Intensity per Unit
Current) versus Peak Current per Segment.

For more information, call your local HP Sales Office or East (201) 265-5000 · Midwest (312) 677-0400 · South (404) 436-6181 · West (213) 877-1282.
Or, write: Hewlett-Packard, 1501 Page Mill Road, Palo Alto, California 94304. In Europe, Post Office Box 85, CH-1217 Meyrin 2, Geneva,
Switzerland. In Japan, YHP, 1-59-1, Yoyogi, Shibuya-Ku, Tokyo, 151.
Printed in U.S.A.

APPENDIX 1-23*

SERIES 1603-02
REFLECTIVE

LIQUID CRYSTAL DISPLAYS

- **Optimum Readability**
- **Single Plane Viewing**
- **MOS Compatibility**
- **No Back Lighting Required**

- **Low Profile**
- **Microwatt Power Consumption**
- **Low Cost**
- **Ideal for High Ambient Light Environment**

The IEE Series 1603-02 is a 3½ decade Liquid Crystal Display with four floating decimals and an overflow plus or minus one (±). A colon is incorporated in the display for additional application-advantages such as clocks, etc.

The IEE Series 1603-02 reflective Liquid Crystal Display consists of a layer of micron-thin liquid crystal material confined between two sheets of glass, one sheet having a clear conductive electrode and the other a reflective coating etched in a segmented pattern to create a digital display. The organic material requires extremely low power in order to be an effective display. Upon generation of an electric field, the liquid layer becomes turbulent and scatters light. This scattering effect (caused by a continuous change in the index of refraction) appears as an optically dense area; by selectively energizing appropriate segments, a digital format is obtained. When the applied field is removed, the liquid crystal material returns to its original quiescent, transparent condition.

Reflective displays eliminate the need for back-lighting, which makes them excellent displays for use in portable equipment or in equipment where low power consumption is a definite consideration.

For optimum life, liquid crystals are operated on A.C. (40-100 Hz) which, coupled with a 15-30 volt range, make them directly compatible with MOS logic. Numerous companies are engaged in the manufacture of standard MOS circuits for use with Liquid Crystal Displays and this list may be obtained from Industrial Electronic Engineers, Inc. (IEE) upon request.

ELECTRICAL SPECIFICATIONS

Operating Voltage: A.C. 40-100 Hz. Typical 24 Volt (Peak to Peak). Maximum 40 Volt. Minimum 10 Volt.

Power Consumption: 20 microwatts per segment (maximum).

Rise Time: 50 milliseconds.

Decay Time: 150 milliseconds.

Contrast Ratio: 15:1 minimum.

Life: 10,000 hrs. minimum at 24 V.A.C.

Operating Temperature: 5°C to 55°C.

Storage Temperature: −10°C to 70°C.

Relative Humidity: 0 to 100%.

MECHANICAL SPECIFICATIONS

Package Size: See diagram.

Character Size: .433″ (11 MM).

Character Width: .260″ (7.6 MM).

Segment Width: .055″ (1.4 MM).

Decimal Point Width: .06″ (1.5 MM).

Decimal Point Height: .08″ (2 MM).

Contacts:

The conductive electrodes on the Series 1603-02 Liquid Crystal Displays are terminated in an edge board configuration having .050″ (1.3 MM) spacing, which allows the use of an edge connector or a spring contact right angle connector to be used in conjunction with printed circuit board patterns.

ORDERING INFORMATION

	Part No.
Liquid Crystal	1603-02
Right Angle Connector	22076-01
Mounting Kit (PC Board with right angle connector attached)	22077-01

AVAILABILITY:

Series 1603-02 displays and optional hardware (connector, PC boards) are available for customer evaluation from shelf stock. For large quantity requirements and/or special designs, consult IEE for information.

*Courtesy of Industrial Electronic Engineers, Inc.

APPENDIX 1-24*

TTL
MSI

TYPES SN5490A, SN5492A, SN5493A, SN54L90, SN54L93,
SN7490A, SN7492A, SN7493A, SN74L90, SN74L93
DECADE, DIVIDE-BY-TWELVE, AND BINARY COUNTERS
BULLETIN NO. DL-S 7211807, DECEMBER 1972

'90A, 'L90 . . . DECADE COUNTERS

'92A . . . DIVIDE-BY-TWELVE
COUNTER

'93A, 'L93 . . . 4-BIT BINARY
COUNTERS

description

Each of these monolithic counters contains four master-slave flip-flops and additional gating to provide a divide-by-two counter and a three-stage binary counter for which the count cycle length is divide-by-five for the '90A and 'L90, divide-by-six for the '92A, and divide-by-eight for the '93A and 'L93.

All of these counters have a gated zero reset and the '90A and 'L90 also have gated set-to-nine inputs for use in BCD nine's complement applications.

To use their maximum count length (decade, divide-by-twelve, or four-bit binary) of these counters, the B input is connected to the Q_A output. The input count pulses are applied to input A and the outputs are as described in the appropriate function table. A symmetrical divide-by-ten count can be obtained from the '90A or 'L90 counters by connecting the Q_D output to the A input and applying the input count to the B input which gives a divide-by-ten square wave at output Q_A.

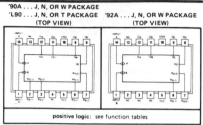

'90A . . . J, N, OR W PACKAGE
'L90 . . . J, N, OR T PACKAGE   '92A . . . J, N, OR W PACKAGE
(TOP VIEW)   (TOP VIEW)

positive logic: see function tables

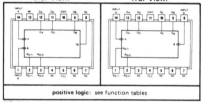

'93A . . . J, N, OR W PACKAGE   'L93 . . . J, N, OR T PACKAGE
(TOP VIEW)   (TOP VIEW)

positive logic: see function tables

NC—No internal connection

TYPES	TYPICAL POWER DISSIPATION
'90A	145 mW
'L90	20 mW
'92A, '93A	130 mW
'L93	16 mW

functional block diagrams

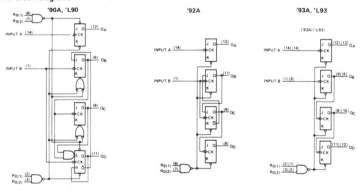

— · · · · dynamic input activated by transition from a high level to a low level.

The J and K inputs shown without connection are for reference only and are functionally at a high level.

*Courtesy of Texas Instruments, Incorporated

TYPES SN5490A, SN5492A, SN5493A, SN54L90, SN54L93, SN7490A, SN7492A, SN7493A, SN74L90, SN74L93 DECADE, DIVIDE-BY-TWELVE, AND BINARY COUNTERS

'90A, 'L90
BCD COUNT SEQUENCE
(See Note A)

COUNT	OUTPUT			
	Q_D	Q_C	Q_B	Q_A
0	L	L	L	L
1	L	L	L	H
2	L	L	H	L
3	L	L	H	H
4	L	H	L	L
5	L	H	L	H
6	L	H	H	L
7	L	H	H	H
8	H	L	L	L
9	H	L	L	H

'90A, 'L90
BI-QUINARY (5-2)
(See Note B)

COUNT	OUTPUT			
	Q_A	Q_D	Q_C	Q_B
0	L	L	L	L
1	L	L	L	H
2	L	L	H	L
3	L	L	H	H
4	L	H	L	L
5	H	L	L	L
6	H	L	L	H
7	H	L	H	L
8	H	L	H	H
9	H	H	L	L

'90A, 'L90
RESET/COUNT FUNCTION TABLE

RESET INPUTS				OUTPUT			
$R_{0(1)}$	$R_{0(2)}$	$R_{9(1)}$	$R_{9(2)}$	Q_D	Q_C	Q_B	Q_A
H	H	L	X	L	L	L	L
H	H	X	L	L	L	L	L
X	X	H	H	H	L	L	H
X	L	X	L	COUNT			
L	X	L	X	COUNT			
L	X	X	L	COUNT			
X	L	L	X	COUNT			

'92A
COUNT SEQUENCE
(See Note C)

COUNT	OUTPUT			
	Q_D	Q_C	Q_B	Q_A
0	L	L	L	L
1	L	L	L	H
2	L	L	H	L
3	L	L	H	H
4	L	H	L	L
5	L	H	L	H
6	H	L	L	L
7	H	L	L	H
8	H	L	H	L
9	H	L	H	H
10	H	H	L	L
11	H	H	L	H

'93A, 'L93
COUNT SEQUENCE
(See Note C)

COUNT	OUTPUT			
	Q_D	Q_C	Q_B	Q_A
0	L	L	L	L
1	L	L	L	H
2	L	L	H	L
3	L	L	H	H
4	L	H	L	L
5	L	H	L	H
6	L	H	H	L
7	L	H	H	H
8	H	L	L	L
9	H	L	L	H
10	H	L	H	L
11	H	L	H	H
12	H	H	L	L
13	H	H	L	H
14	H	H	H	L
15	H	H	H	H

'92A, '93A, 'L93
RESET/COUNT FUNCTION TABLE

RESET INPUTS		OUTPUT			
$R_{0(1)}$	$R_{0(2)}$	Q_D	Q_C	Q_B	Q_A
H	H	L	L	L	L
L	X	COUNT			
X	L	COUNT			

NOTES: A. Output Q_A is connected to input B for BCD count.
B. Output Q_D is connected to input A for bi-quinary count.
C. Output Q_A is connected to input B.
D. H = high level, L = low level, X = irrelevant

schematics of inputs and outputs

'90A, '92A, '93A
EQUIVALENT OF EACH INPUT

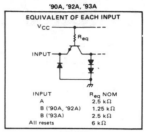

INPUT	R_{eq} NOM
A	2.5 kΩ
B ('90A, '92A)	1.25 kΩ
B ('93A)	2.5 kΩ
All resets	6 kΩ

'L90, 'L93
EQUIVALENT OF EACH INPUT
EXCEPT A AND B OF 'L93

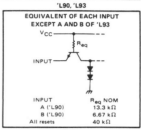

INPUT	R_{eq} NOM
A ('L90)	13.3 kΩ
B ('L90)	6.67 kΩ
All resets	40 kΩ

'L93
EQUIVALENT OF A OR B INPUT

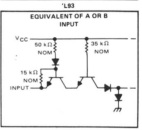

'90A, '92A, '93A, 'L90, 'L93
TYPICAL OF ALL OUTPUTS

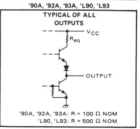

'90A, '92A, '93A: R = 100 Ω NOM
'L90, 'L93: R = 500 Ω NOM

Appendix 2

Standard Resistor and Capacitor Values

APPENDIX 2-1
TYPICAL STANDARD RESISTOR VALUES

Ω	Ω	Ω	$k\Omega$	$k\Omega$	$k\Omega$	$M\Omega$	$M\Omega$
—	10	100	1	10	100	1	10
—	12	120	1.2	12	120	1.2	—
—	15	150	1.5	15	150	1.5	15
—	18	180	1.8	18	180	1.8	—
—	22	220	2.2	22	220	2.2	22
2.7	27	270	2.7	27	270	2.7	—
3.3	33	330	3.3	33	330	3.3	—
3.9	39	390	3.9	39	390	3.9	—
4.7	47	470	4.7	47	470	4.7	—
5.6	56	560	5.6	56	560	5.6	—
6.8	68	680	6.8	68	680	6.8	—
—	82	820	8.2	82	820	—	—

APPENDIX 2-2
TYPICAL STANDARD CAPACITOR VALUES

pF	pF	pF	pF	µF	µF	µF	µF	µF	µF	µF
5	50	500	5000	0.05	0.5	5	50	500	5000	
—	51	510	5100	—	—	—	—	—	—	
—	56	560	5600	0.056	0.56	5.6	56	—	5600	
—	—	—	6000	0.06	—	6	—	—	6000	
—	62	620	6200	—	—	—	—	—	—	
—	68	680	6800	0.068	0.68	6.8	—	—	—	
—	75	750	7500	—	—	—	75	—	—	
—	—	—	8000	—	—	8	80	—	—	
—	82	820	8200	0.082	0.82	8.2	82	—	—	
—	91	910	9100	—	—	—	—	—	—	
10	100	1000		0.01	0.1	1	10	100	1000	10000
—	110	1100		—	—	—	—	—	—	
12	120	1200		0.012	0.12	1.2	—	—	—	
—	130	1300		—	—	—	—	—	—	
15	150	1500		0.015	0.15	1.5	15	150	1500	
—	160	1600		—	—	—	—	—	—	
18	180	1800		0.018	0.18	1.8	18	180	—	
20	200	2000		0.02	0.2	2	20	200	2000	
22	220	2200		—	0.22	2.2	22	—	—	
24	240	2400		—	—	—	—	240	—	
—	250	2500		—	0.25	—	25	250	2500	
27	270	2700		0.027	0.27	2.7	27	270	—	
30	300	3000		0.03	0.3	3	30	300	3000	
33	330	3300		0.033	0.33	3.3	33	330	3300	
36	360	3600		—	—	—	—	—	—	
39	390	3900		0.039	0.39	3.9	39	—	—	
—	—	4000		0.04	—	4	—	400	—	
43	430	4300		—	—	—	—	—	—	
47	470	4700		0.047	0.47	4.7	47	—	—	

Answers to Problems

1-4	0.3 V, 200 pps, 2 ms, 40%, 0.67	**2-23**	0.185 V, 0.405 V
1-5(b)	20%	**2-27**	0.18 V
1-6	5.25 V, 9.52%, 5 μs, 7 μs, 38 μs, 14 330 pps, 1.19, 54.3%	**2-28**	1.8 V, 0.9 V
1-7	0.33 V, 0.77 V	**3-7**	20 kΩ
1-8	100, 6	**3-8**	50 kΩ, −19.3 V
1-9	3.5 μs, 8.4 MHz	**3-13**	8 kΩ, 1.79 mA
1-11	1 MHz, 15.9 Hz	**3-15**	6.8 kΩ
1-12	0.7 μs, 0.63%	**3-17**	1N751, 1 kΩ
1-13	433 Hz, 70 kHz	**3-21**	0.1 μF, 100 kΩ
1-15	314 Hz	**3-24**	1 μF, 47 kΩ
1-16	35 ns, 46 ns	**3-25**	1N757, 1μF, 120 kΩ
2-3	0.754 V	**4-4**	30.3, 182 μA
2-4	0.493 mA	**4-5**	16.2 μA, 114 μA
2-5	263.9 ms	**4-6**	0.228 mW, 1.25 μW, 2.28 mW, 1.25 μW
2-6	2.13 V, 3.62 V	**4-8**	14.999 V, 13.5 V, 1.5 V, 2.026 μs
2-7	830 nA	**4-10**	80.5 pF, 80.5 kHz
2-10	16.7 Hz	**4-12**	20 V, 106 mV
2-11	355 kΩ	**4-16**	1 kΩ, 21.5 kΩ (use 18 kΩ)
2-12	45.5 pF	**4-18**	60.4 kHz
2-13	14.52 μs, 6.6 μs, 33 μs	**4-19**	22 kΩ, 560 kΩ, 17.28 V
2-14	6.54 V	**4-22**	12 kΩ, 680 kΩ, 7500 pF
2-15	9.71 V, 5.29 V	**4-23**	12 kΩ, 10 kΩ, 0.18 μF
2-16	2.94 mA, 1.6 mA	**4-24**	820 Ω, 10 kΩ, 0.47 μF
2-17	5.64 μs, 0.2 V		
2-18	7.8 V, 15.2 μs	**5-4**	1.8 kΩ, 133 kΩ. 1.8 kΩ
2-19	3.12 V, 218 μA	**5-5**	5.6 kΩ, 218 kΩ, 5.6 kΩ

5-6	500 pF, 1.5 kΩ, 20 pF		**8-3**	22 kΩ, 0.02 μF
5-7	1 kΩ, 100 kΩ, 1 kΩ		**8-7**	0.01 μF, 6.8 kΩ, 15 kΩ
5-8	10 kΩ, 200 kΩ, 10 kΩ		**8-8**	151 μs, 76.2%
5-9	(100 pF, 1.5 kΩ, 3 pF), (500 pF, 1.5 kΩ, 20 pF)		**8-10**	33 kΩ, 0.05 μF, 1.2 kΩ, 10 kΩ
5-11	$\pm$15 V, 560 kΩ, 680 kΩ, 0.015 μF		**8-11**	R_A = 2.2 kΩ, R_D = 10 kΩ
5-12	220 kΩ, 180 kΩ, 0.012 μF, 36 μs		**8-13**	120 kΩ, 0.1 μF
5-13	180 kΩ, 150 kΩ, 0.1 μF		**8-15**	610 kΩ, 0.1 μF
5-15	18 kΩ, 72 kΩ, 18 kΩ		**8-16**	(R_A = 100 kΩ, C_C = 0.25 μF, C_A = 0.01 μF), (R_A = 47 kΩ, 100 kΩ, 180 kΩ, 330 kΩ)
5-17	1N746, 220 Ω, 4.7 kΩ, 220 Ω		**8-17**	(33 kΩ, 47 kΩ, 0.1 μF), (82 kΩ, 82 kΩ, 1800 pF)
5-19	3.9 kΩ, 510 pF, 100 Ω		**8-19**	120 kΩ, 3000 pF, (5.8 V to 12 V)
5-20	-19.89 V, 6.63 V		**8-20**	0.05 μF, 820 Ω, 30 kΩ
5-22	150 kΩ, 150 kΩ, 1.5 MΩ, 0.01 μF, 106 Hz			
5-23	1.65 V		**9-2**	18 kΩ, 1 μF, 120 kΩ, 0.15 μF
			9-4	3.9 kΩ, 3.9 kΩ, 2.7 kΩ
6-3	R_E = 4.7 kΩ, R_C = 8.2 kΩ, R_1 = 47 kΩ, R_2 = 47 kΩ		**9-5**	5 kΩ, 1.8 kΩ
6-4	R_{C1} = 3.3 kΩ, R_1 = 56 kΩ		**9-7**	467 Ω, 2.2 kΩ, 5.5 V
6-6	20 pF		**9-8**	14 V, 0.15 μF
6-8	(39 kΩ, 180 kΩ), (39 kΩ, 220 kΩ)		**9-10**	47 kΩ, 0.008 μF, 16.45 V
6-10	2.2 kΩ, 500 Ω, 1.8 kΩ		**9-11**	12 kΩ, 27 kΩ, 390 pF
6-12	(39 kΩ, 100 kΩ), (39 kΩ, 150 kΩ)		**9-13**	1.2 μF, 5.6 kΩ, 30 μF, 270 kΩ, 0.06 μF
6-13	56 kΩ, 270 kΩ, 180 kΩ		**9-15**	0.068 μF, 100 kΩ, 1.8 μF, 3.3 MΩ, 5000 pF
6-16	2.2 kΩ, 1.5 kΩ		**9-17**	(56 kΩ, 150 kΩ), (0.05 μF, 22 kΩ, 4 μF), (100 kΩ)
7-2	3.9 kΩ, 3.9 kΩ, 39 kΩ, 22 kΩ		**9-19**	0.03 μF, 100 kΩ, 100 kΩ, $V_{GS(\text{off})}$ = 7 V, 709
7-3	0.0025 μF		**9-21**	(47 kΩ, 220 kΩ), (22 kΩ, 0.1 μF)
7-4	309 pF		**9-22**	(20 kΩ, 39 kΩ), (10 kΩ, 18 kΩ)
7-5	10 kΩ, 1800 pF		**9-23**	(68 kΩ + 200 kΩ), 220 kΩ, 0.02 μF, 220 kΩ, (39 kΩ + 100 kΩ)
7-8	5.6 kΩ, 150 kΩ, 5.6 kΩ, 0.015 μF		**9-25**	(10 kΩ + 100 kΩ), 120 kΩ, 120 kΩ, 0.02 μF, 5000 pF, 709
7-9	0.004 μF, 680 Ω		**9-26**	100 pF, 330 kΩ, 270 kΩ, (1.8 kΩ + 20 kΩ), 0.15 μF, 709
7-11	1.8 kΩ, 120 kΩ, 1200 pF		**9-27**	0.82 μF, 270 kΩ, 180 kΩ, 100 kΩ, 180 kΩ, (30 kΩ + 2.7 kΩ)
7-14	1 kΩ, 6000 pF, 5.6 kΩ, 10 kΩ, 47 kΩ, 1 kΩ, 1 kΩ			
7-16	68 kΩ, 68 kΩ, 68 kΩ, 0.015 μF			
7-18	5.6 kΩ, 220 Ω, 560 kΩ, 10 pF, 250 pF, 560 kΩ, 15 kΩ			
7-20	180 kΩ, 560 pF, 1 kΩ, 180 kΩ, 180 kΩ, 180 kΩ			

9-29   0.018 μF, $R_1 = R_2 = (100 \text{ k}\Omega + 68$ kΩ), 47 kΩ, 180 kΩ, (27 kΩ + 2.7 kΩ)

9-31   2.7 kΩ, 3.9 kΩ, 3.9 kΩ, 1.3 V, −1.3 V

9-34   ±9 V, 18 kΩ, 33 kΩ, 33 kΩ, 1800 pF, 82 kΩ

9-35   27 kΩ, 27 kΩ, 10 kΩ

9-36   +6 V, −3 V, 8.2 kΩ, 3.9 kΩ, 18 kΩ, 0.027 μF, 82 kΩ

9-38   4700 pF, (2.7 kΩ + 30 kΩ)

9-39   9100 pF, (2.7 kΩ + 30 kΩ), 496 μs, 41 μs

10-2   8.2 kΩ, 0.9 V, 9 V

10-4   1.5 kΩ

10-12   8, 64, 512

11-5 $AB + BC$

11-7 $AB\overline{E} + CD\overline{E}$

11-8 $\overline{ABC}$

11-9 $A(B + \overline{C}) + (\overline{A + D + C})$

12-3   0.2 V, 0.93 V

12-4   0.9 V, 1.4 V

12-6   13

12-7   7

12-13   470 Ω

12-26   330 Ω

12-27   680 Ω

12-29   (R_B = 4.7 kΩ, R_C = 1.5 kΩ), (R_{B5} = 4.7 kΩ, R_{C5} = 1.5 kΩ, R_{B6} = 10 kΩ, R_{C6} = 680 Ω)

12-31 R_B = 27 kΩ, R_C = 18 kΩ

13-5   5.6 kΩ, 56 kΩ, 33 kΩ

13-6   4.7 kΩ, 47 kΩ, 27 kΩ

13-9   56 pF

13-10   120 pF, 174 kHz

13-13   160 pF

13-14   120 pF, 174 kHz

14-12   560 Ω, 39 kΩ

14-29   15, 7

14-33   100 μA

15-2   0.035%, 0.00027%

15-4   0.047%, 0.0001%

15-6   10 kΩ, 47 kΩ, 8.2 kΩ, 0.11%

15-8   0.039 μF, 10.73 μs, 0.08%

15-10   124 kΩ, 62 kΩ, 31 kΩ, 15.5 kΩ, 7.75 kΩ

15-11   31

15-28   6, 15

Index